Study Guide

TO ACCOMPANY

UNIVERSITY PHYSICS

Revised Edition

HARRIS BENSON

Vanier College

John Wiley & Sons, Inc.

New York Chichester Brisbane Toronto Singapore

Copyright © 1996 by Harris Benson.

All rights reserved.

Reproduction or translation of any part of this work beyond that permitted by Sections 107 and 108 of the 1976 United States Copyright Act without the permission of the copyright owner is unlawful. Requests for permission or further information should be addressed to the Permissions Department, John Wiley & Sons, Inc.

ISBN 0-471-14604-8

Printed in the United States of America

10 9 8 7 6 5

PREFACE

This Study Guide is meant to accompany UNIVERSITY PHYSICS, Revised Edition, by Harris Benson, John Wiley and Sons, 1996. It is not a replacement for the textbook. You should read the book before you start a given chapter in the study guide.

The structure of each chapter is as follows:

MAJOR POINTS: These provide you with a brief overview of the central topics in the chapter.

CHAPTER REVIEW: The chapter review includes the most important equations and some discussion of concepts.

EXAMPLES: Several examples are presented throughout the Chapter Review.

SOLUTIONS TO SELECTED TEXT EXERCISES AND PROBLEMS: Solutions to approximately 10% of the exercises and problems in the text are provided.

SELF-TEST: The Self-Test should allow you to judge how well you have grasped the basic ideas in the chapter. Many of the questions have been taken from actual tests. Although the style of questions asked by your own instructor may be different, you should should make an honest attempt to solve these questions under "examination" conditions. Allow yourself about 10 min per question.

ANSWERS TO SELF-TESTS: Solutions to all the Self-Test questions are provided at the end of the Study Guide.

Always keep in mind that the statement "I understand the material but I can't do the problems" is nonsense. An important way in which one demonstrates mastery of a subject is the ability to solve problems. I hope that this Study Guide provides you with enough examples (a total of about 475 solutions) to help you solve the problems set for homework and those that appear in tests. Your comments or suggestions are most welcome. Good luck with your studies.

HARRIS BENSON
Vanier College,
821 Ste. Croix Ave.,
Montreal, H4L 3X9

CONTENTS

Chapter 1

INTRODUCTION

MAJOR POINTS

1. The SI system of base units; the conversion of units
2. The use of significant figures to indicate the precision of data
3. Dimensional analysis
4. Order of magnitude estimates

CHAPTER REVIEW

It would be a good idea for you to read the first two sections of this chapter, What is Physics? and Concepts, Models, and Theories. They should give you some perspective on this subject. You should also browse through Appendix B (Mathematics Review) and Appendix C (Calculus Review) to note what they contain.

UNITS

A measurement of any physical quantity is expressed as a comparison to standards called units. In the SI (System International) the units of mass, length, and time are the kilogram (kg), meter (m), and second (s). Other units are listed in Appendix A of the text. Most physical quantities involve combinations of these base units. For example, the unit of speed is m/s and the unit of density is kg/m^3. Sometimes it is necessary to convert from one set of units to another. The process is illustrated in the next example.

EXAMPLE 1

A nail grows 0.2 mm per week. Express this rate in terms of angstroms (10^{-10} m) per min.

Solution:

When converting from one set of units to another, it is helpful to use factors that equal unity. For example, 1 h/3600 s = 1. First, we note that 1 angstrom is 10^{-7} mm, then

$$0.2 \text{ mm/week} = (0.2 \text{ mm/week})(1 \text{ Å}/10^{-7} \text{ mm})(1 \text{ week}/7 \text{ d})(1 \text{ d}/1440 \text{ min})$$

$$= 198 \text{ Å/min}$$

POWER OF TEN NOTATION

It is usually preferable to express numerical values in terms of scientific (power of ten) notation rather than using zeros. For example,

$$0.000380 \quad = 3.80\text{x}10^{-4}$$

$$1{,}246{,}000 = 1.246 \times 10^6$$

You should be familar with the metric prefixes listed on the inside of the front cover. The following are frequently encountered:

micro	10^{-6}
milli	10^{-3}
centi	10^{-2}
kilo	10^{3}
mega	10^{6}

Note also that several conversion factors are listed on the inside of the back cover.

SIGNIFICANT FIGURES

All measurements have some uncertainty in their value. This uncertainty may be stated explicitly as a range above and below the given value or be implied in the number of significant figures used.
For example,

$$0.03800 = 3.800 \times 10^{-2}$$

has 4 significant figures, whereas

$$1020 = 1.02 \times 10^3 \quad (1.020 \times 10^3\ ?)$$

has at least three significant figures. We would need some additional information to decide whether the zero after the 2 is significant or not. If we are given 1020.0, this definitely has 5 significant figures.

When you make a calculation that uses experimentally obtained values, you must be careful not to state your final result with unwarranted precision. This is easily done by properly limiting the number of significant figures quoted. As a rule of thumb the number of significant figures in the final result should equal that of the quantity with the least number of siginificant figures. For axample,

$$\frac{(1.3)(49.7)}{(5.781)} = (11.18) = 11$$

If this is an intermediate step in a longer calculation, it is usually acceptable to carry along one extra figure, that is, we use 11.2 but round off appropriately at the end.

DIMENSIONS

In dimensional analysis, units are reduced to factors of mass (M), length (L) and time (T). Any equation must be dimensionally consistent; that is, terms on both sides of the equation must have the same dimensions.

EXAMPLE 2

Check the dimensional constistency of the equation $v^2 = 2a\Delta x$ where v is a velocity (m/s), Δx is a displacement (m) and a is an acceleration (m/s^2).

Solution:

The dimensions of the three quantities are

$$[v^2] = L^2\,T^{-2}; \qquad [a] = L\,T^{-2}; \qquad [\Delta x] = L$$

The dimensions of the product

$$[a\,\Delta x] = L^2\,T^{-2}$$

are the same as $[v^2]$, so the equation is dimensionally consistent. The factor of two may or may not be correct (it is correct).

SOLUTIONS TO SELECTED TEXT EXERCISES AND PROBLEMS

Exercise 3

The density of water is about 1 g/cm^3. What is this in SI units?

Solution:

$$1\ g/cm^3 = (1\ g/cm^3)(1\ kg/10^3\ g)(10^6\ cm^3/1\ m^3)$$

$$= 10^3\ kg/m^3$$

Note that each of the factors on the right is equal to unity.

Exercise 11

Specify the number of significant figures in each of the following values:
(a) 23.001 s; (b) $0.500\text{x}10^2$ m; (c) 0.002030 kg; (d) 2700 kg/s.

Solution:

(a) 23.001 s has five S.F.
(b) $5.00\text{x}10^1$ m has three S.F.
(c) $2.03\text{x}10^{-3}$ kg has three S.F. since we cannot be sure about the last zero
(d) 2700 kg/s = $2.7\text{x}10^3$ kg/s has two S.F.

Exercise 25
A watch is advertised as being 99% accurate. Would you buy it?

Solution:
In one day there are 1440 min. One percent of this is 14.4 min, which is not an acceptable error for most people.

Exercise 39
The argument of a trigonometric function must be a dimensionless quantity. If the speed v of a particle of mass m as a function of time t is given by $v = \omega A \sin[(k/m)^{1/2}t]$, find the dimensions of ω and k, given that A is a length.

Solution:
The dimensions of velocity are $[v] = LT^{-1}$, and the sine function is dimensionless, so the dimensions of $[\omega A]$ must be the same as [v]. Since [A] = L, we have $[\omega] = T^{-1}$.
Since the quantity $(k/m)^{1/2}t$ must be dimensionless, and [t] = T, we have that

$$[k/m] = T^{-2}$$

We know that [m] = M, so $[k] = MT^{-2}$.

SELF-TEST

1. Express the following in scientific notation with the appropriate number of significant figures:

$$\frac{(674.9)(1.60\text{x}10^{2})^{1/3}}{(5.8\text{x}10^{-1})}$$

2. A plant grows 0.02 inch per day. Express this in mm/s.

3. The position of a particle as a function of time, t, is given by

$$x = A + Bt + Ct^2$$

What are the dimensions of A, B and C?

Chapter 2

VECTORS

MAJOR POINTS

1. The distinction between scalars and vectors
2. Addition or subtraction of vectors graphically by using the tail-to-tip method
3. (a) Resolving a vector into its rectangular components
   (b) Determining the magnitude and direction of the sum or resultant
4. Unit vector notation
5. (a) The scalar (dot) product of two vectors
   (b) The vector (cross) product of two vectors

CHAPTER REVIEW

SCALARS AND VECTORS

A scalar is a quantity that is specified by a number and a unit. It has only magnitude. Scalars obey the laws of ordinary algebra.

A vector is a quantity that has magnitude and direction and obeys the law of vector addition.

In the text, a vector is printed in boldface, **A**, but is usually written with an arrow on the top: $\vec{A}$. The magnitude of a vector is a (positive) scalar and in the text is printed in italic: $|\mathbf{A}| = A$. In your own writing be careful to distinguish clearly between scalar and vector quantities. You cannot add or equate scalars and vectors, but you can multiply a vector **A** by a scalar c to obtain a vector c**A**.

GRAPHICAL ADDITION OF VECTORS

Tail-to-tip method

In order to add two vectors **A** and **B** graphically we use the **tail-to-tip** method.

1. Draw the coordinate axes
2. Pick a convenient scale, say 1 cm for a 5 m displacement.
3. Draw **A** to scale in the appropriate direction.
4. Place the tail of **B** at the tip of **A**.
5. The **sum** or **resultant**, **R** = **A** + **B**, is drawn from the tail of **A** to the tip of **B**.

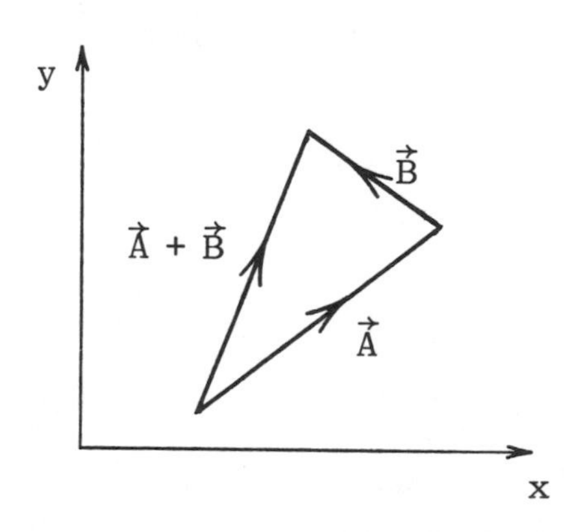

FIGURE 2.1

The process is shown in Fig. 2.1. The order does not matter: **A** + **B** = **B** + **A**. It is clearer to place the arrowheads at the midpoints of the arrows representing the vectors.

Subtraction

One can treat subtraction as a special case of addition. Thus,

$$\mathbf{A} - \mathbf{B} = \mathbf{A} + (-\mathbf{B})$$

Alternatively, we consider the equation,

$$\mathbf{A} = \mathbf{B} + (\mathbf{A} - \mathbf{B})$$

In this case, the vectors are placed tail-to-tail and the vector (**A** - **B**) is drawn from the tip of **B** to the tip of **A**, as shown in Fig. 2.2b. The vector (**A** - **B**) represents the **displacement** from the initial position (the tip of **B**) to the final position (the tip of **A**).

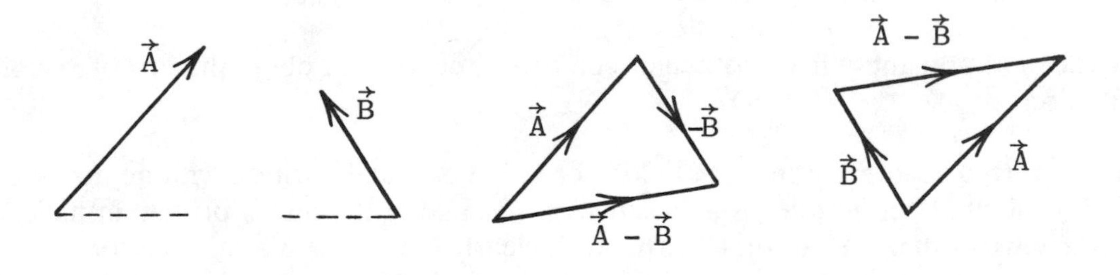

FIGURE 2.2

The tail-to-tip method may be extended to several vectors, as the next example shows.

EXAMPLE 1

Given **A** = 3 m at 60° W of N, **B** = 2 m at 30° S of W, and **C** = 1.5 m E, find (a) **A** + **B**, (b) **A** - **B**, (c) **P** = **B** - **A** - 2**C**.

Solution:

(a) First draw **A** with its tail at the origin, as in Fig. 2.3a. Then draw "miniaxes" at the tip of **A** to help you draw **B** in the proper direction. The sum **A** + **B** is from the tail of **A** to the tip of **B**.
(b) We follow a similar procedure but add -**B** to **A**, see Fig. 2.3b.
(c) If we take the vectors in the order **B**, -**A**, and -2**C**, the resultant **P** crosses **A**, as shown in Fig. 2.3c. If we took the order -**A**, **B**, and -2**C**, we would have obtained an open polygon. (Try it.)

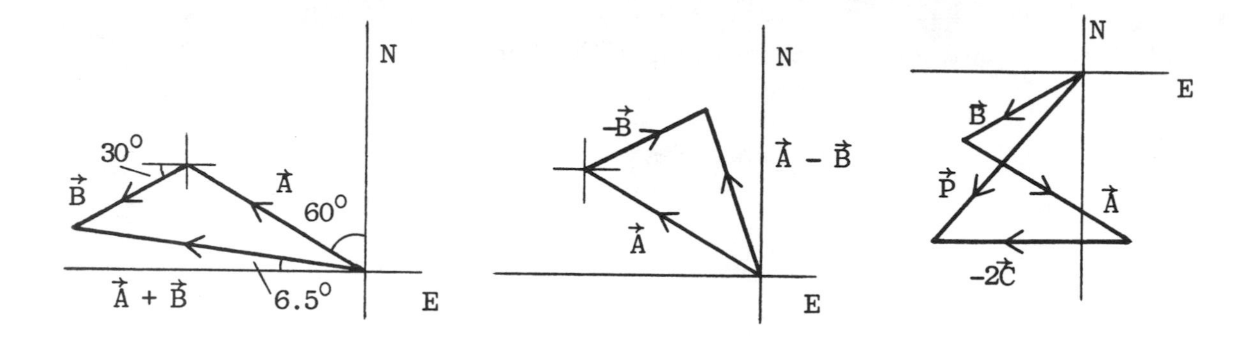

FIGURE 2.3

COMPONENTS

Consider the two-dimensional vector **A** in Fig. 2.4. One can represent **A** either by its magnitude and direction, (A, θ) or by its rectangular components (A_x, A_y). The components of **A** are the "projections" of the vector onto the rectangular (perpendicular) x and y axes:

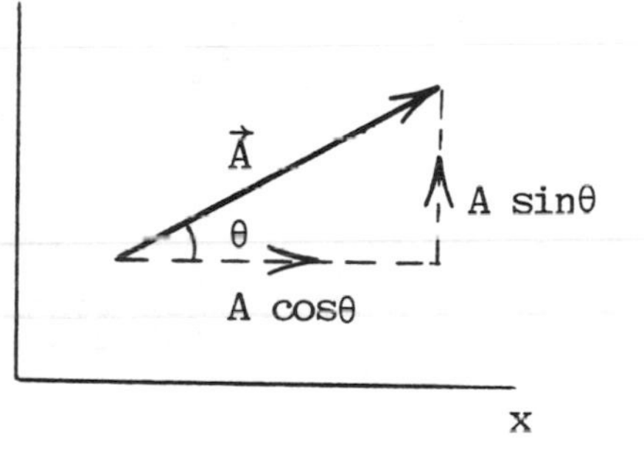

FIGURE 2.4

$$A_x = A \cos\theta$$

$$A_y = A \sin\theta$$

where θ is measured counterclockwise from the +x axis. However, one is often confronted by vectors whose directions are specified relative to the -x axis or the -y axis. It is better to consider each case individually, keeping in mind the following:

The sign of a component is determined by its direction relative to the chosen +x axis or +y axis. Note that the position of the tail of a vector has no bearing on the signs of its components.

EXAMPLE 2
Find the components of the four vectors shown in Fig. 2.5.

Solution:
The components of the vectors are shown in the figure.

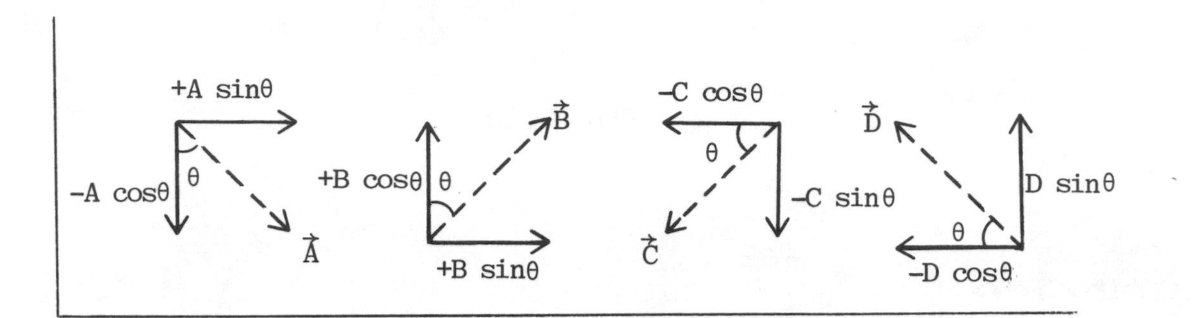

FIGURE 2.5

Addition using components

For two-dimensional vectors, the sum **R** = **A** + **B** is equivalent to two equations in terms of the components:

$$R_x = A_x + B_x; \qquad R_y = A_y + B_y$$

The magnitude and direction of the resultant are given by

$$R = (R_x^2 + R_y^2)^{1/2}; \qquad \tan\boldsymbol{\theta} = R_y/R_x$$

The angle **θ** is measured counterclockwise from the +x axis. The quadrant **θ** is in depends on the individual signs of R_x and R_y.

Unit Vectors

The unit vectors **i**, **j**, and **k**, each of magnitude of unity, serve only to specify a direction in space. They point along the +x, +y and +z axes respectively. A vector **A** is expressed in terms of its components A_x, A_y, and A_z, as follows:

$$\mathbf{A} = A_x\mathbf{i} + A_y\mathbf{j} + A_z\mathbf{k}$$

In order to add (or subtract) two vectors, we add (or subtract) their components.

EXAMPLE 3
Given the vectors **A** = -**i** + 2**j** - 3**k** and **B** = 2**i** - **j** + 5**k**, find
(a) **A** + **B**, (b) **A** - **B**.

Solution:
(a) **A** + **B** = (-1 + 2)**i** + (2 - 1)**j** + (-3 + 5)**k** = **i** + **j** + 2**k**.
(b) **A** - **B** = (-1 - 2)**i** + (2 + 1)**j** + (-3 - 5)**k** = -3**i** + 3**j** - 8**k**.

SCALAR (DOT) PRODUCT

The scalar product of vectors **A** and **B** is defined as

$$\mathbf{A.B} = A\ B\cos\theta$$

where θ is the (smaller) angle between **A** and **B**. Notice that **B.A** = **A.B**, and for the unit vectors:

$$\mathbf{i.i} = \mathbf{j.j} = \mathbf{k.k} = 1; \quad \mathbf{i.j} = \mathbf{i.k} = \mathbf{j.k} = 0$$

In terms of the components,

$$\mathbf{A.B} = A_xB_x + A_yB_y + A_zB_z$$

If **B** = **A**, then **A.A** = $A^2 = A_x^2 + A_y^2 + A_z^2$, which is the Pythagorean theorem in three dimensions.

EXAMPLE 4
Given the vectors **A** = 3**i** + 2**j** - **k** m, and **B** = -2**i** + **j** + 5**k** m, find (a) **A.B**; (b) the angle between **A** and **B**.

Solution:
(a) $\mathbf{A.B} = (3)(-2) + (2)(1) + (-1)(5) = -9\ \text{m}^2$

(b) The magnitudes of the vectors are $A = (9 + 2 + 1)^{1/2} = 3.46$ m, and $B = (4 + 1 + 25)^{1/2} = 5.48$ m. From the definition of the scalar product:

$$\cos\theta = \mathbf{A.B}/AB$$

$$= (-9\ \text{m}^2)/(3.46\ \text{m})(5.48\ \text{m}) = -0.475$$

Thus $\theta = 118^\circ$.

VECTOR (CROSS) PRODUCT

The vector product of two vectors **A** and **B** is defined as

$$\mathbf{A} \times \mathbf{B} = A\,B \sin\theta\,\hat{\mathbf{n}}$$

where the unit vector $\hat{\mathbf{n}}$ is normal (perpendicular) to the plane containing **A** and **B**. Its direction is specified by the right-hand rule or by the bottlecap rule (see Fig. 2.20 in the text). Notice that $\mathbf{B} \times \mathbf{A} = -\mathbf{A} \times \mathbf{B}$ and that if **A** and **B** are parallel, then $\mathbf{A} \times \mathbf{B} = 0$. In particular, for the unit vectors, $\mathbf{i} \times \mathbf{i} = \mathbf{j} \times \mathbf{j} = \mathbf{k} \times \mathbf{k} = 0$, and

$$\mathbf{i} \times \mathbf{j} = \mathbf{k},\ \mathbf{j} \times \mathbf{k} = \mathbf{i},\ \mathbf{k} \times \mathbf{i} = \mathbf{j},\ \mathbf{j} \times \mathbf{i} = -\mathbf{k},\ \mathbf{k} \times \mathbf{j} = -\mathbf{i},\ \mathbf{i} \times \mathbf{k} = -\mathbf{j}$$

You might find a pneumonic helpful: If the unit vectors appear in the cyclic order **i j k i j ...**, then the product is positive.

EXAMPLE 5

Find the vector product of $\mathbf{A} = -2\mathbf{i} + 3\mathbf{j} + 4\mathbf{k}$ m and $\mathbf{B} = \mathbf{i} - 2\mathbf{j} - 5\mathbf{k}$ m.

Solution:

In terms of the components:

$$\mathbf{A} \times \mathbf{B} = (-2\mathbf{i} + 3\mathbf{j} + 4\mathbf{k}) \times (\mathbf{i} - 2\mathbf{j} - 5\mathbf{k})$$

$$= (+4\mathbf{i}\times\mathbf{j} + 10\mathbf{i}\times\mathbf{k}) + (3\mathbf{j}\times\mathbf{i} - 15\mathbf{j}\times\mathbf{k}) + (4\mathbf{k}\times\mathbf{i} - 8\mathbf{k}\times\mathbf{j})$$

$$= (4\mathbf{k} - 10\mathbf{j}) + (-3\mathbf{k} - 15\mathbf{i}) + (4\mathbf{j} + 8\mathbf{i}) = -7\mathbf{i} - 6\mathbf{j} + \mathbf{k}\ \text{m}^2$$

SOLUTIONS TO SELECTED TEXT EXERCISES AND PROBLEMS

Exercise 15.

Given two vectors $\mathbf{A} = 2\mathbf{i} - 3\mathbf{j} + \mathbf{k}$ m and $\mathbf{B} = -\mathbf{i} + 2\mathbf{j} - \mathbf{k}$ m, find: (a) $\mathbf{R} = \mathbf{A} + \mathbf{B}$; (b) R; (c) $\hat{\mathbf{R}}$

Solution:

(a) $\mathbf{R} = \mathbf{i} - \mathbf{j}$ m; (b) $R = (2)^{1/2} = 1.41$ m;
(c) $\hat{\mathbf{R}} = \mathbf{R}/R = 1/(2)^{1/2}\,(\mathbf{i} - \mathbf{j})$ m $= 0.707\mathbf{i} - 0.707\mathbf{j}$ m

Exercise 17.
The vector **A** has a magnitude of 6 m and vector **B** has a magnitude of 4 m. What is the angle between them if the magnitude of their resultant is (a) the maximum possible; (b) the minimum possible; (c) 3 m; and (d) 8 m. Do each part graphically and by components. (Let **A** lie along the x axis.)

Solution:
(a) 0, (b) 180°,
(c) First draw **A** to scale. Then draw an arc of radius 4 m with its center at the tip of **A** and an arc of radius 3 m with its center at the tail of **A**, see Fig. 2.6a. The point of intersection of the arcs allows you to complete the vector triangle. In terms of the components:

$$(6 + 4\cos\theta)^2 + (4\sin\theta)^2 = 9$$

so, $\cos\theta = -0.896$, and $\theta = 154°$;

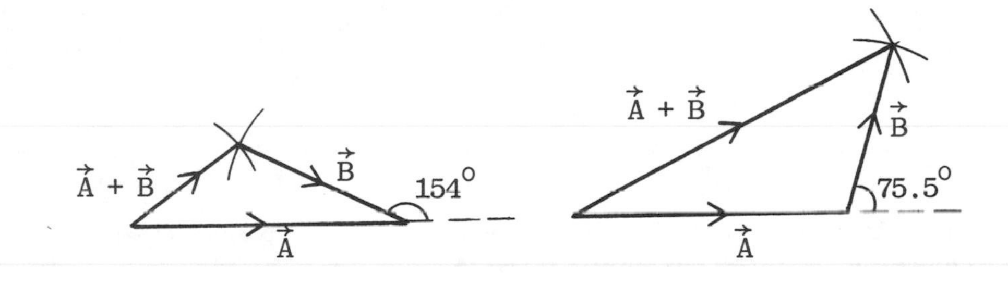

FIGURE 2.6

(d) In this case the arc centered at the tail of **A** has a radius of 8 m, see Fig. 2.6b. In terms of the components:

$$(6 + 4\cos\theta)^2 + (4\sin\theta)^2 = 64$$

so, $\cos\theta = 0.25$, and $\theta = 75.5°$.

Exercise 21.

A ship sails from a point at a distance of 4 km and a bearing of 40° N of E relative to a lighthouse to a point 6 km at 60° N of W. (a) What is its displacement? (b) What is the least distance between them?

Solution:

(a) The initial position of the ship is given by **A** = 3.06**i** + 2.57**j** km; and its final position is **B** = -3**i** + 5.2**j** km, see Fig. 2.7. Its displacement is **B** - **A** = = -6.06**i** + 2.63**j** km, or 6.61 km at 23.5° N of W.

(b) The least distance d between the ship and the lighthouse is the perpendicular distance from the origin to **B** - **A**.

$$d = A\sin(63.5°) = 3.58 \text{ km}.$$

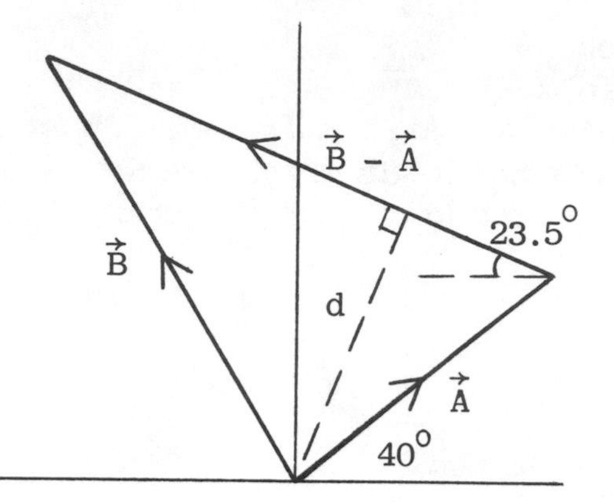

FIGURE 2.7

Exercise 37.

Given that vector **A** = 2**i** + 3**j** m, find the vector **B** of length 5 m that is perpendicular to **A** and lies in the following planes: (a) the xz plane; (b) the xy plane.

Solution:

(a) Since **A** is in the xy plane, we find at once that **B** = ±5**k**;

(b) Since **A** is at 56.3° to the x axis, **B** must be at 56.3° to the y axis, as in Fig. 2.8. This tells us that $B_y/B_x = \pm 2/3$:

$$\mathbf{B} = p(\pm 3\mathbf{i} \mp 2\mathbf{j})$$

The magnitude of this vector is $B = (13)^{1/2}p$, but we know that this should be 5, thus $p = 5/(13)^{1/2}$ and

$$\mathbf{B} = \mp 4.16\mathbf{i} \pm 2.77\mathbf{j} \text{ m}.$$

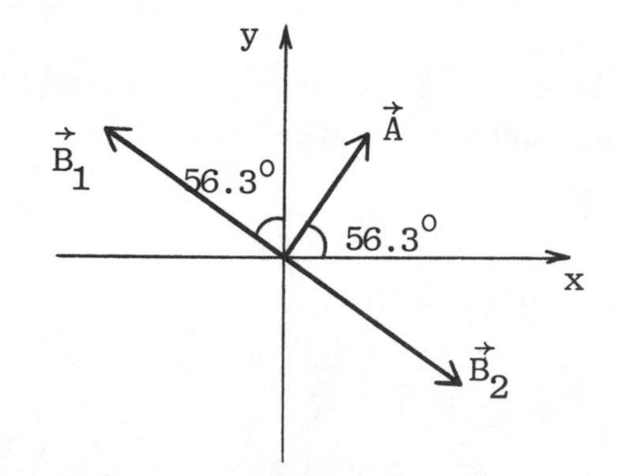

FIGURE 2.8

Exercise 45.
Show that the angles α, β, and γ between a vector **A** and the x, y, and z axes respectively, are given by

$$\cos\alpha = \mathbf{A.i}/A; \qquad \cos\beta = \mathbf{A.j}/A; \qquad \cos\gamma = \mathbf{A.k}/A$$

If **A** = 3**i** + 2**j** + **k**, find the angle between **A** and each axis.

Solution:
We know that

$$\mathbf{A.i} = A_x = A\cos\alpha,$$

thus $\cos\alpha = \mathbf{A.i}/A$, etc.

$\cos\alpha = \mathbf{A.i}/A = 3/(14)^{1/2}$, thus $\alpha = 36.7°$;
$\cos\beta = \mathbf{A.j}/A = 2/(14)^{1/2}$, thus $\beta = 57.7°$;
$\cos\gamma = \mathbf{A.k}/A = 1/(14)^{1/2}$, thus $\gamma = 74.5°$.

Exercise 53.
Vector **A** has a magnitude of 4 m and lies in the xy plane directed at 45° counter-clockwise from the +x axis, whereas **B** has a magnitude of 3 m and lies in the yz plane directed at 30° clockwise form the +z axis, see Fig. 2.9. Find **A x B**.

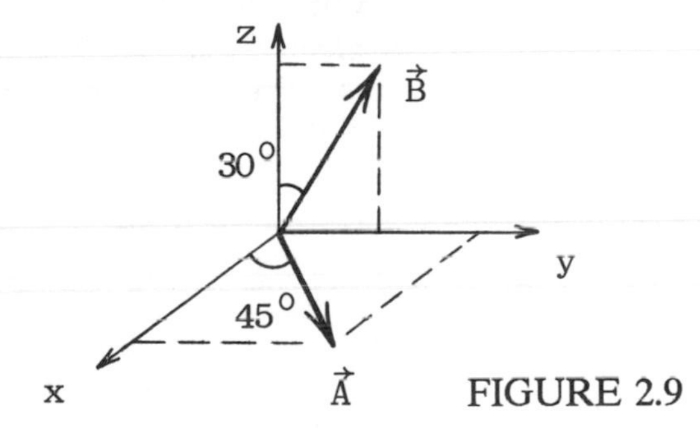

FIGURE 2.9

Solution:
The expression AB sinθ is not helpful since it would be difficult to determine the angle θ between **A** and **B**. Instead, we first find the components, **A** = 2.83**i** + 2.83**j** m, **B** = 1.5**j** + 2.6**k** m. Then,

$$\mathbf{A \times B} = 7.36\mathbf{i} - 7.36\mathbf{j} + 4.25\mathbf{k}\ \text{m}^2.$$

Problem 1.
Find a vector of length 5 m in the xy plane that is perpendicular to **A** = 3**i** + 6**j** - 2**k** m. (Hint: Consider the dot product.)

Solution:
Since **A.B** = 0, we have $3B_x + 6B_y = 0$, i.e. $B_x = -2B_y$, which means

$$\mathbf{B} = p(\pm 2\mathbf{i} \mp \mathbf{j}),$$

where p is a constant. Since $B = p(5)^{1/2} = 5$ m, it follows that $p = (5)^{1/2}$, thus $\mathbf{B} = \pm 4.46\mathbf{i} \mp 2.23\mathbf{j}$ m

Problem 11
A three-dimensional vector **A** has a length of 10 m and makes the angles 65° and 40° with the +x and +z axes, respectively. Find the magnitudes of its Cartesian components.

Solution:
The x and y components are simply

$$A_x = 10 \cos 65° = 4.23 \text{ m}, \quad A_z = 10 \cos 40° = 7.66 \text{ m},$$

From the Pythagorean theorem in three dimensions:

$$A_y = (A^2 - A_x^2 - A_z^2)^{1/2} = 4.84 \text{ m}.$$

SELF-TEST

1. The magnitudes of the two vectors in Fig. 2.10. are A = 2 m and B = 3 m.
   (a) Express **A** and **B** in unit vector notation.
   (b) Find the vector **C** such that **A** - **B** + **C** = 0.
   (c) Draw a vector diagram to illustrate part (b).

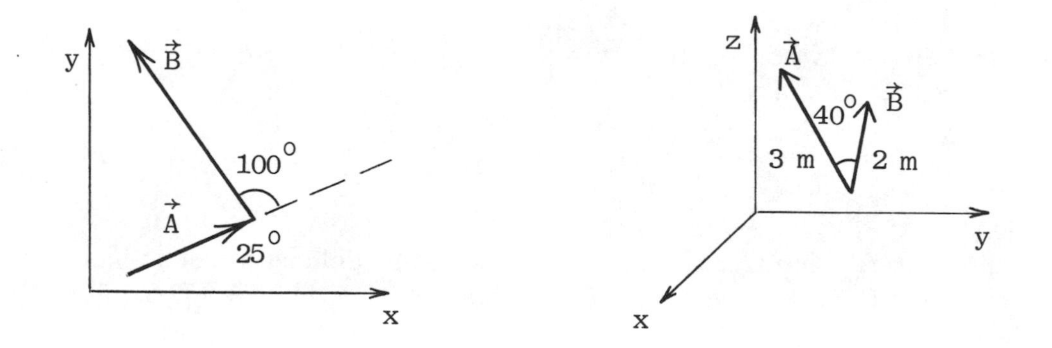

FIGURE 2.10

FIGURE 2.11

2. Vectors **A** and **B** lies in the yz plane as shown in Fig. 2.11. Determine (a) **A.B**; (b) **AxB**.

3. Given **A** = 2**i** - 3**j** + **k**, **B** = -**i** + 2**j** + 4**k**, and **C** = 5**k**, determine each of the following if it is an allowed operation:

   (a) **(A.B)C**; (b) **(A x B).C**; (c) **C(A x B)**; (d) **C x (A.B)**

Chapter 3

ONE-DIMENSIONAL KINEMATICS

MAJOR POINTS

1. The definitions of speed, velocity, and acceleration
2. The distinction between average and instantaneous values
3. (a) The use of tangents to obtain instantaneous values
   (b) The use of areas to obtain displacements and changes in velocity.
4. Vertical free-fall

CHAPTER REVIEW

The subject of kinematics is concerned with the description of the motion of a particle. One-dimensional kinematics is restricted to motion along a line such as the x axis.

POSITION, DISPLACEMENT AND DISTANCE

The position, x, of a particle is specified by a coordinate, a point on an axis. The displacement of a particle is defined as a change in position,

$$\Delta x = x_f - x_i$$

The sign of the displacement indicates its direction relative to the chosen +x axis. The distance traveled by a particle is the length of the actual path taken. Distance is always a positive quantity. In general, its value is not equal to the magnitude of the displacement. For example, if a particle returns to its initial position, its net displacement is zero, but the distance traveled is not zero.

SPEED AND VELOCITY

Both speed and velocity tell us how fast an object has moved, but they have different definitions:

$$\text{Average speed} = \text{Distance traveled/Time interval}$$

This is always a positive quantity, determined by the actual path. In contrast,

$$\text{Average velocity} = \text{Displacement/Time interval}$$

$$v_{av} = \Delta x/\Delta t$$

Average velocity depends on the initial and final positions, not on the length of the path taken. Graphically, the average velocity over some time interval is given by the slope of the line joining the initial and final points on a position-time (x vs t) graph, as shown in Fig. 3.1.

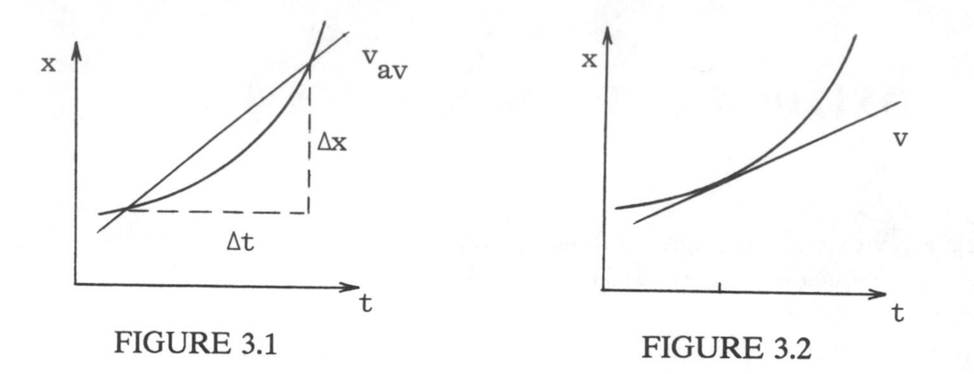

FIGURE 3.1 FIGURE 3.2

Whereas average velocity is defined over a finite time interval, Δt, the instantaneous velocity v of a particle at a given point in time is defined as the rate of change, or derivative of the position with respect to time:

$$v = dx/dt$$

This is called the derivative of x with respect to t and may be determined graphically from the slope of the tangent to the x vs t graph at the given time (see Fig. 3.2).

ACCELERATION

The average acceleration for some time interval Δt and the instantaneous acceleration at some instant are defined in terms of velocity, not in terms of speed:

$$a_{av} = \Delta v/\Delta t; \qquad a = dv/dt$$

The SI unit of acceleration is m/s^2, however you may come across mixed units such as 4 m.p.h/s, which means that the velocity of the body changes by 4 m.p.h. in each second. The sign of the acceleration is determined by its direction relative to the chosen +x axis, not simply by whether the particle has speeded up or slowed down.

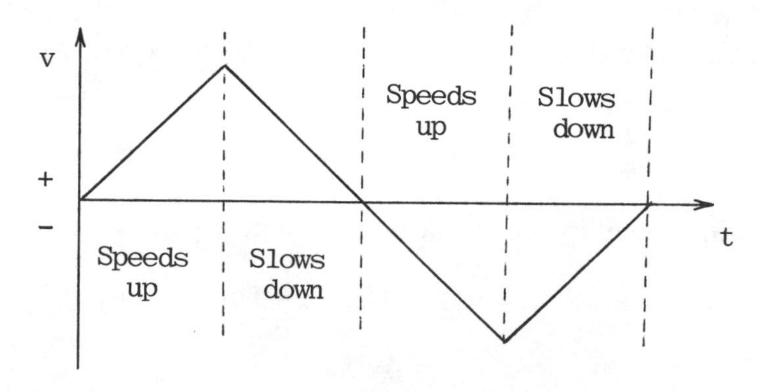

FIGURE 3.3

Figure 3.3 shows a v vs t graph. The acceleration is given by the slope of the lines. When v and a have the same sign they are in the same direction and the particle is speeding up. When v and a have opposite signs, they are in opposite directions and the particle is slowing down. Note in particular that a negative acceleration does not necessarily mean that the particle is slowing down.

Even though you may be comfortable with other expressions, such as v = d/t or a = v/t, it is important that you stop using them. They are misleading. The quantity d is often taken to be distance rather than displacement, and acceleration is defined not in terms of velocity but in terms of the change in velocity.

EQUATIONS OF KINEMATICS (Constant acceleration)

By starting with the definitions of average velocity and average acceleration one can obtain several useful relations if we assume that the acceleration is constant (in magnitude and direction). The term "uniform" is often used instead of "constant".

$$v = v_o + a\,t \tag{3.1}$$

$$x = x_o + v_o\,t + 1/2\,a\,t^2 \tag{3.2}$$

$$v^2 = v_o^2 + 2\,a\,(x - x_o) \tag{3.3}$$

Note carefully the meaning of the symbols: x is a position, not a distance or a displacement and v is a velocity (actually the x component of the velocity), not a speed. Of the five quantities x, v_o, v, a and t you need three in order to solve any problem. The value of the initial position x_o is determined by the choice of origin of the coordinate system.

EXAMPLE 1

Given the v vs t graph of Fig. 3.4a, plot the corresponding a vs t and x vs t graphs. Take the position x = 0 at t = 0.

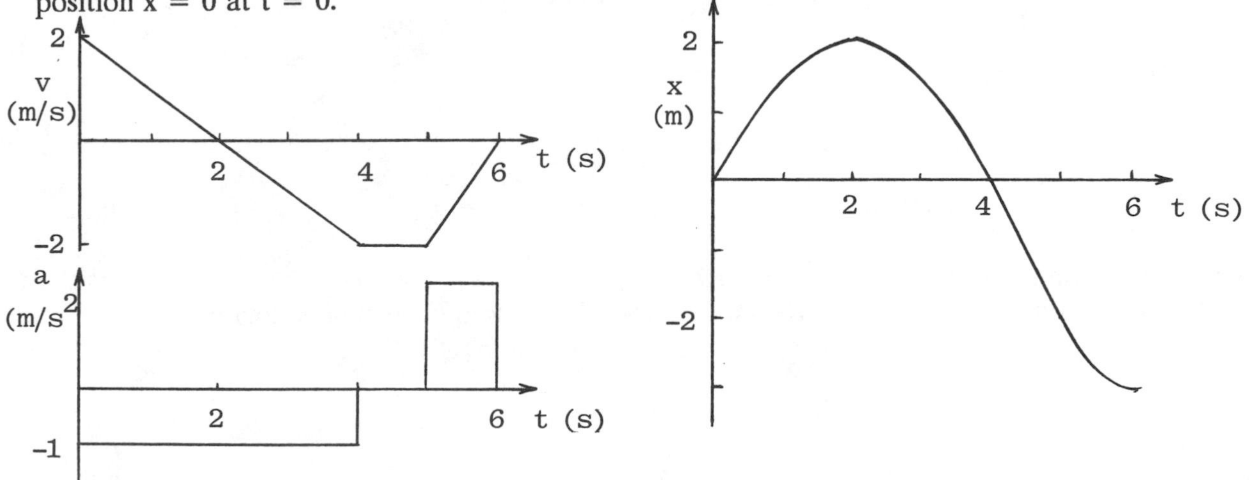

FIGURE 3.4

Solution:

Since $a = dv/dt$, we find the acceleration by taking the slope of the tangent to the v vs t graph. In this case the function is a straight line so we do not need to draw tangents. Between 0 s and 4 s the v vs t graph is a straight line so it has a single slope: -1 m/s^2. Between 6 s and 7 s, the slope of the line is +2 m/s^2. The a vs t graph is drawn in Fig. 3.4b. Note that at $t = 2$ s, the instantaneous velocity $v = 0$, but a is not zero.

Recall that the area between the v vs t graph and the t axis gives the displacement (including the sign) in a given time interval. To plot the x vs t graph, we must determine the displacements in several time intervals. The position at any time is the sum of the displacements up to that time. In Fig. 3.4a we have indicated the areas within 1 s intervals. At $t = 1$ s, the displacement $\Delta x = x_f - x_i = 3$ m. Since $x_i = 0$, we have $x_1 = 3$ m. Between 1 s and 2 s, $\Delta x = 1$ m, so $x_2 = 3 + 1 = 4$ m. Between 2 s and 3 s, $\Delta x = -1$ m, therefore $x_3 = 4 - 1 = 3$ m. Between 3 s and 4 s, $\Delta x = -3$ m, thus $x_4 = 3 - 3 = 0$. Similarly, you should verify that $x_5 = -4$ m and $x_6 = -5$ m.
The x vs t graph is plotted in Fig. 3.4c.

Notice that when the v vs t graph varies linearly, the x vs t graph has a parabolic shape. When v is constant, the x vs t graph is linear. At $t = 0$ and 6 s, when $v = 0$, the slope of the x vs t graph is zero.

EXAMPLE 2

A car accelerates uniformly from 10 m/s to 30 m/s in 10 s. (a) What is the distance traveled? (b) How long does it take to reach the midpoint in space? (c) Where is it at the midpoint in time?

Solution:

(a) Our equations deal with displacement not distance, but in one dimensional motion in a fixed direction, the distance equals the magnitude of the displacement. We choose the origin such that $x_o = 0$.
GIVEN: $v_o = 10$ m/s, $v = 30$ m/s, $t = 10$ s. UNKNOWN: a, x
(a) From Eq. 3.1, the acceleration is

$$a = (v - v_o)/t = (30 \text{ m/s} - 10 \text{ m/s})/(10 \text{ s}) = +2 \text{ m/s}^2$$

To find the displacement Δx, we may use either Eq. 3.2 or 3.3:

$$v^2 = v_o^2 + 2\, a\, \Delta x$$

$$(30)^2 = (10)^2 + 2\,(2)\, \Delta x$$

Thus $\Delta x = 200$ m. (Confirm this by using Eq. 3.2.)
(b) At the midpoint in space $\Delta x = 100$ m. Since we do not know the final velocity, Eq. 3.1 is not suitable. From Eq. 3.2:

$$\Delta x = v_o\, t + 1/2\, a\, t^2$$

$$100 = (10)t + 1/2(2)\, t^2$$

$$t^2 + 10t - 100 = 0$$

Using the quadratic formula,

$$t = [-10 + (10^2 + 4x100)^{1/2}]/2$$

$$= 6.2 \text{ s}, -16.2 \text{ s}.$$

Thus t = 6.2 s.
(c) The midpoint in time is t = 5 s, so

$$\Delta x = v_o\, t + 1/2\, a\, t^2$$

$$= (10)(5) + 1/2\, (2)(5)^2 = 75 \text{ m}.$$

EXAMPLE 3
The driver of a car moving at 30 m/s suddenly sees a truck 60 m ahead moving at 10 m/s in the same direction. The maximum deceleration of the car is 5 m/s². (a) Is there a collision? (b) If not, what is the distance of closest approach?

Solution:
(a) We choose the origin at the initial position of the car, as shown in Fig. 3.5. We first write expression for the positions of the car and truck as functions of time.

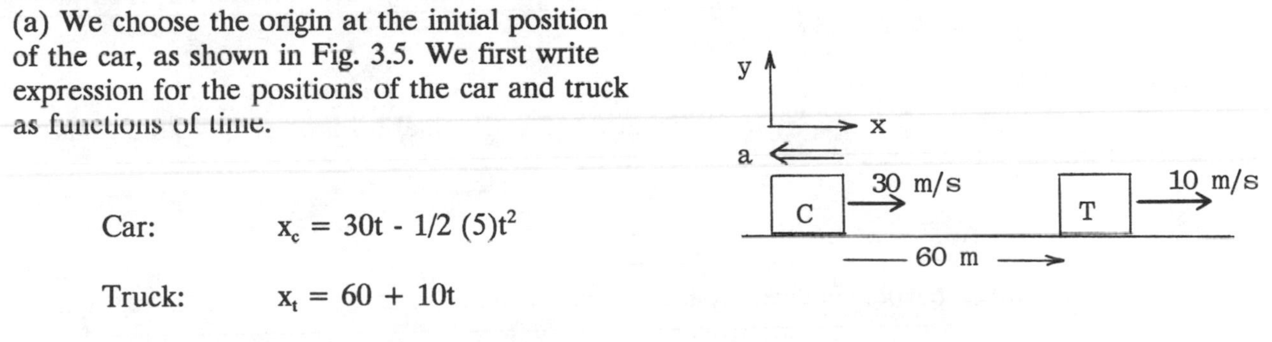

Car: $x_c = 30t - 1/2\, (5)t^2$

Truck: $x_t = 60 + 10t$

FIGURE 3.5

They will collide if $x_c = x_t$. On equating these expressions and dividing by 2.5 we find:

$$t^2 - 8t + 24 = 0$$

Considering the quadratic formula, we note that $b^2 < 4ac$ ($8^2 < 4x24$), which means there is no real solution to the quadratic equation. The car and truck do not collide.
(b) As long as the velocity of the car is greater than that of the truck, the distance between them will continue to diminish. Once the velocity of the car goes below 10 m/s, the separation will increase. Thus, the distance of closest approach occurs when they have the same velocity, 10 m/s.

$v_c = 30 - 5t$; $v_t = 10$

On setting $v_c = v_t$ we find $t = 4$ s.
At this time the positions are:

$$x_c = 30(4) - 2.5(4^2) = 80 \text{ m};$$

$$x_t = 60 + 10(4) = 100 \text{ m}.$$

Thus the minimum separation is 20 m. The x vs t graphs are sketched in Fig. 3.6

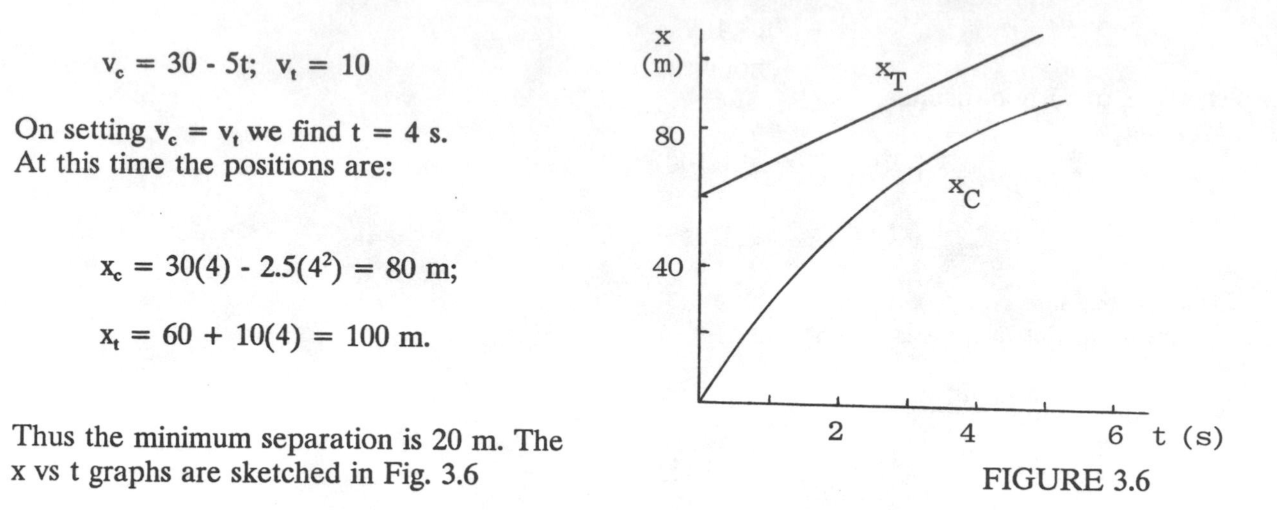

FIGURE 3.6

EXAMPLE 4 (May be omitted if you have just started calculus.)
The position of a particle is given by

$$x = 7 - 2t + 3t^2 \text{ m}$$

Find: (a) The average velocity between 2 s and 3 s. (b) The average acceleration between 2 s and 3 s. (c) The acceleration at 2 s.

Solution:
(a) At 2 s, the initial position is $x_i = 15$ m and at $t = 3$ s, the final position is $x_f = 28$ m. Since $\Delta t = 1$ s, the average velocity is

$$v_{av} = (x_f - x_i)/\Delta t = +13 \text{ m/s}$$

(b) The velocity of the particle is given by the derivative of x:

$$v = dx/dt = -2 + 6t \text{ m/s}$$

At 2 s, $v_i = 10$ m/s and at 3 s, $v_f = 16$ m/s, therefore the average acceleration is

$$a_{av} = (v_f - v_i)/\Delta t = +6 \text{ m/s}^2$$

(c) The acceleration is given by the derivative of v:

$$a = dv/dt = +6 \text{ m/s}^2$$

The acceleration is constant.

VERTICAL FREE FALL

The term free-fall refers to motion solely under the influence of gravity. The motion can be upward, downward or even orbital motion. In the absence of air resistance, the acceleration due to gravity has the same value for all bodies. Near the surface of the earth its magnitude is approximately $g = 9.8\ m/s^2$. (Note that g is a positive quantity.)

It is convenient to switch from x to y when dealing with vertical motion. The sign of the acceleration is determined by the orientation of the axes. It is advisable to stick to just one set of axes, with the y axis pointing upward, as in Fig. 3.7.

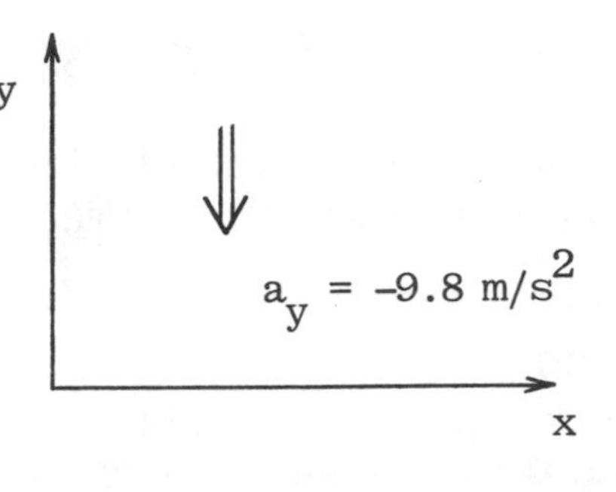

FIGURE 3.7

The acceleration has the same value whether the body is moving upward or downward; its sign does not change. The equations of kinematics are expressed in the text in terms of $a = -g$ where $g = 9.8\ m/s^2$. However, it often happens that students insert $g = -9.8\ m/s^2$ by mistake. To avoid this possibility, we express these equations solely in terms of a.

EXAMPLE 5

A particle is thrown up at 28 m/s from the ground. Find: (a) Its velocity at a height of 20 m. (b) When it is at 12 m. (c) When its speed is 15 m/s.

Solution:

We pick the origin at ground level with the y axis pointing up, therefore $a = -9.8\ m/s^2$.

(a) GIVEN: $v_o = 28$ m/s, $y = 20$ m, $y_o = 0$. UNKNOWN: v, t. From Eq. 3.3:

$$v^2 = v_o^2 + 2\,a\,(y - y_o)$$

$$= (28)^2 + 2(-9.8)(20 - 0) = 392\ m^2/s^2$$

Thus $v = \pm 19.8$ m/s. There are two acceptable solutions, indicating different directions of motion.

(b) GIVEN: $v_o = 28$ m/s, $y = 12$ m. UNKNOWN: v, t. From Eq. 3.2:

$$y = y_o + v_o\,t + 1/2\,a\,t^2$$

$$12 = 0 + (28)t + 1/2(-9.8)t^2$$

or,

$$4.9t^2 - 28t + 12 = 0$$

thus,

$$t = [28 + (28^2 - 4x4.9x12)^{1/2}]/(9.8)$$

$$= 5.24\ s,\ 0.47\ s.$$

Both solutions are acceptable; they correspond to opposite directions of motion.
(c) The question is phrased in terms of speed. This means that the velocity can be $v = +15$ m/s or -15 m/s. From $v = v_o + at$, we find

$$15 = 28 - 9.8t, \qquad \text{so} \qquad t = 1.33 \text{ s}$$

$$-15 = 28 - 9.8t, \qquad \text{so} \qquad t = 4.39 \text{ s}$$

In each of the parts of this example there were two possible answers. You should always read a question with this possibility in mind.

EXAMPLE 6
A person is at the top of a building of height 100 m. Ball A is thrown upward at 5 m/s and ball B is thrown downward at 20 m/s two seconds later. (a) Where and when do the balls meet? (b) What are their velocities when they meet?

Solution:
The coordinate axes and initial values are shown in Fig. 3.8. If ball A is in flight for t seconds, then ball B is in flight for (t - 2) seconds. The positions of the balls as functions of time are given by:

$$y_A = 100 + 5\,t - 4.9\,t^2$$

$$y_B = 100 - 20(t - 2) - 4.9(t - 2)^2$$

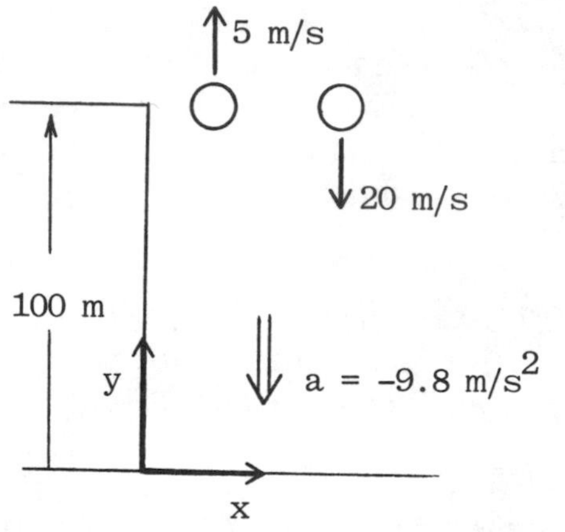

FIGURE 3.8

The balls meet when $y_A = y_B$. This condition leads to $5.4t = 20.4$ or $t = 3.78$ s.
(b) The velocities of the balls are found from $v = v_o + at$:

$$v_A = 5 - 9.8t = -32.0 \text{ m/s}$$

$$v_B = -20 - 9.8(t - 2) = -37.4 \text{ m/s}$$

EXAMPLE 7

A flower pot falls off a balcony and takes 0.12 s to go past a window on a lower floor that is 1.4 m high. From what height above the window sill did the pot fall?

Solution:

Let us say that the balcony is at a distance h above the bottom of the window, as shown in Fig. 3.9. The origin has been chosen at the sill (bottom). The initial position of the pot is therefore $y_o = h$. If it takes time t to reach the top of the window, it takes time (t + 0.12 s) to reach the sill.

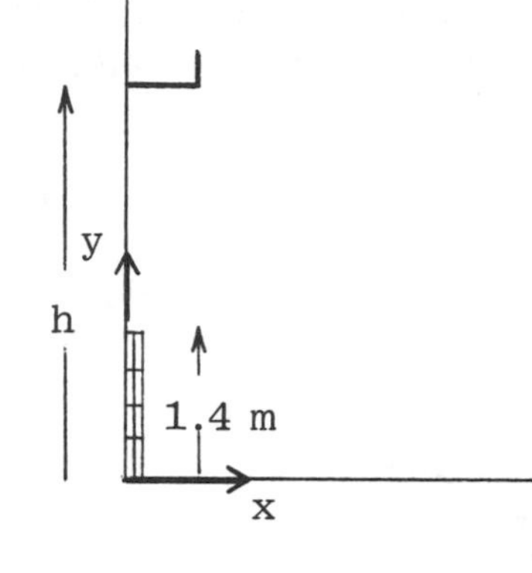

FIGURE 3.9

Since $v_o = 0$, the positions of the pot at the top and the bottom of the window are given by:

$$y_T = 1.4 = h - 4.9t^2 \qquad \text{(i)}$$

$$y_B = 0 = h - 4.9(t + 0.12)^2 \qquad \text{(ii)}$$

Subtracting (ii) from (i) we find:

$$1.4 = 4.9(0.24t + 0.0144)$$

which yields t = 1.13 s. On substituting this into (i) we find $h = 1.4 + 4.9t^2 = 7.65$ m.

SOLUTIONS TO SELECTED TEXT EXERCISES AND PROBLEMS

Exercise 9

Consider a 500-km car race on a 10-km track. Car A finishes the race in 4 h and is 1.5 laps ahead of B at this time. What is B's time for the race?

Solution:

We assume that the speed of car B is constant. Car B covers (500 km - 15 km) = 485 km when A finishes, thus its speed is

$$v_B = 485 \text{ km}/4 \text{ h} = 121.25 \text{ km/h}$$

The time required to complete 500 km is therefore

$$T_B = 500\ \text{km}/v_B = 4.124\ \text{h} = 4\ \text{h}\ 7\ \text{min}\ 25\ \text{s}$$

Exercise 25
From the v versus t graph of Fig. 3.10 find: (a) the time(s) at which the particle is at rest; (b) at what time, if any, the particle reverses the direction of its motion; (c) the average acceleration between 1 and 4 s; (d) the instantaneous acceleration at 3 s.

Solution:
(a) From the graph we see that $v = 0$ at 0 and 3 s;

(b) The velocity changes from negative to positive, indicating a change in direction, at 3 s.

(c) The average acceleration is

$$a_{av} = [4\ \text{m/s} - (-2\ \text{m/s})]/3\ \text{s} = 2\ \text{m/s}^2$$

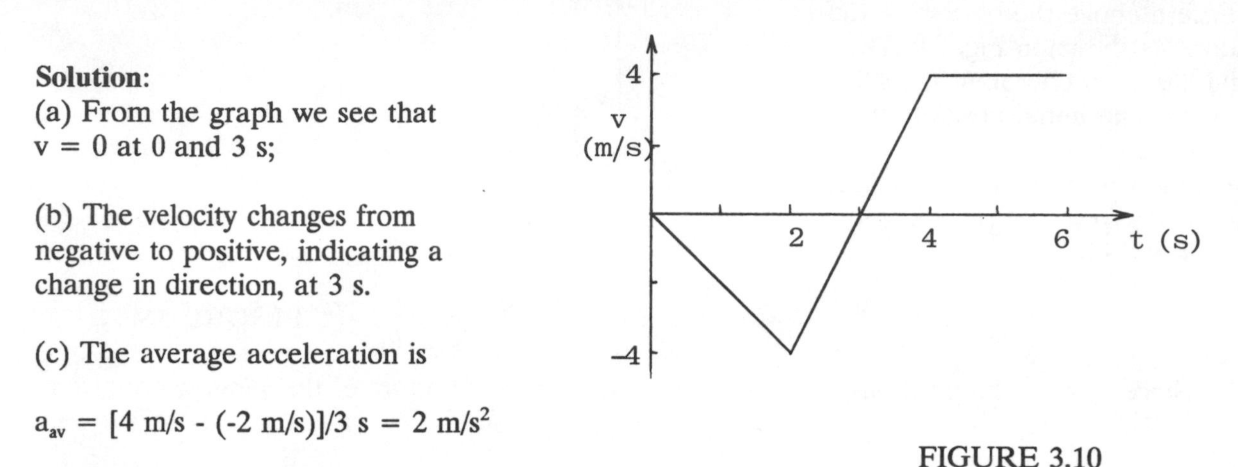

FIGURE 3.10

(d) The fact that the instantaneous velocity is zero at 3 s does not mean that the instantaneous acceleration is zero at this time. The slope of the line is $a = 4\ \text{m/s}^2$.

Exercise 29
Use the v versus t graph of Fig. 3.11 to estimate: (a) the average velocity for the first 6 s; (b) the average speed for the first 6 s.

Solution:
(a) We need to find the net displacement for the first 6 s. This is done by adding the areas between the function and the t axis - keeping in mind that areas below the axis are negative. The net displacement for 6 s is 15 m. Thus the average velocity is 15 m/6 s = +2.5 m/s.

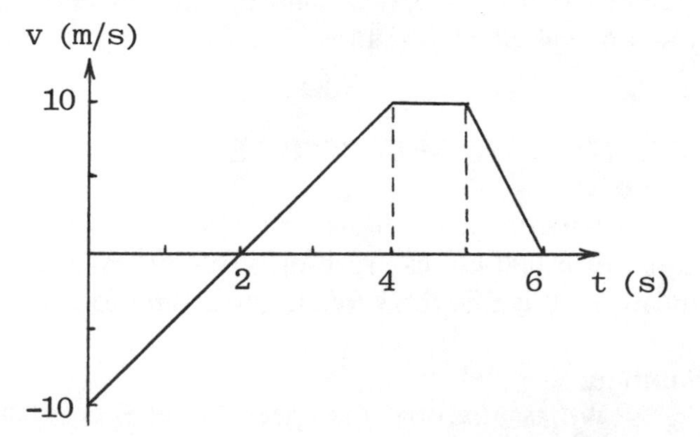

(b) To find the average speed we need the distance traveled. In this case we add up the areas but take all of them to be positive. The total distance traveled is 35 m, thus the average speed is 35 m/6 s = 5.83 m/s.

FIGURE 3.11

Exercise 45
The driver of a truck moving at 30 m/s suddenly notices a moose 70 m straight ahead. If the driver's reaction time is 0.5 s and the maximum deceleration is 8 m/s^2, can he avoid hitting the moose without steering to one side?

Solution:

In 0.5 s, the truck moves 15 m. We choose the origin at this point, so the initial position of the moose is (70 m - 15 m) = 55 m. Let us see how far the truck travels before coming to a stop. Since the final velocity is zero and $a = -8$ m/s^2, then $v^2 = v_o^2 + 2a\,\Delta x$ leads to

$$\Delta x = -v_o^2/2a = (30 \text{ m/s})^2/2(8 \text{ m/s}^2) = 56.25 \text{ m}.$$

The moose would be hit. The driver needs a further 1.25 m.

Exercise 69
An object is fired vertically up from the ground. Find its maximum height and the time of flight given that it loses 30% of its initial speed in 1.8 s and is moving upward.

Solution:

Since the ball loses 30% of its initial speed, it has 70% remaining, that is $v = 0.7v_o$. From $v = v_o + at$, we have

$$0.7v_o = v_o - (9.8 \text{ m/s}^2)(1.8 \text{ s})$$

thus, $v_o = 58.8$ m/s. At the maximum height H, we know $v = 0$. From $v^2 = v_o^2 + 2a(H - 0)$, with $a = -g$, we have

$$H = v_o^2/2g = 176 \text{ m}$$

Since velocity on landing is $v = -v_o$, from $v = v_o + at = v_o - gt$, we deduce that the time of flight is $t = 2v_o/g = 12$ s

Problem 13
A car starts from rest and accelerates uniformly for 200 m. It moves at constant speed for 160 m and then decelerates to rest in 50 m. The whole trip takes 33 s. How long did it move at constant speed? (Hint: Draw the v versus t graph.)

Solution:

Let us say that that car accelerates for a period t_1 and travels at constant speed for a period t_2. The deceleration is four times larger than the acceleration so the final deceleration period is $t_1/4$. The v vs t graph is shown in Fig. 3.12. The maximum velocity is v.

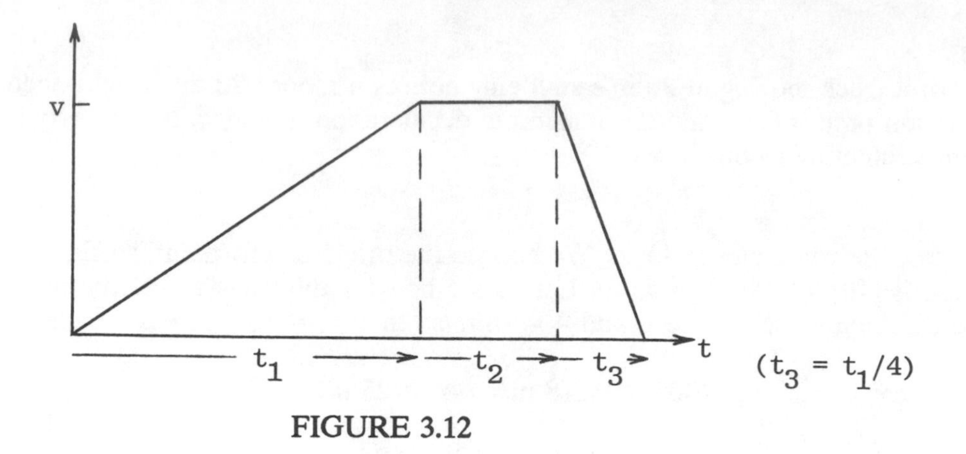

FIGURE 3.12

In the first time interval t_1 the displacement is the area of the triangle,

$$(vt_1)/2 = 200 \text{ m}$$

which gives us $v = 400/t_1$. The time t_2 during which the speed is constant is $t_2 = (33 - t_1 - t_1/4)$. The area of the rectangle is

$$v(33 - 5t_1/4) = 160 \text{ m}$$

On substituting $v = 400/t_1$ into this equation we find $t_1 = 20$ s, and then $t_2 = 8$ s.

Problem 23
A climber can estimate the height of a cliff by dropping a stone and noting the time at which he hears the impact on the ground. Suppose this time is 2.5 s. Find the height of the cliff under the following conditions: (a) by assuming the speed of sound is large enough to be ignored; (b) by taking the speed of sound to be 330 m/s.

Solution:

(a) We choose the origin at the bottom of the cliff and the axes as shown in Fig. 3.13. With $y_o = H$ and $y = 0$, the equation $y = y_o + v_o t + 1/2\, at^2$ gives us

$$0 = H - 4.9t^2, \text{ so } H = 30.6 \text{ m}$$

(b) If t_1 is the time taken by the stone to fall and t_2 is the time for the sound to travel back up, then

$$H = 4.9t_1^2; \quad \text{and} \quad H/330 = t_2$$

where $t_1 + t_2 = 2.5$ s. Equating these expressions for H:

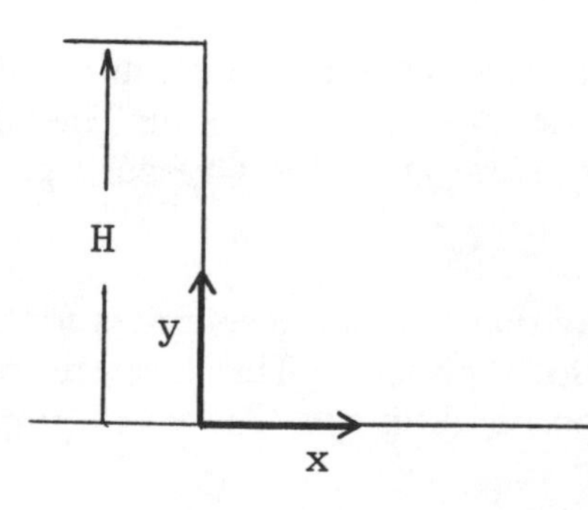

FIGURE 3.13

$$330(2.5 - t_1) = 4.9t_1^2$$

Solving the quadratic leads to $t_1 = 2.41$ s and then H = 28.5 m.

SELF-TEST

1. Consider the v-t graph shown Fig. 3.14. Find: (a) the instantaneous acceleration at 3 s; (b) the position at 6 s. Take x = 0 at t = 0. (c) The average speed between 2 s and 6 s. (d) During what interval(s) is the body speeding up?

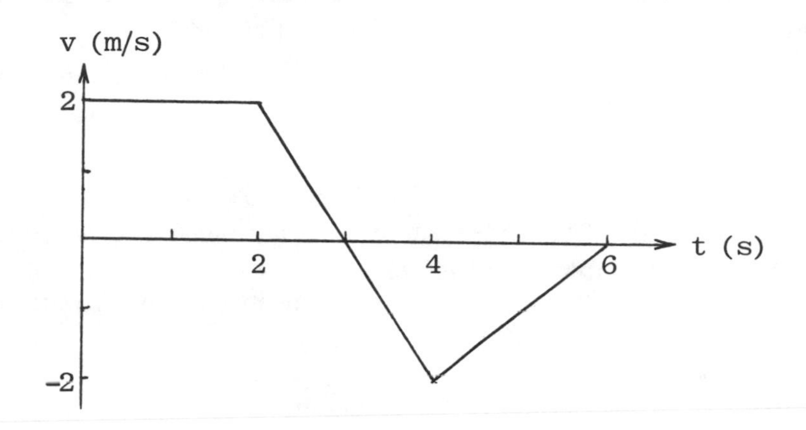

FIGURE 3.14

2. A car and a truck are initially at rest with the truck 14 m ahead of the car. At t = 0 s, the car accelerates at 4 m/s². At t = 1 s, the truck accelerates at 2 m/s² in the same direction.
   (a) Where and when does the car catch the truck?
   (b) What are their velocities at the meeting point?

3. A ball thrown vertically up from the roof of a building 40 m high lands 4 s later. (a) When is it 25 m above the ground? (b) What is the maximum height? (c) What is the velocity on landing? Use 10 m/s²

Chapter 4

INERTIA AND TWO-DIMENSIONAL MOTION

MAJOR POINTS

1. The concept of inertia; Newton's first law
2. In projectile motion, the horizontal and vertical motions are independent.
3. In uniform circular motion there is an inward, centripetal, acceleration
4. Determining the velocity of one body relative to another

CHAPTER REVIEW

INERTIA

The inertia of a body is its tendency to resist any change in its velocity. Thus a body at rest tends to stay at rest, whereas a moving body tends to maintain its velocity. Its resists speeding up, slowing down or changing the direction of its velocity. According to Newton's first law,

A body free of external influence will maintain a constant velocity.

A reference frame in which Newton's first law is valid is called an inertial reference frame. Any frame that moves at constant velocity relative to a known inertial frame is also an inertial frame. An accelerated frame, such as a rotating platform or a bus coming to a stop, is a noninertial frame.

PROJECTILE MOTION:

Projectile motion near the surface of the earth consists of two independent motions: A horizontal motion at constant speed and a vertical motion subject to the acceleration due to gravity. Figure 4.1 in the text illustrates the fact that the horizontal components of the displacements in each time interval are equal. Air resistance is ignored. With the y axis vertically upward we have

$$a_y = -9.8 \text{ m/s}^2; \qquad a_x = 0$$

The equations of kinematics are

$$x = v_{ox}\, t$$

$$v_y = v_{oy} + a\, t$$

$$y = y_o + v_{oy}\, t + 1/2\, a\, t^2$$

$$v_y^2 = v_{oy}^2 + 2\, a\, (y - y_o)$$

Note that the components v_{ox} and v_{oy}, rather than v_o, appear in these equations. If you write these equations in terms of a = -g, you must be careful that g = 9.8 m/s^2, a positive quantity.

If a ball is thrown at angle θ above the horizontal the components of the initial velocity are

$$v_{ox} = v_o \cos\theta; \qquad v_{oy} = v_o \sin\theta$$

Do not write $v_{ox} \cos\theta$. If θ is below the horizontal, then $v_{oy} = -v_o \sin\theta$.

EXAMPLE 1

A bullet is fired horizontally at 300 m/s at a target 180 m away at the same horizontal level. (a) How far below the target does the bullet strike? (b) At what angle to the horizontal should the bullet be fired?

Solution:

(a) We take the origin at the initial position of the bullet as in Fig. 4.1. We are given v_{ox} = 300 m/s and v_{oy} = 0. The horizontal and vertical positions at a later time t are given by

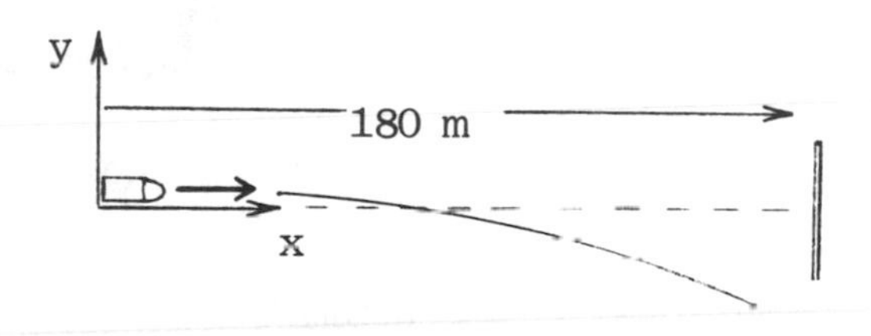

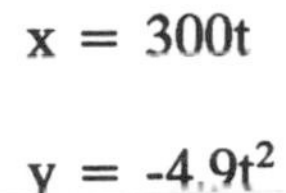

$$x = 300t$$

$$y = -4.9t^2$$

The bullet strikes the target at

$$t = (180 \text{ m})/(300 \text{ m/s}) = 0.6 \text{ s}$$

FIGURE 4.1

Thus its vertical position is $y = -4.9(0.6)^2 = -1.76$ m.

(b) In order for the bullet to strike the target, we must have y = 0 when x = 180 m. If θ is the initial angle above the horizontal, then

$$x = 300 \cos\theta t; \qquad y = 300 \sin\theta t - 4.9t^2$$

We have two equations in two unknowns, θ and t. Since x = 180 m, we substitute t = x/(300 cosθ) = 0.6/cosθ and y = 0 into the equation for y:

$$0 = \sin\theta/\cos\theta \qquad - 4.9(0.6/\cos\theta)^2$$

After removing a factor of cosθ and using the identity sin2θ = 2sinθcosθ, we find sin2θ = 0.882, which means θ = 61.9°.

EXAMPLE 2
A projectile is fired from ground level at 35° to the horizontal. At a later time its horizontal displacement is 40 m and its height is 20 m. Find: (a) The initial speed; (b) The velocity at the given point.

Solution:
Do not make the unwarranted assumption that we have been given the maximum height of the projectile. Start by writing expressions for the horizontal and vertical positions:

$$x = v_o \cos\theta\, t \qquad \text{(i)}$$

$$y = v_o \sin\theta\, t - 4.9t^2 \qquad \text{(ii)}$$

We are given x = 40 m, y = 20 m and θ = 35°. We substitute $t = 40/(v_o \cos\theta)$ into the equation for y:

$$20 = 40 \tan\theta - 4.9(40/v_o \cos\theta)^2$$

Solving this we find v_o = 38.2 m/s.
(b) From (i) we find the time taken to reach the given point:
$t = 40/(38.2 \cos 35°) = 1.28$ s. The components of the velocity at this time are

$$v_x = v_{ox} = v_o \cos 35° = 31.3 \text{ m/s}$$

$$v_y = v_{oy} + a\, t = v_o \sin 35° - 9.8\, t = 9.37 \text{ m/s}$$

Finally, $\mathbf{v} = 31.3\mathbf{i} + 9.4\mathbf{j}$ m/s.

UNIFORM CIRCULAR MOTION

Acceleration is defined in terms of the rate of change of velocity, not of speed. When a particle moves in a circle at constant speed its velocity is constantly changing because the direction of this vector is changing.
Thus the particle is accelerating even though its speed is constant.
The centripetal (inward) acceleration (Fig. 4.2) is

$$a = v^2/r$$

The period of the circular motion is the time required to complete one circle:

$$T = 2\pi r/v$$

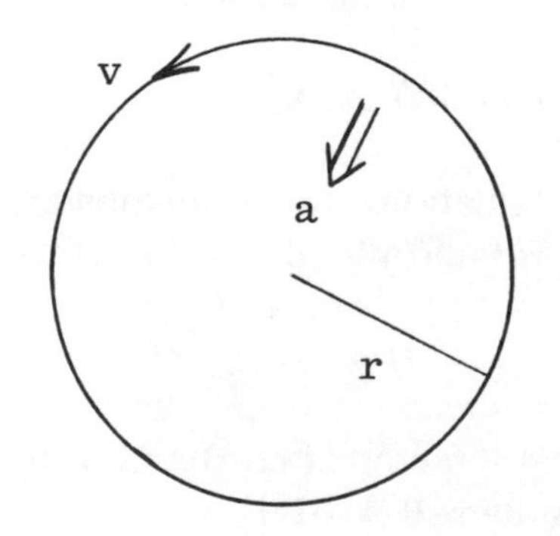

FIGURE 4.2

EXAMPLE 3
A stone moves at constant speed in a circle of radius 1.8 m. It makes 2.5 revolutions per second. What is its centripetal acceleration?

Solution:
The period of the motion is $T = 1/2.5$ s $= 0.4$ s, and the circumference is $2\pi r = 11.3$ m, thus the speed is

$$v = 2\pi r/T = 28.5 \text{ m/s}$$

The centripetal (inward) acceleration is

$$a = v^2/r = (28.5 \text{ m/s})^2/(1.8 \text{ m}) = 443 \text{ m/s}^2$$

RELATIVE MOTION

We sometimes need to find the velocity of one body relative to another that is also moving relative to some frame, such as the earth. The relationship between the velocities is clarified by the use of double subscripts:

$\mathbf{v}_{AB}$	=	$\mathbf{v}_{AC}$	+	$\mathbf{v}_{CB}$
Velocity of A relative to B		Velocity of A relative to C		Velocity of C relative to B

Note carefully the order of the subscripts. We assume that the reference frames are inertial (there is no acceleration or rotation of the axes) and that the rectangular axes are parallel. In general,

$$\mathbf{v}_{BA} = -\,\mathbf{v}_{AB}$$

You should begin each problem using the double subscripts, then switch to single-letter notation if you wish.

EXAMPLE 4
A pilot has to reach a destination 200 km northeast. The airspeed of the plane is 180 km/h. If there is a wind blowing at 40 km/h due south. (a) What is the required heading of the plane? (b) How long does the trip take?

Solution:
In such problems it is important to start with the double subscript notation. We use P for the plane, A for the air and G for the ground. We choose east and north as the x and y axes. We are given $|\mathbf{v}_{PA}| = 180$ km/h, $\mathbf{v}_{AG} = -40\mathbf{j}$ km/h and the direction of $\mathbf{v}_{PG}$ as 45° to the x axis. The relationship between these velocities is

$$\mathbf{v}_{PG} = \mathbf{v}_{PA} + \mathbf{v}_{AG}$$

The vector triangle is shown in Fig. 4.3. The simplest approach to finding the angle θ is to use the law of sines:

$$40/\sin\theta = 180/\sin 135°$$

Thus, $\sin\theta = 0.222$ and $\theta = 12.8°$. The required heading is 57.8° N of E. (The component method leads to a messy quadratic.)

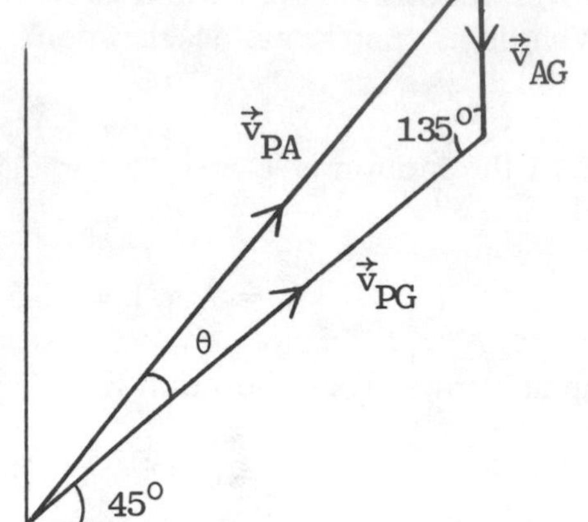

FIGURE 4.3

(b) To find the time we need the magnitude of $\mathbf{v}_{PG}$, the velocity of the plane relative to the ground. We now know the third angle in the triangle is 180 - 147.8 = 32.2°. From the law of sines:

$$180/\sin 45° = v_{PG}/\sin 32.2°$$

so, $v_{PG} =$ 136 km/h. The time taken is t = (200 km/136 kmph) = 1.47 h.

SOLUTIONS TO SELECTED TEXT EXERCISES AND PROBLEMS

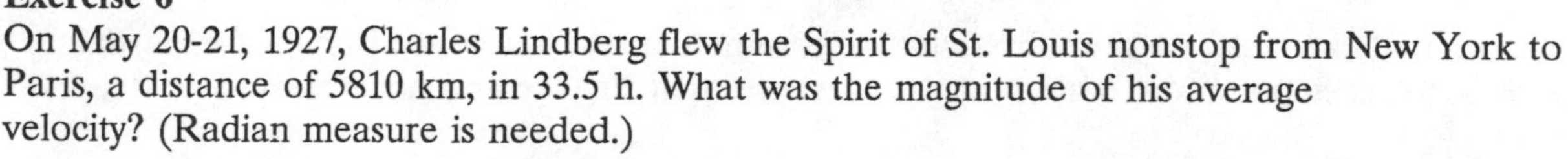

Exercise 6
On May 20-21, 1927, Charles Lindberg flew the Spirit of St. Louis nonstop from New York to Paris, a distance of 5810 km, in 33.5 h. What was the magnitude of his average velocity? (Radian measure is needed.)

Solution:

The 5810 km distance is along the curved surface of the earth. We need to find the displacement $\Delta\mathbf{r} = \mathbf{r}_2 - \mathbf{r}_1$, where $r_2 = r_1$ is the radius of the earth; see Fig. 4.4. The angular displacement is

$$\Delta\theta = s/r = 5810 \text{ km}/6370 \text{ km} = 0.912 \text{ rad}$$

The magnitude of the displacement is the length of the straight line joining the initial and final points:

$$\Delta r = 2r\tan(\Delta\theta/2) = 5611 \text{ km}$$

FIGURE 4.4

The magnitude of the average velocity is $v_{av} = \Delta r/\Delta t = 5611 \text{ km}/33.5 \text{ h} = 167$ km/h

Exercise 17
A basketball is thrown at 45° above the horizontal. The hoop is located 4 m away horizontally at a height of 0.8 m above the point of release. What is the required initial speed?

Solution:

With the axes as shown in Fig. 4.5, the coordinates of the hoop are given by

$$x = 4 \text{ m} = v_o \cos 45° \, t$$

$$y = 0.8 \text{ m} = v_o \sin 45° \, t - 4.9t^2$$

Substitute $t = 4/(v_o \cos 45°)$ into the second equation to find $v_o = 7$ m/s.

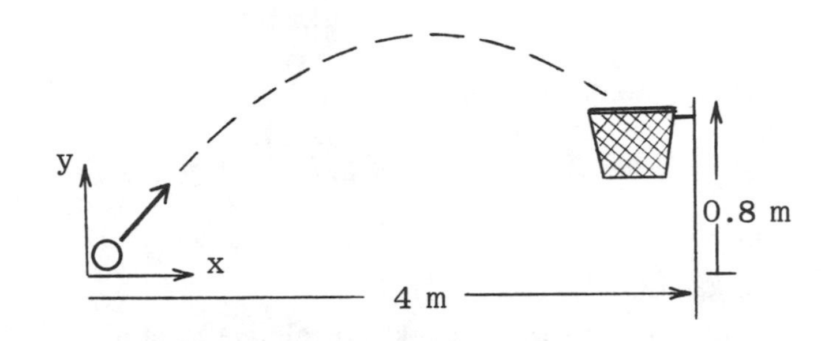

FIGURE 4.5

Exercise 31
A medieval catapult, as in Fig. 4.39 (in the text), could project a 75-kg stone at 50 m/s at 30° above the horizontal. Suppose the target is a fortress wall of height 12 m at a horizontal distance of 200 m. (a) Would the stone hit the wall? (b) If so, at what height, and (c) at what angle?

Solution:
(a) and (b). With the origin at the catapult, the equations for x and y are

$$x = 50 \cos 30° \, t \qquad \text{(i)}$$

$$y = 50 \sin 30° \, t - 4.9 \, t^2 \qquad \text{(ii)}$$

With $x = 200$ m in (i) we find $t = 4.62$ s. Using this time in (ii) we find $y = 10.9$ m. Since 10.9 m is less than 12 m, the stone would hit the wall.
(c) To find the angle of impact we need the direction of motion, which is given by the velocity.

$$v_x = v_{ox} = 50 \cos 30°; \qquad v_y = 50 \sin 30° - gt$$

Thus,

$$\text{Tan}\alpha = v_y/v_x = -20.3/43.3$$

which leads to $\alpha = 25.1°$ below the horizontal.

Exercise 45
A particle travels in a circular path of circumference 8 m and makes 5 revolutions per second. What is its centripetal acceleration?

Solution:
The period is 1/5 s = 0.2 s, and $2\pi r$ = 8 m, so the speed is

$$v = 2\pi r/T = 40 \text{ m/s}$$

We find r = 1.27 m, so

$$a = v^2/r = 1260 \text{ m/s}^2$$

Exercise 49
A ship is sailing west at 5 km/h relative to land. To people on the ship, a balloon appears to move away horizontally at 10 km/h at 37° S of E. What is the wind velocity relative to land?

Solution:
Let us use B for the balloon, S for the ship and L for land. The velocities are related according to (see Fig. 4.6)

$$\mathbf{v}_{BL} = \mathbf{v}_{BS} + \mathbf{v}_{SL}$$

The balloon actually moves with the wind, so we switch to simpler notation

$$\mathbf{W} = \mathbf{B} + \mathbf{S}$$

The components of this equations are

$$W_x = B_x + S_x = +8 - 5 = 3 \text{ m/s}$$

$$W_y = B_y + S_y = -6 + 0 = -6 \text{ m/s}$$

so, $\mathbf{W} = 3\mathbf{i} - 6\mathbf{j}$, or 6.72 m/s at 63.5° S of E.

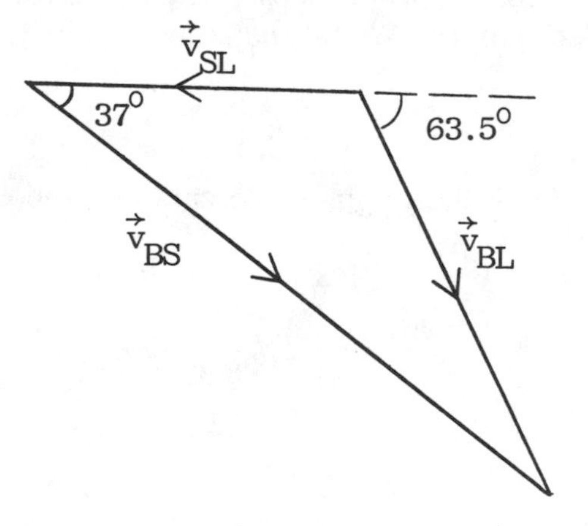

FIGURE 4.6

Problem 11
A cannon on the ground can fire a ball at 50 m/s at 53° above the horizontal. The target is 60 m above the ground. At what horizontal distance from the target should the cannon be located?

Solution:

We choose the origin at the cannon (Fig. 4.7). The components of the initial velocity are $v_{ox} = 50 \cos 53° = 30$ m/s and $v_{oy} = 50 \sin 53° = 40$ m/s. The coordinates at time t are given by

$$x = 30\, t$$

$$y = 40\, t - 4.9\, t^2$$

Use $t = x/30$ and $y = 60$ to obtain

$$4.9x^2 - 1200x + 54{,}000 = 0$$

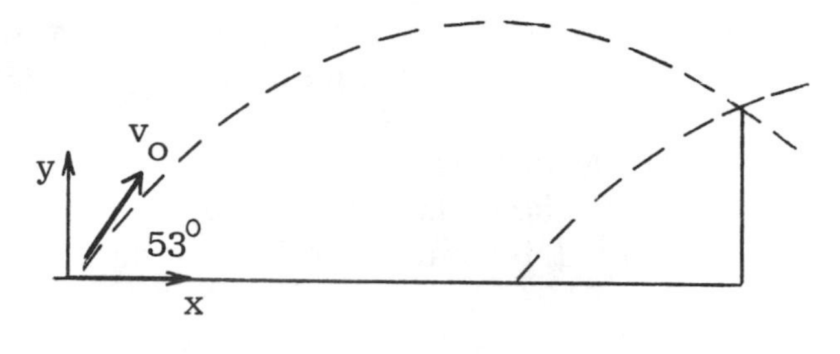

FIGURE 4.7

This yields x = 59.4 m, 185.5 m. Both solutions are acceptable and the trajectories are illustrated in the figure.

Problem 17

A car goes out of control and slides off a steep embankment of height h at θ to the horizontal (see Fig. 4.45). It lands in a ditch at a distance R from the base. Show that the speed at which the car left the slope was

$$v_o = (R/\cos\theta)\, [g/2(h - R\tan\theta)]^{1/2}$$

Solution:

The axes are shown in Fig. 4.8. The x and y coordinates of the landing spot are x = R and y = 0, so

$$R = v_o \cos\theta\, t$$

$$0 = h - v_o \sin\theta\, t - 1/2\, gt^2$$

Substituting for t from the x equation into the y equation we find

$$h = R \tan\theta + (g/2)(R/v_o \cos\theta)^2$$

This can be rearranged to yield

$$v_o = (R/\cos\theta)[g/2(h - R\tan\theta)]^{1/2}$$

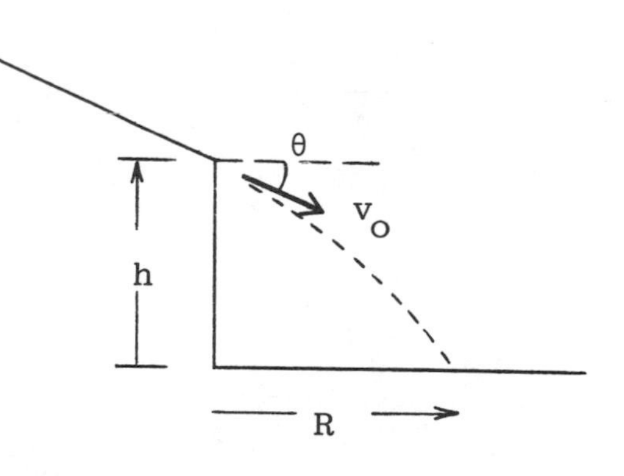

FIGURE 4.8

SELF-TEST

1. A stone is thrown from the top of a building of height 44 m with a speed v_o at angle θ below the horizontal. It lands in 2 s at a point 32 m from the base. (a) Find v_o and θ. (b) At what angle to the horizontal does it hit the ground?

2. A particle moving in a circle at a constant speed of 2 m/s has an acceleration of 8 m/s^2. How many revolutions does it complete in 5 s?

3. A pilot sets course at 30° N of E and measures his airspeed to be 160 km/h. By observing landmarks he realizes that his true course is 50° N of E and his speed relative to the ground is 150 km/h. What is the wind velocity?

Chapter 5

DYNAMICS 1

MAJOR POINTS

1. The concepts of force and mass
2. Newton's second law.
3. The distinction between mass, weight, and apparent weight
4. Newton's third law; the identification of action-reaction pairs.
5. The use of free-body diagrams to analyze the forces acting on a body.

CHAPTER REVIEW

Dynamics is a branch of mechanics that relates the acceleration of a body to the net force acting on it. We have an intuitive understanding of force as a either a push or a pull. We need to define this quantity more precisely. From Newton's first law we might infer that "force" causes acceleration. The aim of dynamics is to find the acceleration of a body given all the forces acting on it.

MASS AND WEIGHT

The mass of a body is a measure of its inertia, that is, its resistance to a change in velocity. The mass of a body is independent of its location; it has the same value here as in outer space.

The weight of an object is the net gravitational force acting on it. At the surface of a planet of mass M and radius R, the gravitational force acting on a particle of mass m is given by Newton's law fo gravitation

$$F = G\,m\,M/R^2$$

This is usually written in the form

$$\mathbf{W} = m\,\mathbf{g}$$

where **g** is the gravitational force per unit mass, measured in N/kg. Since **W** = m**g** has the same form as **F** = m**a** and N/kg reduces to m/s^2, it is customary, although not quite correct, to say that **g** is the "acceleration due to gravity". (This issue is discussed briefly in Ch. 6 and more fully in Ch. 13 of the text.)

<u>Distinction between mass and weight</u>

As we noted above, the mass of a body is an intrinsic property of that body, independent of its location. The weight of the body, on the other hand depends on the local value of **g**, which depends on various factors.

Suppose you were on the moon. Since the value of **g** at the surface of the moon is about

1/6 th the value on earth, you would find it much easier to lift or to hold a ball. However, the mass of the ball would be the same as it is on earth. Thus you would experience the same resistance if you try to thrown the ball or to kick it.

NEWTON'S SECOND LAW

Newton's second law of motion is

$$\Sigma \mathbf{F} = m \mathbf{a}$$

Expressed in the form $\mathbf{a} = \Sigma \mathbf{F}/m$ it says:

> The acceleration of a body of mass m is directly proportional to the net force, $\Sigma\mathbf{F}$, acting on it and inversely proportional to its mass.

Newton's second law is valid only in an inertial frame (one in which the first law is valid). The fictitious forces that appear in (accelerated) noninertial frames are discussed in Section 6.8 of the text. Usually you will deal with two rectangular components of this law

$$\Sigma F_x = m a_x ; \qquad \Sigma F_y = m a_y$$

Note that the acceleration is in the same direction as the net force. The direction of motion of a particle (given by its velocity) may be in any direction.

NEWTON'S THIRD LAW

Figure 5.1 shows two objects A and B. We take $\mathbf{F}_{AB}$ to be the force exerted on A by B and $\mathbf{F}_{BA}$ to be the force exerted on B by A. According to Newton's third law:

$$\mathbf{F}_{AB} = - \mathbf{F}_{BA}$$

The forces are equal and opposite and act on different bodies.

FIGURE 5.1

FREE BODY DIAGRAM

An important tool in the solution of mechanics problems is the "free-body diagram" (FBD). In this diagram we isolate the body in whose acceleration we are interested and draw all the external forces acting ON it. Refer to the Problem Solving Guide on p. 87 of the text.

HELPFUL HINTS

1. Use a sharp pencil and a ruler. A neat and carefully drawn FBD is half the problem done.

2. Choose the axes such that the acceleration, given or assumed, is in the positive direction of either the x or the y axis. Do not rotate these axes when drawing the FBD.

3. Watch the direction of the forces. For example, the weight W = mg, of a body is vertically downward, not at some arbitrary angle. The normal force N exerted by a surface is just that: normal, which means perpendicular, to the surface.

4. It may help you to identify the forces acting on a body if you were to imagine yourself in its place. Furthermore, you should be able to identify the body (table, incline, the earth, a rope etc.) responsible for a given force. Keep in mind that **ma** is not to be included in the FBD; it is the resultant of the all the forces.

Equilibrium

The term equilibrium means balance. A body is in translational equilibrium if the net force on it is zero, that is, $\Sigma \mathbf{F} = 0$. The body may be at rest (static equilibrium) or moving at constant velocity (dynamic equilibrium). Rotational equilibrium is discussed in Ch. 12.

When a body is thrown vertically up, it comes to rest momentarily at the highest point. Is it in equilibrium at this point? Students sometimes think that the force exerted by the hand somehow stays with the ball. As the ball rises, it is assumed that both gravity and the "force of the hand" (which is supposed to diminish with height) act on the ball, becoming equal at the highest point. This is not so! The ball is subject only to the force of gravity (and of course air resistance). At the highest point the velocity of the ball is instantaneously zero, but its acceleration is still 9.8 m/s^2 vertically downward. Since there is only one force acting on it, the ball is not in equilibrium.

PROBLEM-SOLVING GUIDE

1. Identify the forces acting on each block and draw the vectors on the diagram.
2. Show the direction of the acceleration with a double-stem arrow. Pick suitable axes for each block. (The acceleration should be along either the +x axis or the +y axis.
3. Draw a free-body diagram. If a force lies along one of the axes draw it as a solid arrow and indicate the sign. If a force does not lie along an axis draw it as a dashed arrow, but draw the components as solid arrows.
4. Label each component, for example: $+F \sin\theta$, or $-F \cos\theta$.

Tension

The tension in a rope is the force exerted by one section on an adjacent section or on a support, such as a wall. Unless otherwise indicated, we assume that strings and ropes are massless and the tension is uniform throughout.

EXAMPLE 1

Two blocks with masses $m_1 = 3$ kg and $m_2 = 5$ kg are connected by a light rope; see Fig. 5.2. The horizontal surface is frictionless. Find (a) the acceleration, and (b) the tension in the rope.

Solution:

The forces acting on each block are indicated in Fig. 5.2. Block m_2 accelerates downward and m_1 to the left. Note that $a_1 = a_2 = a$. The axes for each block are oriented so that the acceleration points along one positive axis. By inspecting the free-body diagrams, we write Newton's second law in component form:

$(m_1) \quad \Sigma F_x = T = m_1 a \qquad$ (i)

$(m_1) \quad \Sigma F_y = N - m_1 g = 0 \qquad$ (ii)

$(m_2) \quad \Sigma F_y = m_2 g - T = m_2 a \qquad$ (iii)

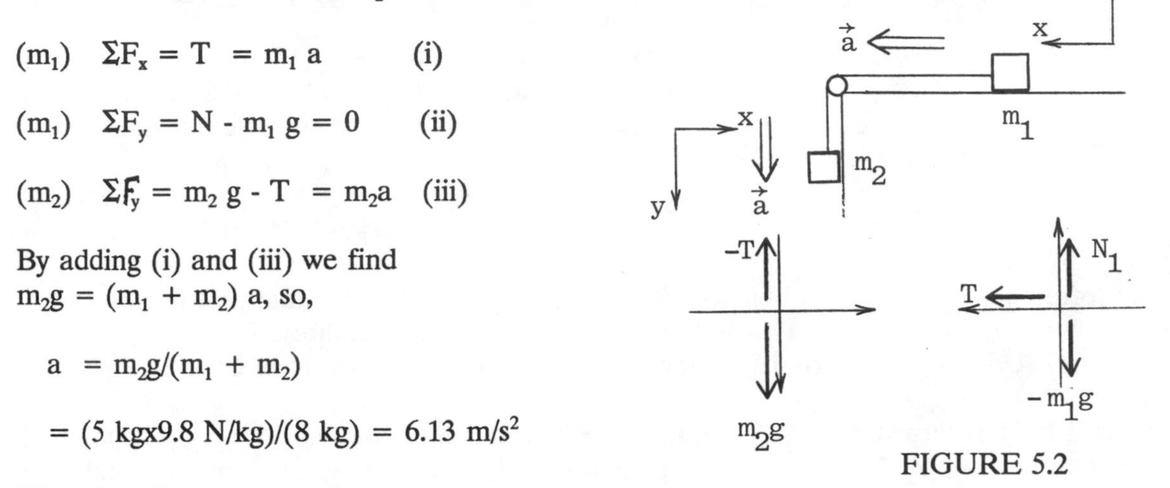

FIGURE 5.2

By adding (i) and (iii) we find
$m_2 g = (m_1 + m_2)\,a$, so,

$$a = m_2 g/(m_1 + m_2)$$

$$= (5 \text{ kg} \times 9.8 \text{ N/kg})/(8 \text{ kg}) = 6.13 \text{ m/s}^2$$

Note that if $m_2 >> m_1$, then $a \approx g$.

EXAMPLE 2

Two blocks with masses $m_1 = 3$ kg and $m_2 = 5$ kg are connected by a light string. Block m_2 hangs vertically below m_1, as in Fig. 5.3. Find the force F_o and the tension in the string given that the blocks accelerate downward at 3 m/s^2.

Solution:

It might help to "isolate" each block by drawing a dashed circle around it. The tension acts upward on m_2 and downward on m_1.

$m_2 g - T = m_2 a \qquad$ (i)

$T + m_1 g - F_o = m_1 a \qquad$ (ii)

Adding (i) and (ii) we find

$F_o = (m_1 + m_2)(g - a) = 56$ N.

From either (i) and (ii) we find T = 25 N.

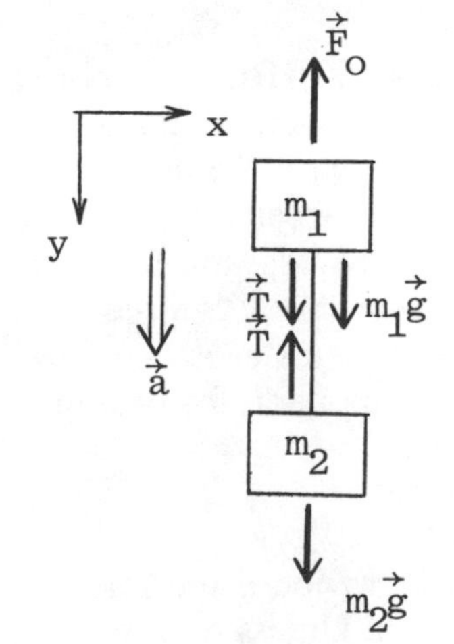

FIGURE 5.3

EXAMPLE 3

A 2 kg block is on a frictionless 20° incline as shown in Fig. 5.4a. A force F = 6 N acts at α = 40° to the incline. Determine (a) the acceleration, and (b) the normal force acting on the block.

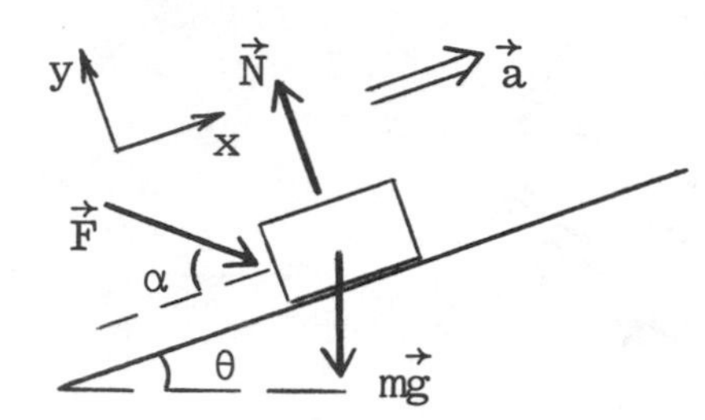

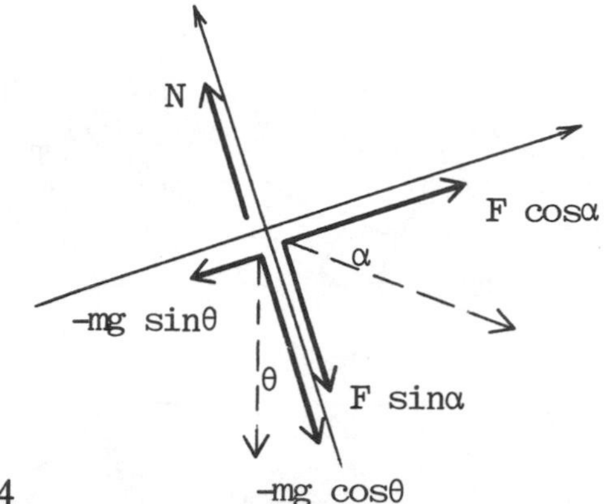

FIGURE 5.4

Solution:

The direction of the acceleration of the block is not clear. We assume it is up the incline and choose the axes accordingly, with the x axis along the acceleration. Next, we start the free-body diagram (FBD) by drawing the axes again (as thin lines) accurately parallel to the tilted axes in the sketch. Do not rotate the axes to be horizontal and vertical.

Draw the normal force **N** accurately perpendicular to the incline and **mg** accurately vertical. A sloppy drawing will make your life that much harder. Since **N** is along the y axis, draw it as a solid arrow on the FBD, Fig. 5.4b. Get into the habit of drawing the forces just off the axes; this helps to ensure you do not forget a force or a component. Since **mg** is vertical, draw it as a dashed arrow and draw its components as solid arrows. In the end, we count only the solid arrows that lie along the axes. The force **F** should be drawn as a dashed arrow with its tail near the origin, not with its tip pointing to the origin. Draw the components as solid arrows. In all cases make sure the arrows in the FBD are drawn with a ruler accurately parallel to the vectors in the sketch.

Now comes the important part: Determine the components of the forces that do not lie along the axes. The sign of each component is determined by its direction relative to the axis. In finding the magnitudes (i.e. whether to use cosθ or sinθ) you might draw thin lines to complete the rectangle.

The FBD helps us write Newton's second law in component form:

$$\Sigma F_x = F \cos\alpha - m g \sin\theta = m a \qquad \text{(i)}$$

$$\Sigma F_y = N - F \sin\alpha - m g \cos\theta = 0 \qquad \text{(ii)}$$

Equation (ii) immediately leads to N = 22.3 N. (Check it.) Then from (i),

$$(6 \text{ N})\cos 40° - (2 \text{ kg})(9.8 \text{ N/kg})\sin 20° = 2 a$$

Thus, $a = -1.05 \text{ m/s}^2$.

EXAMPLE 4
Two blocks with masses $m_1 = 2.4$ kg and $m_2 = 3.6$ kg are connected by a massless rope that hangs over a pulley as in Fig. 5.5. What is the acceleration of the blocks? Take $\theta = 30°$.

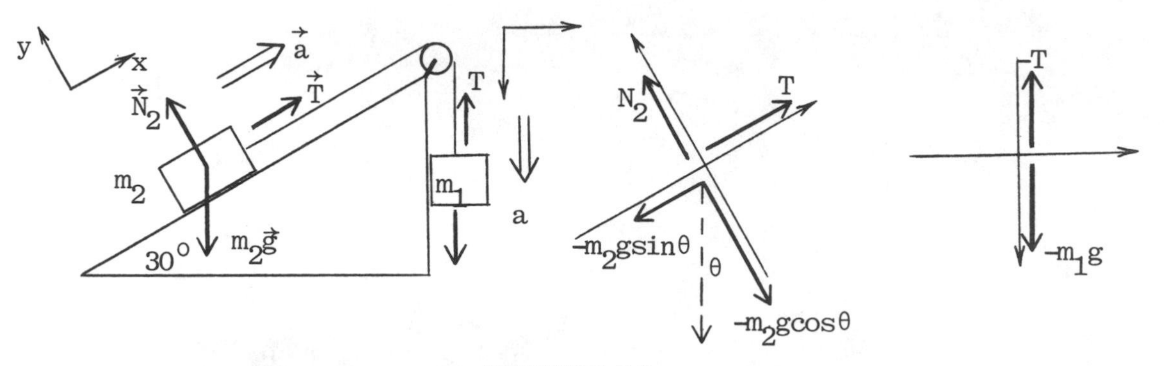

FIGURE 5.5

Solution:

Initially we do not know the direction of the acceleration so we assume that m_1 accelerates downward and then make the appropriate choices for the two sets of axes. The forces acting on each block are shown in the sketch and the corresponding FBD's are also shown. Note that there is only one tension T: Do not label the forces exerted by the ropes T_1 and T_2.

Using the FBD we write Newton's second law just for the lines along which the blocks are free to move.

$$(m_1) \qquad \Sigma F_x = m_1 g - T = m_1 a \qquad \text{(i)}$$

$$(m_2) \qquad \Sigma F_y = T - m_2 g \sin\theta = m_2 a \qquad \text{(ii)}$$

On adding (i) and (ii) we obtain $(m_1 g - m_2 g \sin\theta) = (m_1 + m_2)a$ which gives us

$$a = (m_1 g - m_2 g \sin\theta)/(m_1 + m_2)$$

When you obtain such an algebraic expression it is a good idea to first check that it is dimensionally correct. We know $[a] = LT^{-2}$ $[mg \sin\theta] = (M)(LT^{-2})$; $[m_1 + m_2] = M$. Clearly $[mg \sin\theta/(m_1 + m_2)] = LT^{-2}$as it should.

Next, you should try extreme values for some of the quantities. For example, $m_1 = 0$ or $m_2 = 0$, and $\theta = 0$ or $90°$. With the given values, $a = 0.98$ m/s^2.

SOLUTIONS TO SELECTED TEXT EXERCISES AND PROBLEMS

Exercise 15

A person lowers his 50-kg torso by 15 cm and jumps vertically. If the torso rises 40 cm above normal, find the force (assumed to be constant) exerted on the torso at the hip joint.

Solution:

In order for the torso to rise 0.4 m into the air, the initial speed v required is given by

$$0 = v^2 - 2g(0.4 \text{ m})$$

so $v^2 = 7.84 \text{ m}^2/\text{s}^2$. This v is the final speed for the period during which the feet are in contact with the ground. The initial velocity is zero, and the acceleration of the torso occurs over a distance of 0.15 m, thus

$$v^2 = 2a(0.15)$$

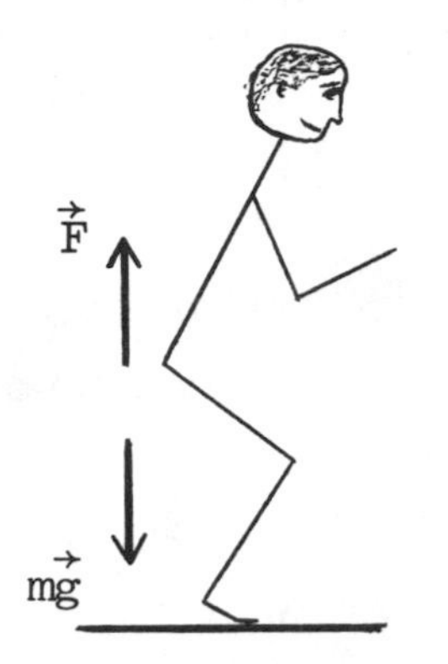

FIGURE 5.6

and $a = 26.1 \text{ m/s}^2$. Using Fig. 5.6 we see that the upward force F exerted on the torso is given by

$$F - mg = ma$$

so, $F = m(g + a) = 1795$ N.

Exercise 25

A Polaris missile has a mass of $1.4\text{x}10^4$ kg and its engine has a thrust of $2\text{x}10^4$ N. If its engine fires in the vertical direction for 1 min, to what vertical height would it rise in the absence of air resistance?

Solution:

The forces acting on the missile are the thrust, T, upward and its weight, mg, downward, see Fig. 5.7. Thus,

$$T - mg = ma$$

This leads to $a = 4.49 \text{ m/s}^2$. Assuming that the rocket starts at rest, its vertical displacement while the engine is on, is

$$\Delta y_1 = 1/2\ at^2 = 1/2(4.49)(60^2) = 8.08 \text{ km}$$

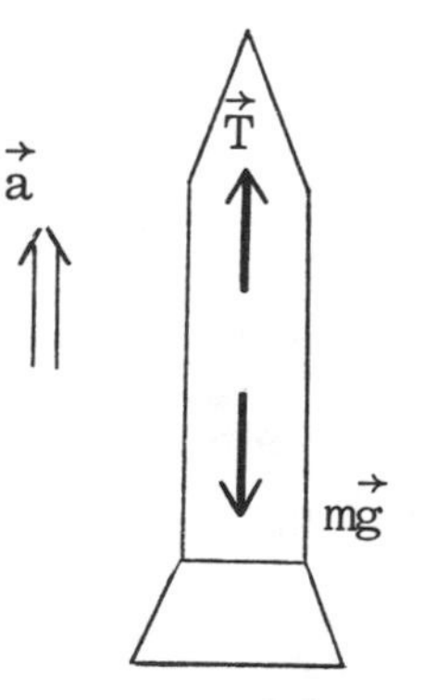

FIGURE 5.7

When the engine stops, the velocity is $v = at = 269.4$ m/s and from then on the rocket is then in free-fall. The added height in free-fall is given by

$$0 = v^2 - 2g\ \Delta y_2$$

thus $\Delta y_2 = v^2/2g = 3.70$ km. The total height is $\Delta y_1 + \Delta y_2 = 11.8$ km. This result can only be approximate since we have ignored the fact that g decreases with altitude and the effects of air resistance.

Exercise 29

A hot-air balloon of mass M descends with an acceleration a ($< g$). How much ballast should be thrown out for it to accelerate upward with the same acceleration? Assume that the buoyant force is unchanged.

Solution:

The forces on the balloon are shown in Fig. 5.8, where F_B is the buoyant force exerted by the air, and m is the mass of the ballast thrown out.

(a down) $\quad Mg - F_B = Ma$

(a up) $\quad F_B - (M - m)g = (M - m)a$

Adding these equations we find $mg = (2M - m)a$, thus,

$$m = 2Ma/(a + g)$$

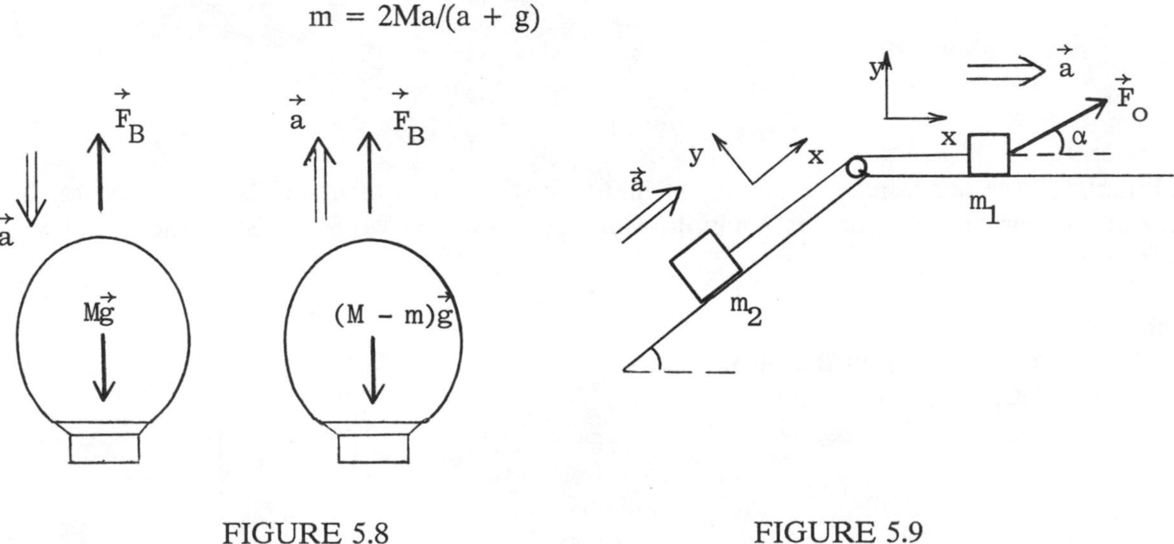

FIGURE 5.8

FIGURE 5.9

Exercise 43

Two blocks with masses $m_1 = 3$ kg and $m_2 = 5$ kg are connected by a light rope and slide on a frictionless surface as in Fig. 5.9. A force $F_o = 10$ N acts on m_2 at $\alpha = 20°$ to the horizontal. Find the acceleration of the system and the tension in the rope.

Solution:

We do not know the direction of the acceleration. We assume it is in the direction shown and draw the axes accordingly.

$$F_o \cos\alpha - T = m_1 a$$

$$T - m_2 g \sin\theta = m_2 a$$

Adding these equations:

$$F_o \cos\alpha - m_2 g \sin\theta = (m_1 + m_2)a$$

Using the given values we find $a = -0.66\ \mathrm{m/s^2}$, and then $T = 12.7$ N.

Exercise 47

A rocket has a net acceleration of 2.4 $\mathrm{m/s^2}$ directed at 60° above the horizontal. What is the apparent weight of an 80-kg astronaut near the earth's surface?

Solution:

The astronaut feels horizontal and vertical forces. We do not know the direction of **N** but the direction of the acceleration is given. In this problem is more convenient to use axes that are horizontal and vertical, as shown in Fig. 5.10. The components of the second law are

$$N_x = ma \cos\theta$$

$$N_y - mg = ma \sin\theta$$

With the given values we find $N_x = 96$ N and $N_y = 950$ N. The magnitude is

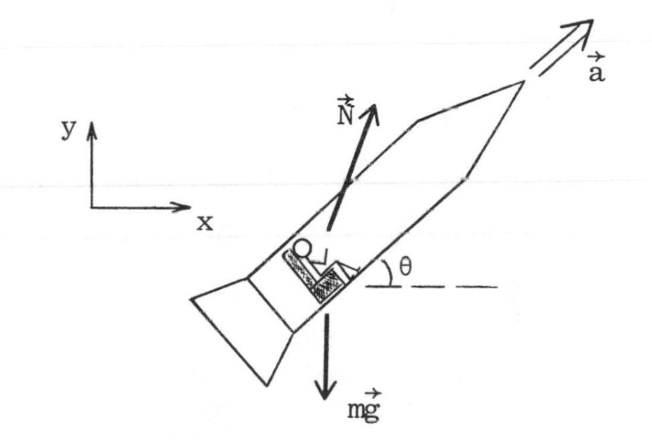

FIGURE 5.10

$$N = (950^2 + 96^2)^{1/2} = 955 \text{ N}$$

The direction is given by $\tan\theta = N_x/N_y$. We find $\theta = 5.77°$ to the vertical. The apparent weight is directed opposite to **N**.

Problem 1

A painter of mass $M = 75$ kg stands an a platform of mass $m = 15$ kg. He pulls on a rope that passes around a pulley, as in Fig. 5.11. Find the tension in the rope given that (a) he is at rest, or (b) he accelerates upward at 0.4 $\mathrm{m/s^2}$. (c) If the maximum tension the rope can withstand is 700 N, what happens when he ties the rope to a hook on the wall?

Solution:

Our system is the painter and his cage. The forces at his feet are internal to the system and thus can be ignored. His hands pull down on the rope therefore the rope pulls up on him. The net force vertically up is 2T. [If you wrap a rope around your waist and pull on the rope that is hung over a branch, you will find it much easier to hold yourself up than if you were to simply grab onto a vertical rope.]

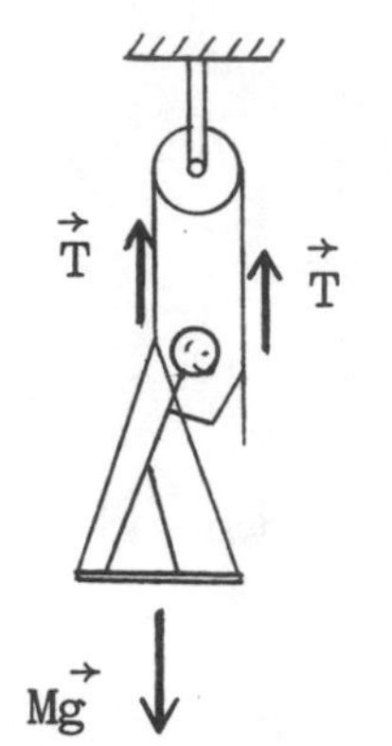

(a) $$2T - Mg = 0$$

where M = 90 kg, so T = 441 N.

(b) $$2T - Mg = 90a$$

so, T = 459 N.

FIGURE 5.11

(c) When free end of the rope is attached to a support, there is only one tension force acting on the system (at the top of the cage). Thus, T = 90g > 700 N, which means that the rope breaks.

Problem 3

Two blocks of masses m_1 = 2 kg and m_2 = 5 kg hang over a massless pulley as in Fig. 5.12. A force F_o = 100 N acting at the axis of the pulley accelerates the system upward. Find (a) the acceleration of each mass; (b) the tension in the rope.

Solution:

Since the pulley is massless, $F_o - 2T = 0$, so $T = F_o/2 = 50$ N. Let us say that A is the upward acceleration of the pulley and a' is the magnitude of the acceleration of the blocks relative to the pulley.

$$5g - T = 5(a' - A) \qquad \text{(i)}$$

$$T - 2g = 2(A + a') \qquad \text{(ii)}$$

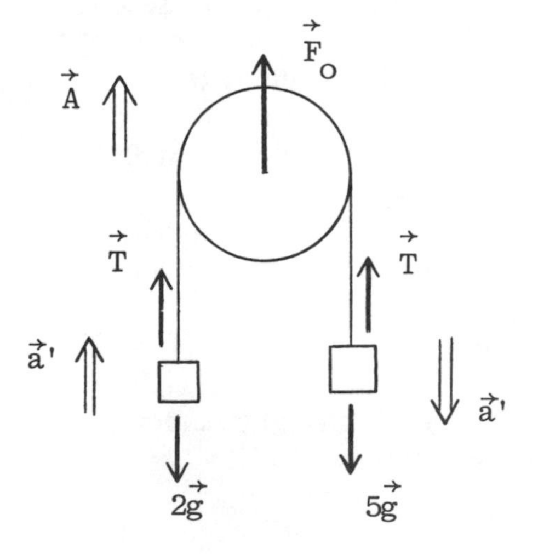

To eliminate A, multiply (i) by 2, and (ii) by 5, and then add. This leads to 3T = 20a', and so a' = 150/20 = 7.5 m/s^2. Either (i) or (ii) then yields A = 7.7 m/s^2.

$$a_2 = (A + a') = 15.2 \text{ m/s}^2 \text{ upward,}$$

FIGURE 5.12

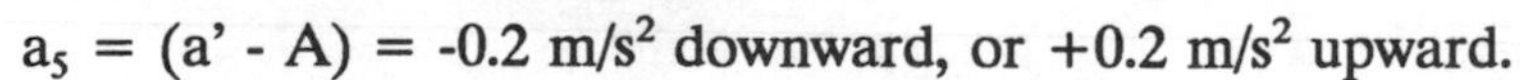

$$a_5 = (a' - A) = -0.2 \text{ m/s}^2 \text{ downward, or } +0.2 \text{ m/s}^2 \text{ upward.}$$

SELF-TEST

1. Two blocks, $m_1 = 7$ kg and $m_2 = 4$ kg, are on a wedge and connected by a massless rope that hangs over a pulley as shown in Fig. 5.13. Determine the acceleration of the blocks and the tension in the rope.

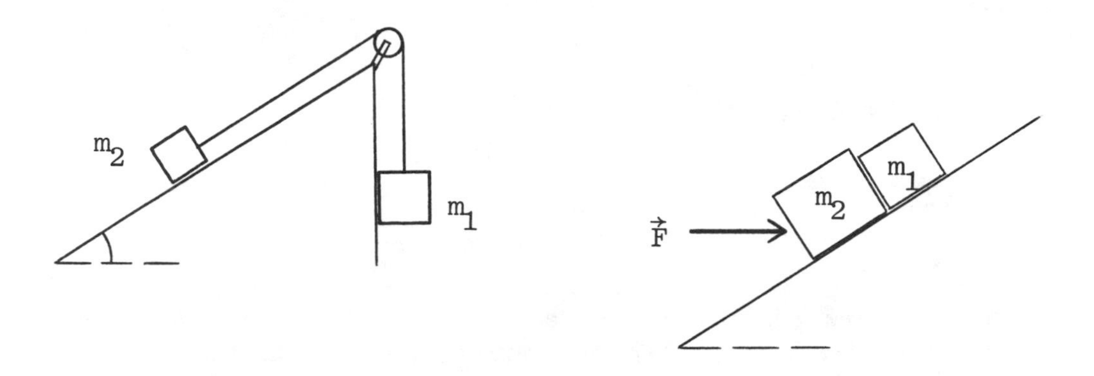

FIGURE 5.13 FIGURE 5.14

2. Two blocks with masses $m_1 = 2$ kg and $m_2 = 3$ kg are on a frictionless incline set at 30° to the horizontal, see Fig. 5.14. They are pushed by a horizontal force F = 60 N. Find (a) the acceleration of the blocks, (b) the force exerted by m_2 on m_1.

3. A 60 kg woman is in an elevator that is moving down at 4 m/s and slowing down at 1.5 m/s^2. (a) What is her apparent weight? (b) According to Newton's third law, what is the magnitude of the reaction to her true weight? On what body does this force act?

Chapter 6

DYNAMICS II

MAJOR POINTS
1. The distinction between static and kinetic friction
2. The dynamics of circular motion
3. Satellite orbits; Kepler's third law

CHAPTER REVIEW

FRICTION

Friction is a force that opposes the relative motion of two surfaces in contact. If the two bodies are at rest relative to each other, the friction is static whereas if they are sliding relative to each other the friction is kinetic. You should note that friction does not always "oppose the motion" of an object. In fact it is the force of friction that moves a car or a walking person forward. The driven wheel or the shoe pushes backward on the road and the frictional force opposes this "relative motion". Thus the frictional force exerted by the road on the wheel or the shoe is in the forward direction. The force of friction between dry, unlubricated surfaces is related to the normal (perpendicular) force. Thus, the force of kinetic friction is given by

$$f_k = \mu_k N$$

where μ_k is the coefficient of kinetic friction. The force of static friction depends on the applied force; only the maximum value can be related to the normal force:

$$f_{s(max)} = \mu_s N$$

EXAMPLE 1

A block slides on a rough incline for which $\mu_k = 0.25$, see Fig. 6.1a. Find its final velocity after it slides 1 m along the incline given that its initial velocity is 4 m/s (a) up the incline, (b) down the incline.

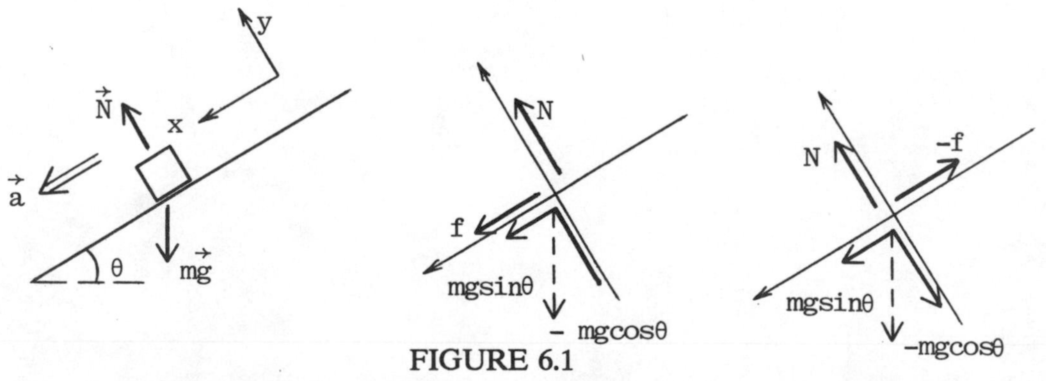

FIGURE 6.1

Solution:
The direction of the acceleration is clearly down the incline. The direction of the force of friction is opposite to the velocity of the block. The forces acting on the block and the FBD are shown in Fig. 6.1. Since the motion is up the incline, the frictional force is down the incline. Newton's second law in component form is:

$$\Sigma F_x = mg \sin\theta + f = m a \qquad \text{(i)}$$

$$\Sigma F_y = N - mg \cos\theta = 0 \qquad \text{(ii)}$$

We use $N = mg \cos\theta$ and substitute $f_k = \mu_k N$ into (i) to obtain

$$mg \sin\theta + \mu_k(mg \cos\theta) = m a$$

Thus,

$$a = g(\sin\theta + \mu_k \cos\theta)$$

$$= (9.8 \text{ N/kg})(0.6 + 0.25 \times 0.8) = 7.84 \text{ m/s}^2.$$

The final velocity may be found with the equation of kinematics. With the present axes, $a = +7.84 \text{ m/s}^2$, but $\Delta x = -1$ m, so

$$v^2 = v_o^2 + 2 a \Delta x$$

$$= (4 \text{ m/s})^2 - 2(7.84 \text{ m/s}^2)(1 \text{ m}) = 0.32 \text{ m}^2/\text{s}^2$$

which yields $v = 0.56$ m/s.

(b) If the motion is down the incline, the frictional force is up along the incline. The components of the second law become

$$\Sigma F_x = mg\sin\theta - f = ma$$

$$\Sigma F_y = N - \mu_k mg\cos\theta = 0$$

From these we find

$$a = g(\sin\theta - \mu_k \cos\theta) = 3.92 \text{ m/s}^2$$

The direction of the force of kinetic friction is opposite to the velocity of the body relative to the surface; but the frictional force may be in the same direction as the acceleration or opposite to it. The final velocity is given by

$$v^2 = (4 \text{ m/s})^2 + 2(3.92 \text{ m/s}^2)(1 \text{ m}) = 23.8 \text{ m}^2/\text{s}^2$$

Thus, $v = 4.9$ m/s.

EXAMPLE 2
A block of mass m = 1.2 kg is held against a rough wall (μ_s = 0.2) by a force F directed at an angle α = 10° above the horizontal, as shown in Fig. 6.2a. What is the minimum value of F for the block to remain stationary?

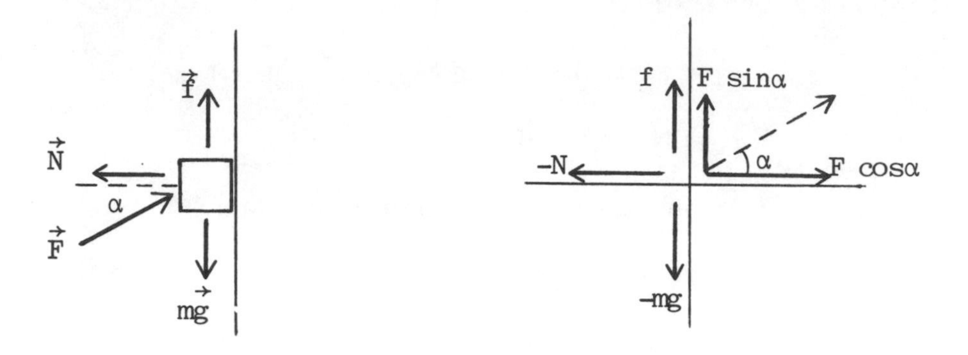

FIGURE 6.2

Solution:
Since we are looking for the minimum value of F, the tendency of the block is to slide downward, so the force of friction is directed upward. Furthermore, the force of static friction will have is maximum value. From the FBD in Fig. 6.2b, we see that the component form of the second law is:

$$\Sigma F_x = F\cos\alpha - N = 0 \qquad \text{(i)}$$

$$\Sigma F_y = f - mg + F\sin\alpha = 0 \qquad \text{(ii)}$$

The maximum possible value of the static frictional force is $f_{s(max)} = \mu_s N$, where from (i) $N = F\cos\alpha$. Substituting this into (ii) we find

$$\mu_s(F\cos\alpha) - mg + F\sin\alpha = 0$$

Thus,

$$F = mg/(\mu_s \cos\alpha + \sin\alpha) = 31.7 \text{ N}$$

From (ii) we may infer that if $F\sin\alpha$ were greater than mg, the force of friction would be directed downward.

CIRCULAR MOTION

When a particle moves with speed v in a circular path of radius r, it experiences a centripetal (inward) acceleration. This is because the direction of the velocity is changing even though the speed is constant. The centripetal force needed to produce this acceleration (see Fig. 6.3) is

$$F = m\,v^2/r$$

The centripetal force may be provided by the tension in a rope, friction, the force of gravity, the normal force exerted by a surface, etc., or by a combination of such forces.

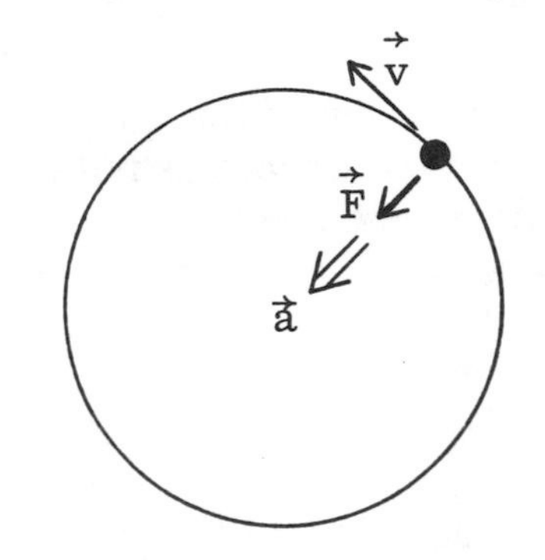

FIGURE 6.3

EXAMPLE 3

A person is on a ride that takes him in a vertical circle of radius 8 m. At the highest point the person is upside down and his apparent weight is one-half his true weight. What is the speed at this point?

Solution:

As Fig. 6.4 shows, at the highest point the acceleration is directed downward and the normal force N exerted by the floor of the vehicle on the feet of the person is also directed downward. Thus,

$$\Sigma F_y = N + mg = mv^2/r$$

The quantity N is equal to the apparent weight, whereas mg is the true weight. We are given that N = 0.5mg, so

$$1.5mg = mv^2/r$$

which means $v = (1.5rg)^{1/2} = 10.8$ m/s.

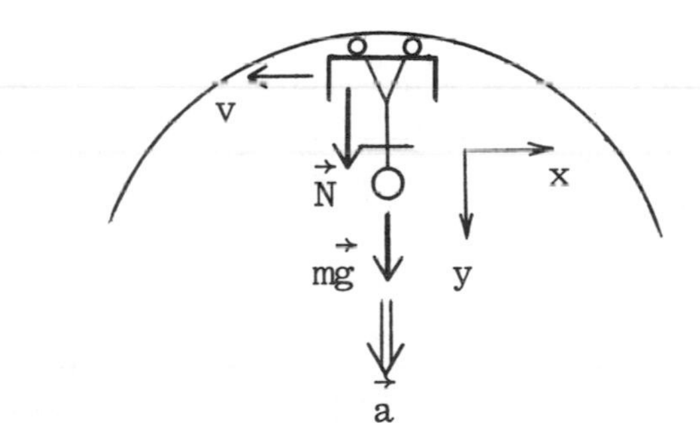

FIGURE 6.4

EXAMPLE 4
A car travels at 12 m/s around a flat curve of radius 40 m. What is the minimum coefficient of friction required?

Solution:

The acceleration of the car is directed toward the center. We draw a side view of the system in Fig. 6.5. A top view is not useful since the normal force and the weight could not be drawn. We choose the direction of the x axis to be along the acceleration. You might be tempted to draw the frictional force directed outward (for whatever reason). Then you must ask yourself where the necessary centripetal force comes from. You won't find anything. In fact, it is the inward frictional force that provides the inward centripetal force. The components of the second law are

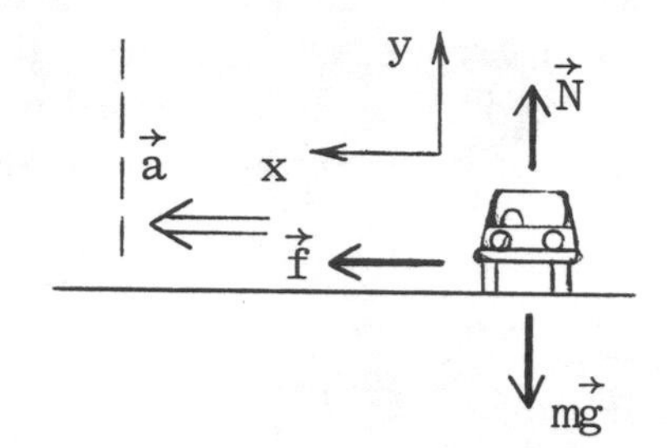

FIGURE 6.5

$$\Sigma F_x = f = mv^2/r$$

$$\Sigma F_y = N - mg = 0$$

Using $f_{s(max)} = \mu_s N = \mu_s (mg)$, we find

$$\mu_s = v^2/rg = 0.367$$

This is the minimum coefficient of static friction required to prevent the car from slipping sideways.

ORBITS

Consider a satellite of mass m in a stable circular orbit of radius r about the sun or a planet of mass M. The centripetal force is provided by gravitational force given by Newton's law of gravitation:

$$G\, m\, M/r^2 = m\, v^2/r$$

From this equation we obtain the orbital speed,

$$v = (G\, M/r)^{1/2}$$

The period T of the orbit is the time taken for one revolution,

$$T = \frac{2\pi r}{v} = \frac{2\,\pi}{\sqrt{GM}}\, r^{3/2}$$

Squaring both sides we obtain Kepler's thrid law of planetary motion:

$$T^2 = \kappa r^3$$

where $\kappa = 4\pi^2/GM$ depends only on the mass of the central body.

EXAMPLE 5
Use data on the orbit of the earth to find (a) the period of Venus given that the radius of its orbit is 1.08×10^{11} m, and (b) the mass of the sun.

Solution:
(a) According to Kepler's third law, the period T of a planet and the radius r of its orbit are related by

$$T^2 = \kappa r^3$$

where $\kappa = 4\pi^2/GM$ depends only on the mass of the sun. Thus,

$$(T_V / T_E)^2 = (r_V / r_E)^3$$

Since $T_E = 1$ year and $r_E = 1.49 \times 10^{11}$ m, we find $T_V = 0.617$ y.
(b) From Kepler's third law, the mass of the sun is given by

$$M = 4\pi^2 r^3/GT^2$$

$$= 4\pi^2(1.49 \times 10^{11})^3/(6.67 \times 10^{-11}\ \text{N.m}^2/\text{kg}^2)(3.156 \times 10^7\ \text{s})^2$$

$$= 1.97 \times 10^{30}\ \text{kg}.$$

SOLUTIONS TO SELECTED TEXT EXERCISES AND PROBLEMS

Exercise 9
A 2.5 kg block is on a 53° incline for which $\mu_k = 0.25$ and $\mu_s = 0.5$. Find its acceleration given that: (a) it is initially at rest; (b) it is moving up the slope; (c) it is moving down the slope.

Solution:
(a) Since the block tends to slide downward, the force of friction is up along the incline. The forces and axes are shown in Fig. 6.6a. The maximum available force of static friction is $f_{s(max)} = \mu_s N$, where in this case
$N = mg\cos\theta$. We must first find out whether the block even starts to move. The net downward force along the incline is

$$\Sigma F_x = mg\sin\theta - \mu_s(mg\cos\theta) = 12.2\ \text{N}$$

Since $\Sigma F_x > 0$, the body does start to move and the friction becomes kinetic.

$$\Sigma F_x = mg\sin\theta - \mu_k(mg\cos\theta) = ma$$

$$19.6 - \mu_k(14.7) = 2.5a$$

Thus $a = 6.37\ m/s^2$, downward

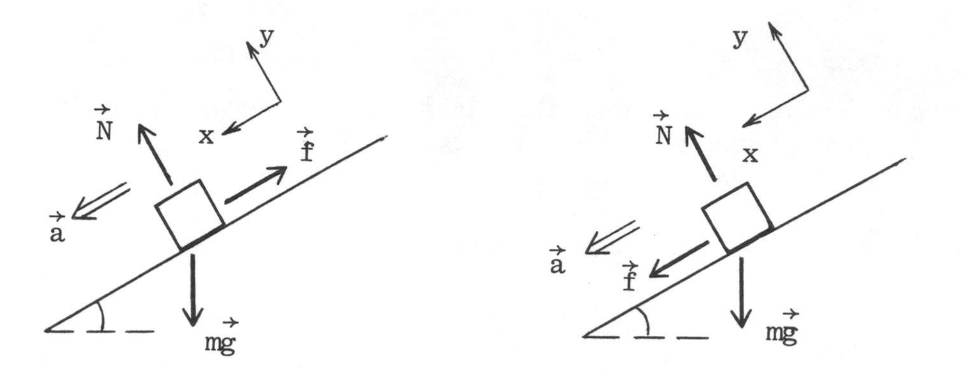

FIGURE 6.6

(b) If the body moves up the slope, the friction is directed downward, as in Fig. 6.6b. In this case,

$$\Sigma F_x = mg\sin\theta + \mu_k(mg\ \cos\theta) = ma$$

thus $a = 9.31\ m/s^2$ (downward)

(c) This case is similar to part (a), so $a = 6.37\ m/s^2$ (downward)

Exercise 17
A block of mass M is subject to the force F shown in Fig. 6.7a. The friction coefficients are μ_k and μ_s. (a) What minimum value of F will start the block moving from rest? (b) Show that if $\mu_s = \cot\theta$, the block cannot be moved for any value of F. (c) To what situation does the condition $\mu_k = \cot\theta$ apply?

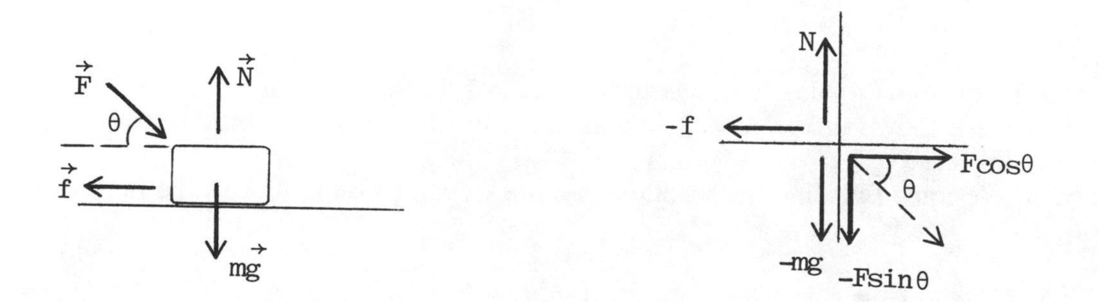

FIGURE 6.7

Solution:

The free-body diagram is shown in Fig. 6.7b.

$$\Sigma F_x = F\cos\theta - \mu N = ma \qquad \text{(i)}$$

$$\Sigma F_y = N - F\sin\theta - mg = 0 \qquad \text{(ii)}$$

When $N = F\sin\theta + mg$ from (ii) is substituted into (i), we find

$$\Sigma F_x = F(\cos\theta - \mu\sin\theta) - \mu mg = ma$$

or,
$$\Sigma F_x = F(\cot\theta - \mu)\sin\theta - \mu mg = ma \qquad \text{(iii)}$$

(a) From (iii) we need $F > \mu_s mg/(\cot\theta - \mu)\sin\theta$ for the block to start moving.
(b) If $\mu_s = \cot\theta$, then the expression for F in part (a) becomes infinite.
(c) If $\mu_k = \cot\theta$, then from (iii) we find $a = -\mu_k g$. In this case, the added frictional force caused by the component of F normal to the plane is just equal to the horizontal component of F. Assuming that the block is initially moving, the acceleration would be the same as if there were no F present.

Exercise 27
A child pulls a 3.6 kg sled at 25° to slope that is at 15° to the horizontal, as in Fig. 6.8a. The sled moves at constant velocity when the tension is 16 N. What is the acceleration of the sled if the rope is released?

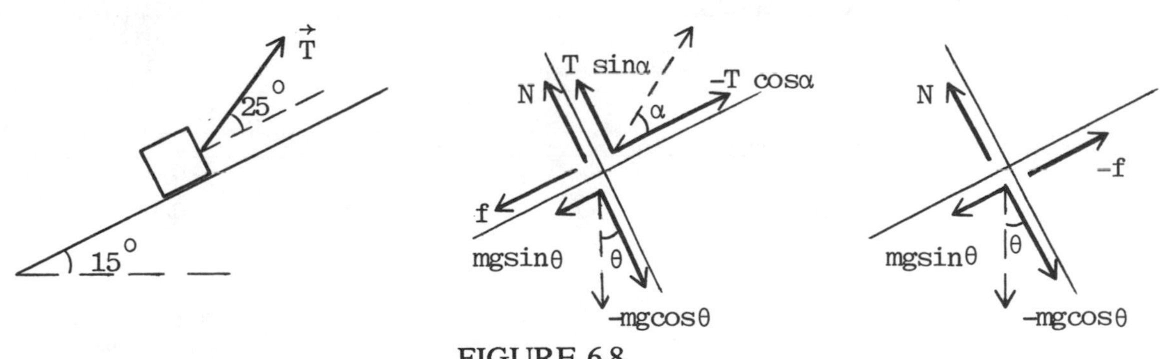

FIGURE 6.8

Solution:

The forces on the sled as it is being dragged up are shown in Fig. 6.8b. We use the initial condition, a = 0, to find μ_k.

$$\Sigma F_x = mg\sin\theta + f - T\cos\alpha = 0 \qquad \text{(i)}$$

$$\Sigma F_y = N + T\sin\alpha - mg\cos\theta = 0 \qquad \text{(ii)}$$

From (ii) $N = mg\cos 15° - T\sin 25° = 27.3$ N. From (i) we find

$$\mu N = T\cos 25° - mg\sin 15°$$

thus $\mu = 0.197$. When $T = 0$, $f = \mu N = \mu(mg\cos\theta)$ is up the incline. From the free-body diagram, Fig. 6.8c, we see that

$$\Sigma F_x = mg\sin 15° - \mu mg\cos 15° = ma$$

which yields $a = 0.67$ m/s^2, downward.

Exercise 33

A small block is placed inside a cylinder of radius R = 40 cm that rotates with a period of 2 s about a horizontal axis as in Fig. 6.9. Show that the maximum angle θ reached by the block before its starts to slip is given by

$$g\sin\theta = \mu_s(g\cos\theta + v^2/R)$$

where $\mu_s = 0.75$ is the coefficient of static friction and v is the speed of the block. (b) Determine θ. (Hint: Use $\sin^2\theta = 1 - \cos^2\theta$ and solve a quadratic in $\cos\theta$.)

Solution:

(a) Just before the block starts to slip it is in uniform circular motion with $a_r = v^2/r$ and $a_t = 0$. The frictional froce is up along the rim. The forces and axes are shown in the figure.

$$\Sigma F_x = mg\sin\theta - \mu N = 0 \quad \text{(i)}$$

$$\Sigma F_y = N - mg\cos\theta = mv^2/R \quad \text{(ii)}$$

Substitute for N from (ii) into (i) to obtain $g\sin\theta = \mu(g\cos\theta + v^2/R)$.

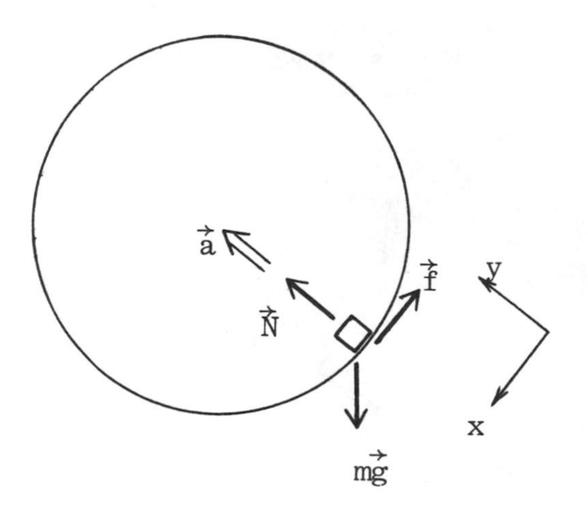

FIGURE 6.9

(b) When the given values are substituted into the equation we find, $\sin\theta = 0.75\cos\theta + 0.302$. Use $\sin\theta = (1 - \cos^2\theta)^{1/2}$ to obtain a quadratic equation in $\cos\theta$ (which we write as "c" for simplicity):

$$1.563c^2 + 0.453c - 0.909 = 0$$

The solution is $\cos\theta = 0.631$, which leads to $\theta = 50.9°$.

Exercise 51
The moon Io of Jupiter is in a circular orbit of radius $4.22\text{x}10^5$ km with a period of 1.77 d. (a) The period of another moon, Europa, is 3.55 d. What is the radius of its orbit? (b) What is the mass of Jupiter?

Solution:
(a) According to Kepler's third law the period and radius of an orbiting satellite, planet or moon are related by $T^2 = \kappa r^3$, where κ depends on the mass of the central body. Thus, for two moon orbiting Jupiter, we have

$$(r_2/r_1)^3 = (T_2/T_1)^2$$

This leads to $r_2^3 = (3.55/1.77)^2(4.22\text{x}10^8 \text{ m})^3$, so $r_2 = 6.71 \text{ x } 10^5$ km.

(b) According to Kepler's third law $T^2 = 4\pi^2 r^3/GM$, where M is the mass of the central body thus

$$M = 4\pi^2 r^3/GT^2$$

The given period is T = (1.77 d)(24 h)(3600 s/h) = $1.53\text{x}10^5$ s. When this and $r = 4.22\text{x}10^8$ m are used, we find $M = 1.90\text{x}10^{27}$ kg.

Exercise 55
In November 1984 the space shuttle Discovery was placed in circular orbit at an altitude of 315 km in order to catch up with the disabled Westar 6 satelllite in circular orbit at an altitude of 360 km. Suppose that these two objects were initially on opposite sides of the earth. How many orbits would the satellite have made for the shuttle to be beneath the satellite (that is, along a radial line and closer to the earth)? Take R_E = 6370 km.

Solution:
As Fig. 6.10 shows, the radius of the orbit is $r = R_E + h$, where R_E is the radius of the earth and h is the altitude. If "1" refers to the space shuttle and "2" refers to Westar, then r_1 = 6685 km and r_2 = 6730 km.
From Kepler's third law, the periods and radii of the two orbits are related by

$$(T_1/T_2)^2 = (r_1/r_2)^3$$

$$= (6685/6730)^3$$

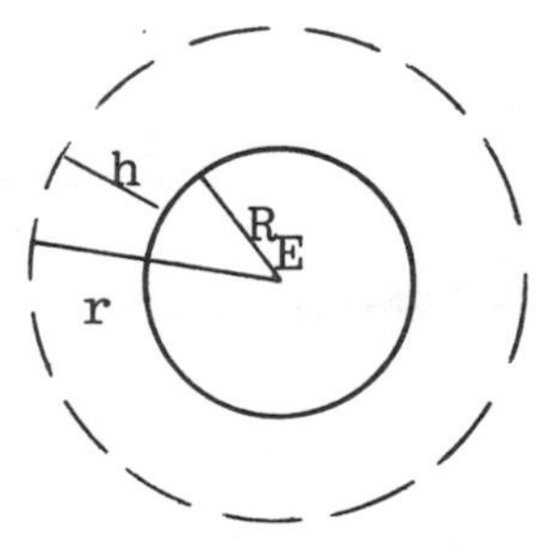

This leads to $T_1/T_2 = 0.99$. If Westar makes n orbits then the shuttle must make n + 1/2 orbits to be radially below Westar. Equating the total times for these orbits, we have

$$nT_2 = (n + 0.5)T_1$$

FIGURE 6.10

This means $n/(n + 0.5) = 0.99$, which leads to $n = 49.5$ orbits. In practice, the space shuttle would rendezvous fairly quickly by switching to another (elliptical) orbit.

Problem 5

A toy car of mass m can travel at a fixed speed. It moves in a circle on a horizontal table. The centripetal force is provided by a string attached to a block of mass M that hangs as shown in Fig. 6.11. The coefficent of static friction is μ. Show that the ratio of the maximum radius to the minimum radius possible is

$$(M + \mu m)/(M - \mu m)$$

Solution:

We are given that v and the tension $T = Mg$ are constant. At the larger radius R_1, the tension is not sufficient to provide the centripetal force so the friction is directed inward as in Fig. 6.11a, thus

$$T + \mu mg = mv^2/R_1 \qquad \text{(i)}$$

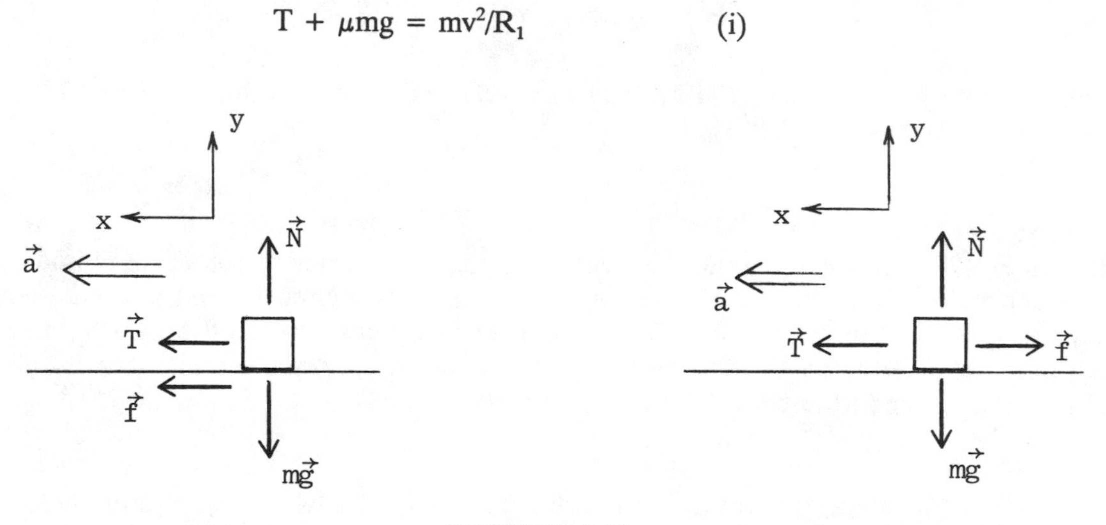

FIGURE 6.11

At the smaller radius R_2, the tension is too large and so the frictional force is outward, as in Fig. 6.11b:

$$T - \mu mg = mv^2/R_2 \qquad \text{(ii)}$$

Since $T = Mg$, the ratio (ii)/(i) yields

$$R_1/R_2 = (M + \mu m)/(M - \mu m)$$

Problem 11
A block rests on an incline that is at an angle θ to the horizontal, as in Fig. 6.12. The coefficients of friction are μ_s and μ_k. (a) Show that the block starts to slide when $\mu_s = \tan\theta_s$, where θ_s is the maximum value of θ. (b) Show that if θ is slightly greater than θ_s, the time taken to slide a distance d along the incline is given by

$$t^2 = 2d/[g\cos\theta_s(\mu_s - \mu_k)]$$

(c) Could one find μ_k in terms of θ? If so, how?

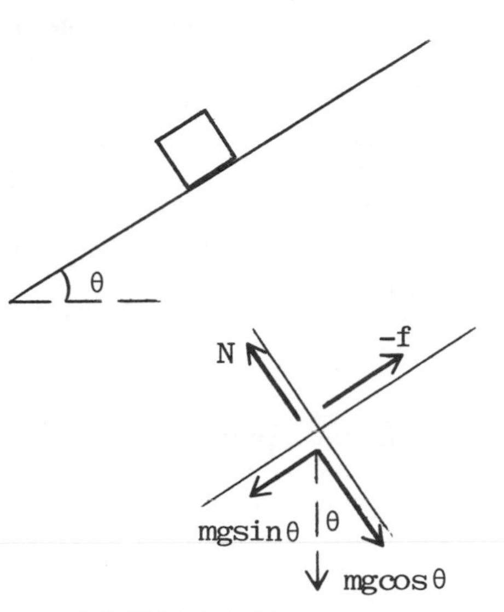

FIGURE 6.12

Solution:
(a) The forces and the axes are shown in Fig. 6.12. The net force down the incline is

$$\Sigma F_x = mg\sin\theta - \mu_s(mg\cos\theta)$$

The block will slide only if $\Sigma F_x > 0$.
If $\Sigma F_x = 0$, we find $\tan\theta_s = \mu_s$.

(b) Once the block is sliding, we have

$$\Sigma F_x = mg\sin\theta - \mu_k(mg\cos\theta) = ma \quad \text{(i)}$$

Since θ is close to θ_s, we have

$$a = g(\sin\theta_s - \mu_k\cos\theta_s) = g\cos\theta_s(\mu_s - \mu_k)$$

The distance moved along the incline is $d = 1/2\, at^2$, so $t^2 = 2d/a$, and

$$t^2 = 2d/[g\cos\theta_s(\mu_s - \mu_k)]$$

(c) When $a = 0$ in (i), we have $\tan\theta_k = \mu_k$. The block will slide at constant speed down the incline.

SELF-TEST

1. A block of mass m = 2 kg is held at rest on a rough incline (θ = 40°) by a force F directed along the incline. If the coefficient of static friction is μ_s = 0.6, what are the maximum and minimum values of F for which the block will stay at rest? Note that $\tan\theta > \mu_s$.

2. In a conical pendulum, a bob of mass m = 3 kg rotates in a horizontal circle at the end of a string of length L = 2 m that makes an angle θ = 37° with the vertical, as in Fig. 6.13. Find the tension in the string and the speed of the bob.

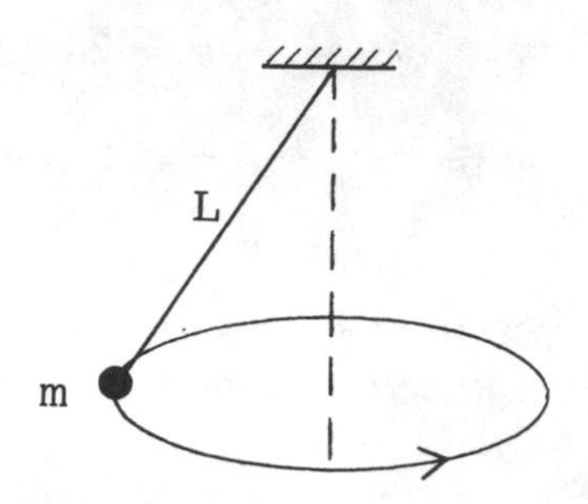

FIGURE 6.13

3. Two moons are in circular orbit about a planet of mass M and radius R. The altitudes of the moons are h_A = R and h_B = 2R. Determine the ratio of (a) the speeds, v_B/v_A; (b) the periods T_B/T_A.

Chapter 7

WORK AND ENERGY

MAJOR POINTS

1. (a) The definition of work done by a constant force
   (b) Work as the area under the F versus x curve
2. Kinetic energy: The energy on object has by virtue of its motion
3. The work energy theorem: $W_{NET} = \Delta K$
4. The definition of power

CHAPTER REVIEW

WORK

The work done by a constant force **F** when its point of application undergoes a displacement **s** is

$$W = \mathbf{F}.\mathbf{s} = F\ s\ \cos\theta$$

where θ is the angle between **F** and **s**, as in Fig. 7.1. The SI unit of work is the joule (J). Work is done only by the component of the force along the direction of the displacement. In unit vector notation $\mathbf{F} = F_x\,\mathbf{i} + F_y\,\mathbf{j} + F_z\,\mathbf{k}$ and $\mathbf{s} = \Delta x\mathbf{i} + \Delta y\mathbf{j} + \Delta z\mathbf{k}$, then

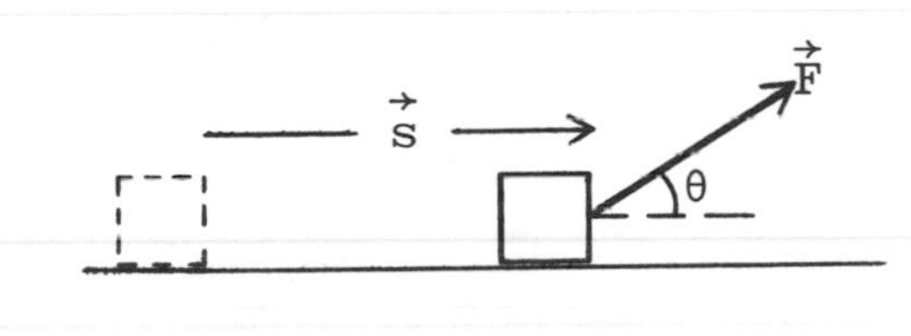

FIGURE 7.1

$$W = F_x\,\Delta x + F_y\,\Delta y + F_z\,\Delta z$$

Work is a scalar quantity that can be positive or negative. Thus if **F** is perpendicular to **s** the work done is zero. If several forces act on a body the net work done is simply:

$$W_{NET} = \mathbf{F}_{NET}\,.\,\mathbf{s}$$

Graphically work done by a force in the x direction, say, may be represented by the area under the F versus x curve. For an infinitesimal displacement we treat the force as being constant during the displacement, and the work done by F is

$$dW = F\ dx$$

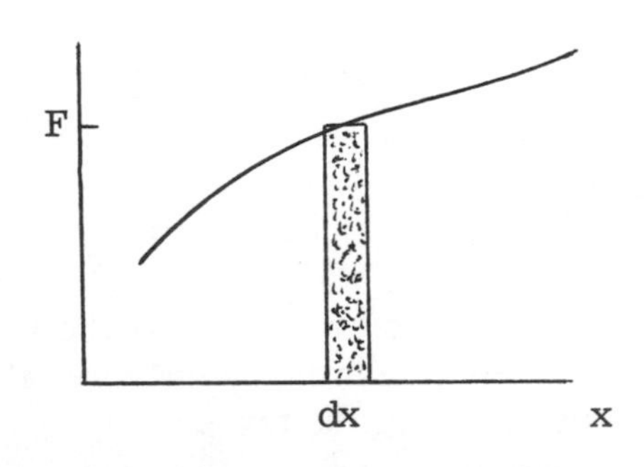

FIGURE 7.2

This is the area of the thin rectangle

in Fig. 7.2. In general, the work done by a force **F** along a path from point A to point B is

$$W = \int \mathbf{F}.d\mathbf{s}$$

$$= \int F_x\, dx + \int F_y\, dy + \int F_z\, dz$$

The sign of each integral depends on the sign of the component of **F** and the direction in which the path is taken (which is indicated by the order of the limits).

EXAMPLE 1
A force **F** = 2**i** - 3**j** + **k** N acts on a particle as it moves from $\mathbf{r}_1$ = 3**i** - 5**k** m to $\mathbf{r}_2$ = 2**j** + **k** m. What is the work done by the force on the particle?

Solution:
The displacement of the particle is **s** = $\mathbf{r}_2 - \mathbf{r}_1$ = -3**i** + 2**j** + 4**k** m.
The work done is

$$W = \mathbf{F} \cdot \mathbf{s}$$

$$= (2\mathbf{i} - 3\mathbf{j} + \mathbf{k}).(-3\mathbf{i} + 2\mathbf{j} + 4\mathbf{k}) = -6 - 6 + 4 = -8 \text{ J}$$

EXAMPLE 2
A force varies with position according to **F**(x, y) = 3x**i** - y^2**j** N. What is the work done by this force from (-2 m, -1 m) to (1 m, 2 m)?

Solution:

$$W = \int F_x\, dx + \int F_y\, dy = \int_{-2}^{1} 3x\, dx - \int_{-1}^{2} y^2\, dy$$

$$= \left| 3x^2/2 \right| - \left| y^3/3 \right| = -4.5 - 3 = -7.5 \text{ J}$$

WORK DONE BY FRICTION
When a block is dragged along a surface the force of friction **f** is opposite to the displacement if the block, thus the work done by friction is negative

$$W_f = \mathbf{f} \cdot \mathbf{s} = f\, s \cos(180^\circ) = -\, f\, s$$

This work is done by friction on the block and on the surface, both of which become warmer. Although in many instances the work done by friction is negative, this is not always true (see p. 126 in the text).

WORK DONE BY THE (CONSTANT) FORCE OF GRAVITY

Near the surface of the earth the force of gravity may be taken to be constant. In this case, the work done by the force $\mathbf{mg} = -\,mg\,\mathbf{j}$ in a displacement $\mathbf{s} = \Delta x\mathbf{i} + \Delta y\mathbf{j} + \Delta z\mathbf{k}$ is

$$W_g = \mathbf{mg} \cdot \mathbf{s} = (-\,mg\,\mathbf{j}) \cdot (\Delta x\mathbf{i} + \Delta y\mathbf{j} + \Delta z\mathbf{k})$$

$$W_g = -\,m\,g\,\Delta y = -\,m\,g\,(y_f - y_i)$$

An important feature of the work done by gravity is that it depends only on the initial and final vertical coordinates, not on the path taken. Note carefully the negative sign in this equation and that y is the vertical coordinate. If a body moves downward, then $\Delta y < 0$ and $W_g > 0$.

WORK DONE BY A SPRING

The force exerted by a spring is given by Hooke's law

$$F_{sp} = -\,k\,x$$

where the spring constant k is measured in N/m, and x is the displacement from the equilibrium position. The work done by this non-constant force may be found from the area under the F versus x graph. Alternatively we could use integration:

$$W_{sp} = \int F_x\,dx = \int (-\,kx)\,dx$$

$$= \left| -kx^2/2 \right|_{x_i}^{x_f}$$

$$W_{sp} = -\,1/2\,k\,(x_f^2 - x_i^2)$$

Note the presence of the negative sign. The work done by a spring depends only on the initial and final points, not on the path taken. You should be aware that Hooke's law is sometimes expressed in terms of the external force required to produce an extension: $F = +kx$.

WORK-ENERGY THEOREM

According to the work-energy theorem, the net work done on a body is equal to the change in its kinetic energy:

$$W_{NET} = \Delta K = K_f - K_i$$

where K, the kinetic energy, is defined as

$$K = 1/2\,mv^2$$

Kinetic energy is energy a body has by virtue of its motion. The work-energy theorem can be derived from Newton's second law $F = ma$. When using dynamics, we say that a force acting on a body causes it to accelerate. According to the work-energy theorem, when work is done on a body its kinetic energy changes. This approach turns out to be quite simple and powerful and furthermore, it leads to the fundamental principle of the conservation of energy (Ch. 8). The use of the concepts of work and energy is basically easier because these quantities are scalars, whereas force is a vector.

EXAMPLE 3

A child pulls a block of mass m = 2 kg up a slope inclined at $\theta = 30°$ to the horizontal, as in Fig. 7.3. The rope is at $\alpha = 20°$ to the incline and the tension in the rope is T = 30 N. The coefficient of friction is $\mu_k = 0.45$. What is the speed of the block after it has moved a distance d = 0.5 m along the incline? The spring (k = 12 N/m) is initially unextended and the block is at rest.

Solution:

The work done by the tension in the rope is

$$W_T = Td \cos\alpha = 14.1 \text{ J}$$

Since $\Delta y = +d \sin\theta$, the work done by gravity is

$$W_g = -mg\Delta y = -mgd \sin\theta = -4.9 \text{ J}$$

The work done by the spring is negative. With $x_i = 0$ and $x_f = d$:

$$W_{sp} = -k(x_f^2 - x_i^2)/2$$

$$= -1/2\ kd^2 = -1.5 \text{ J}$$

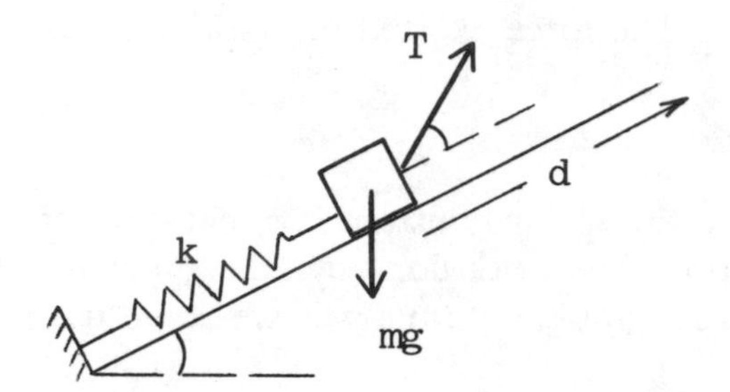

FIGURE 7.3

The normal force is $N = mg \cos\theta - T \sin\alpha = 6.71$ N (draw a free-body diagram if you do not see this). Since $f = \mu N$, the work done by friction is

$$W_f = -fd = -\mu\ N\ d = -1.5 \text{ J}$$

Since the block starts at rest, the change in kinetic energy is simply

$$\Delta K = 1/2\ mv^2 - 0$$

According to the work-energy theorem, $W_{NET} = \Delta K$, thus

$$W_T + W_g + W_{sp} + W_f = \Delta K$$

When the calculated values are inserted we find v = 2.49 m/s.

POWER

If an amount of work ΔW is done by some agent in a time interval Δt, the average power delivered by the agent is defined as

$$P_{av} = \Delta W/\Delta t$$

The SI unit of power is Joule/sec = Watt (W). The instantaneous power is the rate at which work is being done

$$P = dW/dt$$

Since W = **F.s**, we also have that P = dW/dt = **F**.ds/dt, that is

$$P = \mathbf{F} \cdot \mathbf{v}$$

Power may also be defined in terms of the rate at which one form of energy is being tranformed into another, thus

$$P = dE/dt$$

EXAMPLE 4

An athlete in training gives a spring (k = 50 N/m) an extension of 40 cm in 0.6 s. What is his average power output?

Solution:

The work done by the person is $W_{ext} = -W_{sp}$, so

$$W_{ext} = 1/2\ kx^2 = (25\ \text{N/m})(0.4)^2 = 4\ \text{J}$$

The average power output is

$$P_{av} = \Delta W/\Delta t = 4\ \text{J}/0.6\ \text{s}$$

$$= 6.67\ \text{W}$$

SOLUTIONS TO SELECTED TEXT EXERCISES AND PROBLEMS

Exercise 11

An inclined conveyor belt lowers a block of mass M = 20 kg at a constant speed v = 3 m/s, as shown in Fig. 7.4. (a) What is the work done by the motor on the block when the block moves 2 m? (b) What would be the work done by the motor to raise the block through the same distance at constant speed?

Solution:

The tension in the belt is determined by the frictional force between the block and the belt. If there is no slipping, $f = mg \sin\theta$. The speed of the block is not relevant since it is constant.

(a) The tension force is opposite to the displacement so

$$W_{motor} = -mgs \sin\theta = -134 \text{ J}$$

(b) In this case, the tension is in the same direction as the displacement, so $W_{motor} = +134$ J

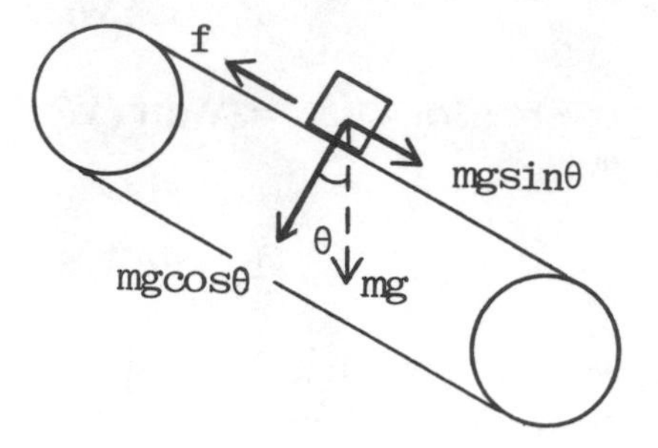

FIGURE 7.4

Exercise 15

The variation of a force with position is depicted in Fig. 7.5. Find the work from (a) $x = 0$ to $x = -A$, (b) $x = +A$ to $x = 0$.

Solution:

In each case we must find the area between the force function and the x axis. The sign of the work depends on the signs of both F_x and the displacement Δx.

(a) Here $F_x < 0$ and $\Delta x < 0$, so $W > 0$. The area of the triangle is

$$W = +F_oA/2$$

(b) Here $F_x > 0$ and $\Delta x < 0$, so $W < 0$. The area of the triangle is

$$W = -F_oA/2$$

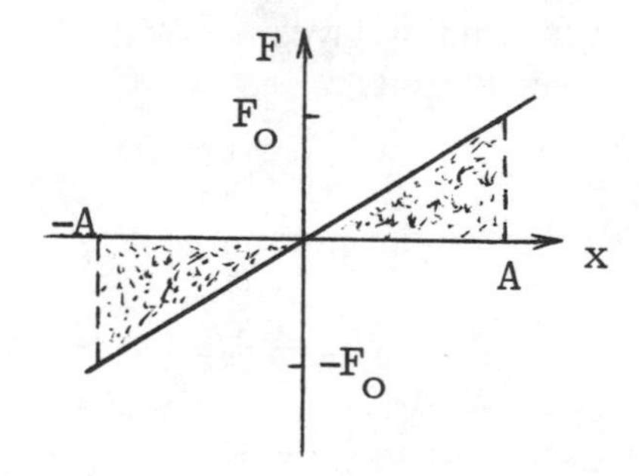

FIGURE 7.5

This example shows that areas above the x axis are not necessarily positive and that areas below the x axis are not necessarily negative.

Exercise 41

A force $F = 24$ N acts at 60° above the horizontal on a 3-kg block that is attached to a spring whose stiffness constant is $k = 20$ N/m (see Fig. 7.6). Assmume $\mu_k = 0.1$. The system starts at rest with the spring unextended. If the block moves 40 cm, find the work done (a) by F, (b) by friction, (c) by the spring. (d) What is the final speed of the block?

Solution:

(a) $W_F = Fs\cos\theta = 4.8$ J

(b) Since $N - mg + F\sin\theta = 0$, we have $N = 8.62$ N and $f = \mu N = 0.86$ N. Thus,

$$W_f = -fs = -0.34 \text{ J}$$

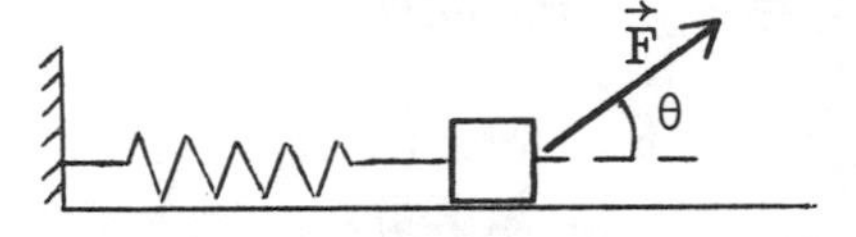

(c) $W_{sp} = -1/2\, k(x_f^2 - x_i^2)$

$$= -1/2\,(20)(0.4)^2 = -1.6 \text{ J}$$

(d) $W_{NET} = W_F + W_f + W_{sp} = 2.86$ J and $\Delta K = 1/2\, mv^2$. From $W_{NET} = \Delta K$ we find $v = 1.38$ m/s.

FIGURE 7.6

Exercise 53
A champion cyclist can sustain an output of 0.5 hp for 10 min. How far could he travel at constant speed if the net retarding force is 18.5 N?

Solution:
Since 1 hp = 746 W, his power output is P = 373 W. At constant speed, the output of the cyclist equals the loss to the retarding force, thus

$$P = fv = (18.5 \text{ N})v$$

We find v = 20.2 m/s. At this speed his displacement in 10 min is

$$\Delta x = v\Delta t = (20.2 \text{ m/s})(600 \text{ s}) = 12.1 \text{ km}$$

Exercise 59
The heat of combustion of gasoline is $3.4\text{x}10^7$ J/L. The fuel economy of a car is rated at 12 km/L at 100 km/h. If the mechanical output at this speed is 25 hp, what is the efficiency (power output/power input) of the motor?

Solution:
We need to calculate the power input, measured in J/s, delivered by the combustion of the gasoline. By carefully watching the units of the three quantities involved, we find

$$(3.4\text{x}10^7 \text{ J/L})(100 \text{ km}/3600 \text{ s})(1 \text{ L}/12 \text{ km}) = 7.87\text{x}10^4 \text{ W}$$

The power output is 25 hp = $1.87\text{x}10^4$ W, thus

$$\text{Efficiency} = 1.87\text{x}10^4 \text{ W}/7.87\text{x}10^4 \text{ W} = 23.7\%$$

Problem 7
A person lifts a 25-kg crate hanging over a pulley by walking horizontally, as shown in Fig. 7.7. As the person walks 2 m, the angle of the rope to the horizontal changes from 45° to 30°. How much work does the person do if the crate rises at constant speed?

Solution:
The fact that the speed is constant tells us that the tension in the rope is fixed at T = mg. The work done is the product of the tension and the height through which the block is raised. Thus we need the change in length,

$$\Delta \ell = \ell_2 - \ell_1$$

of the rope held by the person, as indicated in Fig. 7.7. From the law of sines:

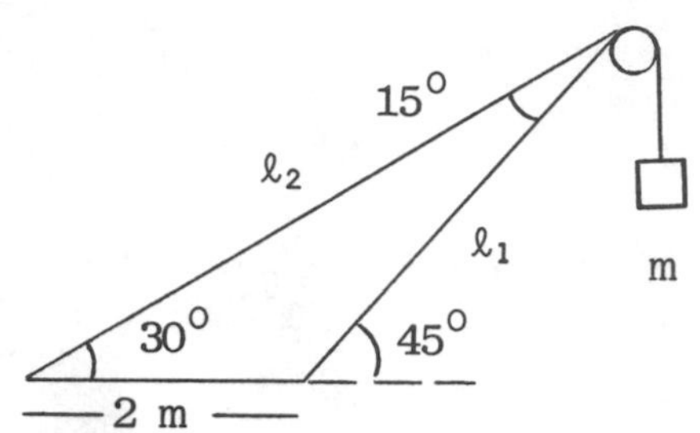

FIGURE 7.7

$$2/\sin 15° = \ell_1/\sin 30° = \ell_2/\sin 45°$$

thus, $\ell_1 = 3.86$ m and $\ell_2 = 5.46$ m, and so $\Delta\ell = 1.6$ m. The work done is

$$W = mg\,\Delta\ell = 392 \text{ J}$$

Problem 9
The retarding force on a 1100 kg car is given by $f = 200 + 0.8v^2$, where v is in m/s. (a) What is the power required at the wheels at 20 m/s on a horizontal road? (b) What is the "road power" needed to travel up a 5° incline at 0.5 m/s^2 when v = 20 m/s? (c) Given that only 15% of the energy supplied by the combustion of fuel is deliviered to the wheels, what is the "fuel input power" (in hp) required for part (b)?

Solution:
(a) The retarding force at 20 m/s is $f = 200 + 0.8(20)^2 = 520$ N. At constant speed the driving force F at the wheels equals f. The power required to travel at constant speed on a horizontal road is

$$P = Fv = (520 \text{ N})(20 \text{ m/s})$$

$$= 10.4 \text{ kW} = 13.9 \text{ hp}$$

(b) Figure 7.8 shows the car accelerating up the incline. The driving force at the wheels is given by Newton's second law:

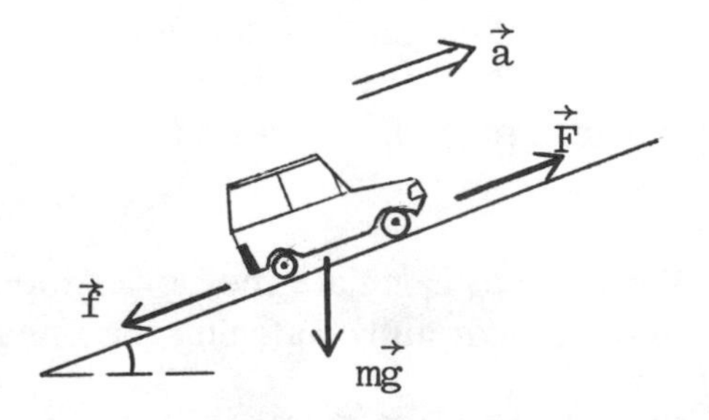

FIGURE 7.8

$$F - f - mg\sin\theta = ma$$

thus,

$$F = f + mg\sin\theta + ma = 2010\ \text{N}$$

The power required at the wheels is therefore

$$P = Fv = (2010\ \text{N})(20\ \text{m/s}) = 40.2\ \text{kW} = 53.9\ \text{hp}$$

(c) FIP = P/Efficiency = 268 kW = 359 hp.

SELF-TEST

1. A horizontal force F = 60 N acts on a block of mass m = 2.5 kg attached to a spring (k = 15 N/m) on a rough incline (μ_k = 0.12 and θ = 30°), see Fig. 7.9. The block starts at rest with the spring unextended and is moved 80 cm up the incline.
   Find: (a) The work done by each of the forces acting on the block; (b) The change in kinetic energy of the block; (c) The instantaneous power delivered by the force F at the end of this displacement.

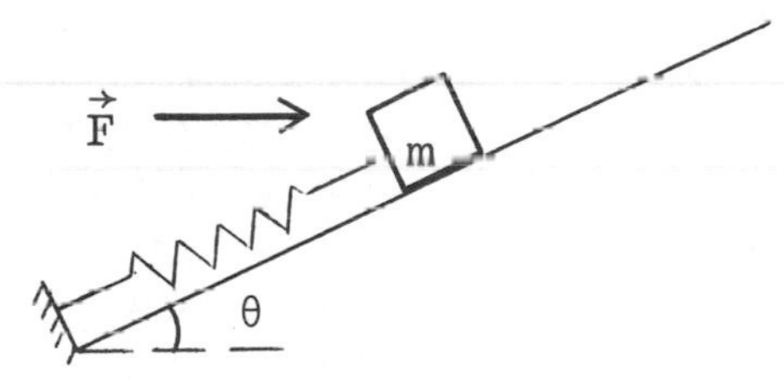

FIGURE 7.9

2. (a) Car A has twice the mass of car B but initially they have the same kinetic energy. What is the ratio of their speeds, v_B/v_A? (b) When both cars slow down by 4 m/s, the kinetic energy of car A is one half that of B. Determine the initial speeds v_A and v_B.

3. A particle is acted on by a force $\mathbf{F} = 5/x^2\ \mathbf{i}$. What is the work done by this force from x = 2 m to x = 5 m?

Chapter 8

CONSERVATION OF ENERGY

MAJOR POINTS

1. The distinction between conservative and nonconservative forces
2. (a) The concept of potential energy
   (b) The relation between potential energy and conservative forces
3. (a) Gravitational potential energy near the earth's surface
   (b) The potential energy of a spring
4. (a) The conservation of mechanical energy
   (b) Modification of the conservation law if work is done by nonconservative forces.
5. Gravitational potential energy taking into account Newton's law of gravitation

CHAPTER REVIEW

CONSERVATIVE FORCES

One can define a conservative force in either of the following ways:

(1) The work done by a conservative force depends only on the initial and final points, not on the path taken.
(2) The work done by a conservative force is zero around any closed loop.

We noted in Ch. 7 that the work done by the force of gravity and by the spring force depend only on the initial and final points, thus these two forces are conservative. In contrast, the work done by friction does depend on the length of the path and so it is a non-conservative force. Similarly, the force you exert with your hand is a nonconservative force because its magnitude can vary arbitrarily. A conservative force is a function only of position. Fluid resistance depends on the speed of a body relative to the fluid, and so it is a nonconservative force.

POTENTIAL ENERGY

Potential energy is energy shared by two interacting particles by virtue of their positions. When you lift up a ball from the ground, the work done by you is stored as potential energy in ball-earth system. If the kinetic energy of the ball is unchanged, then all the work appears as a change in potential energy:

(Constant speed) $$W_{EXT} = +\Delta U = U_f - U_i$$

Note that it is the change in U that appears; one can assign the initial potential energy U_i any convenient value. If $U_i = 0$, we see that

The potential energy of a system is the external work needed to bring the particles from the $U = 0$ configuration to the given positions with no change in kinetic energy.

Potential energy can be defined only for a conservative force because the work done by such a force depends only only the initial and final points, not the path taken. A scalar potential energy can be associated with a conservative force, but not with a nonconservative force. The change in potential energy between two points is defined as the negative of the work done by the conservative force:

$$W_c = -\Delta U = -(U_f - U_i)$$

There is no reference to an external agent and the speed of a particle does not have to be constant. In general, the conservative force F_c varies with position, so

$$U_f - U_i = -\int_i^f F_c \cdot ds$$

If the potential energy function is known, one can determine a component of the associated conservative force along the displacement ds:

$$F_s = -\partial U/\partial s$$

where often s is along the x, y, z, or r axis.

EXAMPLE 1
(a) A conservative force is given by $F_x = 3x^2 - 4x + 5$ N. Find the change in potential energy from $x = 1$ m to $x = 3$ m. (b) A potential energy function is given by $U = 5x^2y^3 - 3xy^2$. Find the force as a function of position.

Solution:
(a)

$$U_f - U_i = -\int F_x\, dx = -\int (3x^2 - 4x + 5)\, dx$$

$$= -\left| x^3 - 2x^2 + 5x \right|_1^3 = -20 \text{ J}$$

(b) The components of the force are

$$F_x = -\partial U/\partial x = 10xy^3 - 3y^2$$

$$F_y = -\partial U/\partial y = 15x^2y^2 - 6xy$$

<u>Gravitational potential energy near the earth's surface.</u>
Near the earth's surface, g is approximately constant. The gravitational potential energy of a particle of mass m at a vertical position y is

$$U_g = mgy$$

Note that $U_g = 0$ at $y = 0$. This function depends only on the vertical position.

Spring Potential Energy

The work done to expand or to compress a spring by x is stored as elastic potential energy in the spring:

$$U_{sp} = 1/2\ kx^2$$

where k is the spring constant is measured in N/m.

CONSERVATION OF MECHANICAL ENERGY

The mechanical energy of a system is the sum of its kinetic and potential energies. In general,

$$E = K + U_g + U_{sp}$$

If there is no work done on an system by any external force or any internal nonconservative force, the mechanical energy of the system is constant:

$$\Delta E = 0; \qquad \text{or} \qquad E_f = E_i$$

If you use $E_f = E_i$, it is necessary to specify the points at which $U_g = 0$. ($U_{sp} = 0$ when the extension or compression is zero.). If you use $\Delta E = \Delta K + \Delta U = 0$, you do not need to set the $U_g = 0$ level, however, care must be taken with the signs of the changes ΔK and ΔU.

MECHANICAL ENERGY AND NONSERVATIVE FORCES

If work is done by nonconservative forces, then the conservation of energy equation is modified:

$$\Delta E = E_f - E_i = W_{NC}$$

That is, the change in mechanical energy of the system is equal to the work done by noncsonservative forces.

EXAMPLE 2

A block of mass $m = 1.5$ kg starts at rest and slides 1.2 m down a frictionless incline ($\theta = 30°$) and then compresses a horizontal spring ($k = 24$ N/m), see Fig. 8.1. (a) Find the maximum compression of the spring given that it is initially at its natural length. (b) What is the speed of the block at the bottom of the incline?

Solution:

(a) The initial energy is the potential energy of the block

$$E_i = mgy = mgd\sin\theta = 8.82 \text{ J}$$

We have set the $U_g = 0$ level at the bottom of the incline. The final energy is the potential energy of the spring, which is compressed by a distance x:

$$E_f = 1/2\, kx^2 = 12x^2$$

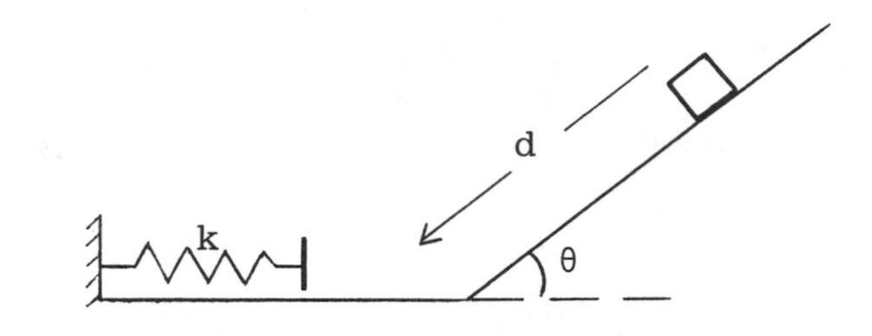

FIGURE 8.1

From the conservation of energy $E_f = E_i$:

$$1/2\, kx^2 = mgd\sin\theta$$

We find x = 0.86 m.

(b) The initial energy is still $E_i = mgd\sin\theta$. At the bottom of the incline the block has only kinetic energy, so $E_f = 1/2\, mv^2$. From $E_f = E_i$, we have

$$mgd\sin\theta = 1/2\, mv^2$$

This leads to v = 3.43 m/s. Note that we did not need to find this v in (a).

EXAMPLE 3

Two blocks, $m_1 = 1.5$ kg and $m_2 = 4.5$ kg, hang on either side of a pulley as shown in Fig. 8.2. Block m_2 is on a rough incline ($\mu_k = 0.1$) which is at angle $\theta = 50°$ to the horizontal. Block m_1 hangs vertically and is connected to a vertical spring (k = 8 N/m). The system starts at rest with the spring unextended. What is the speed of m_2 after it slides 80 cm down the incline?

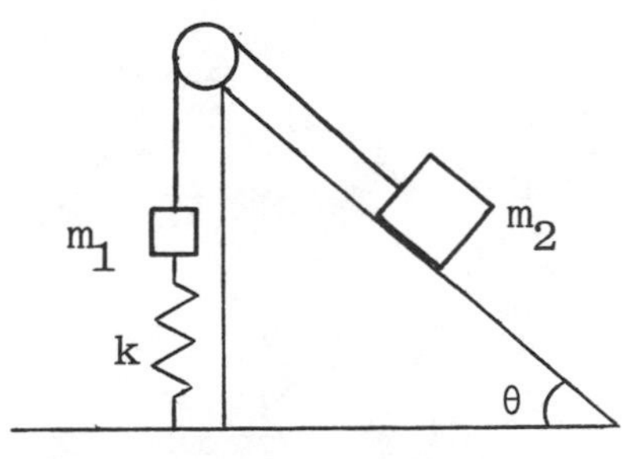

FIGURE 8.2

Solution:
We do not know the initial positions of the blocks, hence it is easier to use the form $\Delta E = W_{NC}$. An increase in any quantity means that the change is positive. Although we have been asked only for the speed of m_2, one must not forget that its is connected to m_1. Thus,

$$\Delta K = 1/2\ (m_1 + m_2)v^2 = 3v^2$$

The potential energy of m_1 increases and that of m_2 decreases. When m_2 slides a distance d:

$$\Delta U_g = m_1gd - m_2gd \sin\theta = -15.3 \text{ J}$$

The potential energy of the spring increases, so

$$\Delta U_{sp} = 1/2\ kd^2 = 2.56 \text{ J}$$

Since the normal force is $N = m_2g \cos\theta$, the nonconservative work done by friction is

$$W_{NC} = -fd = -\mu(m_2g \cos\theta)d = -2.26 \text{ J}$$

The modified form of the conservation of energy principle is:

$$\Delta K + \Delta U_g + \Delta U_{sp} = W_{NC}$$

Using the numerical values above, we find v = 1.87 m/s.

EXAMPLE 4
A roller coaster car starts at a height of 20 m and completes a loop-the-loop of radius 7 m. What is the force exerted by the car on a 60 kg person at the highest point in the loop?

Solution:
This problem requires two distinct methods. First we must use the conservation of energy to find the speed of the car at the highest point in the loop. Then we use Newton's second law to find the force exerted by the car on the person.

With the choice of $U_g = 0$ at the lowest point, the initial and final energies are (see Fig. 8.3)

$$E_i = mgH$$

$$E_f = 1/2\ mv^2 + mg(2R)$$

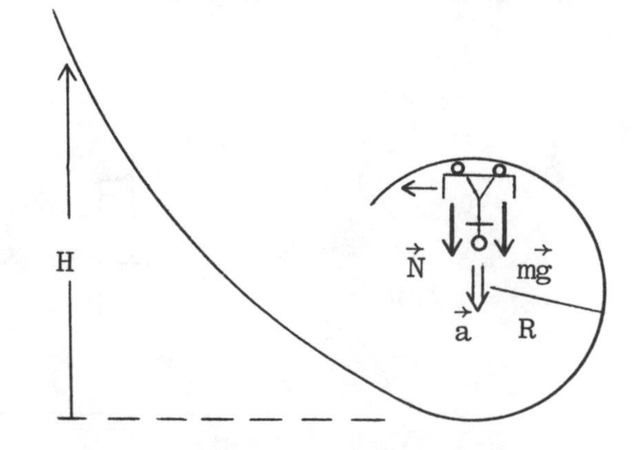

FIGURE 8.3

From $E_f = E_i$, we find

$$v^2 = 2g(H - 2R)$$

thus v = 11.7 m/s. At the highest point the force exerted by the car on the person is downward, so from Newton's second law

$$\Sigma F_y = N + mg = mv^2/R$$

so,

$$N = m(v^2/R - g)$$

$$= 585 \text{ N}$$

Note that if $v^2 = Rg$, $N = 0$ and the person would be apparently weightless at the highest point.

GRAVITATIONAL POTENTIAL ENERGY

The expression $U_g = mgy$ is valid only near the earth's surface where g is taken to be constant, independent of height. In general, one must take into account Newton's law of gravitation, $F = GmM/r^2$. When this is done, one finds the potential energy function for a particle of mass m and a planet of mass M is

$$U = -GmM/r$$

where r is the distance to the center of the planet. The relationship between this expression and $U_g = mgh$ is discussed in Example 8.10 in the text.

The mechanical energy of a rocket or a satellite at a distance r from the center of the planet takes the form

$$E = 1/2\ mv^2 - GmM/r$$

By applying Newton's second law, F = ma, that is $GmM/r^2 = mv^2/r$, on finds that the energy of a satellite in circular orbit is

(Orbit) $$E = -GmM/2r$$

An important concept is that of the escape speed, v_{esc}. If a rocket is fired at the earth's surface with v_{esc}, it will reach $r = \infty$ with zero speed and will not return to earth. At the surface of the earth (or other planet), of radius R, the energy is

$$E_i = 1/2\ mv^2 - GmM/R$$

To find the minimum escape speed, we need $E_f = 0$ at infinity. From $E_f = E_i$ we find that

$$v_{esc} = (2GM/R)^{1/2}$$

EXAMPLE 5
A rocket fired vertically reaches altitude of 2000 km with a speed of 5 km/s. What was its initial speed?

Solution:
The initial energy at $r = R$ and the final energy at $r = R + h$ are

$$E_i = 1/2\ mv_i^2 - GmM/R$$

$$E_f = 1/2\ mv_f^2 - GmM/(R + h)$$

From $E_i = E_f$ we find

$$v_i^2 = v_f^2 + GM[1/R - 1/(R + h)]$$

Next we use the values $R = 6.37\text{x}10^6$ m, $h = 2\text{x}10^6$ m, $M = 6\text{x}10^{24}$ kg, and $v_f = 5\text{x}10^3$ m/s to find $v_i = 6.33$ km/s.

EXAMPLE 6
The acceleration due to gravity at the surface of the moon is one-sixth that on earth. Given that the escape speed from the surface of the earth is 11.2 km/s, what is the escape speed from the surface of the moon? Ignore the rotational motion.

Solution:
The gravitational force at the surface of the moon of radius R may be written $mg_o = GmM/R^2$, where

$$g_o = GM/R^2$$

where g_o is the value at the surface. The escape speed $(2GM/R)^{1/2}$ can thus be written as

$$v_{esc} = (2g_oR)^{1/2}$$

At the surface of the moon $g_o = 9.8/6 = 1.63$ N/kg or 1.63 m/s^2 and $R = 1.74\text{x}10^6$ m (see the front cover), so we find $v_{esc} = 2.4$ km/s.

EXAMPLE 7
What is the escape speed from a small moon of radius 300 km and density 4.5 g/cm^3?

Solution:
The mass $M = \rho V = \rho 4\pi R^3/3$, where $\rho = 4500$ kg/m^3 is the density. The escape speed is

$$v_{esc} = (2GM/R)^{1/2}$$

$$= (8\pi\rho GR^2/3)^{1/2} = 476 \text{ m/s}$$

SOLUTIONS TO SELECTED TEXT EXERCISES AND PROBLEMS

Exercise 5

A simple pendulum has a length of 75 cm and a bob of mass 0.6 kg. When the string is at 30° to the vertical, the bob has a speed of 2 m/s. (a) What is the maximum speed of the bob? (b) What is the maximum angle to the vertical?

Solution:

We pick the $U_g = 0$ level at the lowest point in the swing, see Fig. 8.4. The vertical height of the bob is $y = L - L\cos\theta$, thus with the given values

$E = 1/2\ mv^2 + mgL(1 - \cos\theta) = 1.79$ J

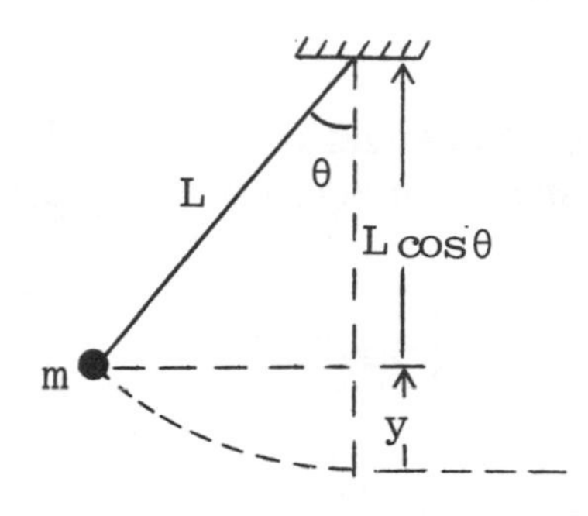

FIGURE 8.4

(a) At the lowest point,

$$E = 1/2\ mv_{max}^2 = 1.79 \text{ J}$$

thus, $v_{max} = 2.44$ m/s

(b) At the highest point,

$$E = mgL(1 - \cos\theta_{max}) = 1.79 \text{ J}$$

thus, $\theta_{max} = 53.6°$

Exercise 9

A 500-g block is dropped from a height of 60 cm above the top of a vertical spring whose stiffness constant is k = 120 N/m (see Fig. 8.5). Find the maximum compression. (You will need to solve a quadratic equation.)

Solution:

We choose the $U_g = 0$ level at the lowest point, where the compression is x, see Fig. 8.5. The initial and final energies are

$$E_i = mg\,(x + 0.6)$$

$$E_f = 1/2\ kx^2$$

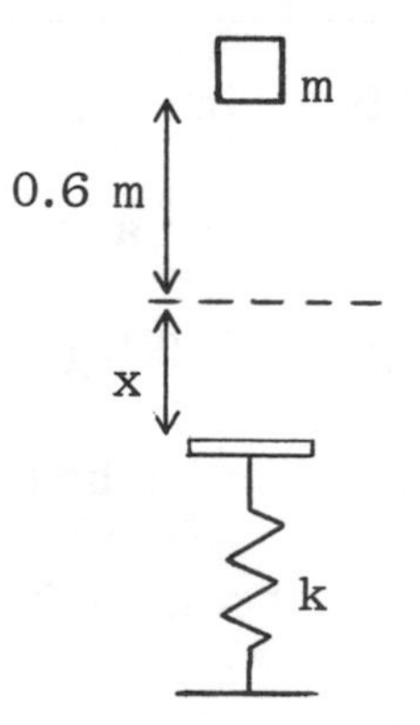

FIGURE 8.5

Note that we do not need to calculate any kinetic energy. On equating $E_f = E_i$ we obtain a quadratic equation

$$60x^2 - 4.9x - 2.94 = 0$$

The solution is x = 0.266 m.

Exercise 37
Given the force $F_x = ax/(b^2 + x^2)^{3/2}$, find the potential energy function. Take $U = 0$ at $x = \infty$.

Solution:
A change in potential energy is given by

$$U_f - U_i = -\int F_x\, dx$$

In the present case,

$$U(x) - U(\infty) = -\int_{\infty}^{x} ax\, dx/(b^2 + x^2)^{3/2}$$

where $U(\infty) = 0$. You could use the substitution $u = b^2 + x^2$, which means $du = 2xdx$ proceed or refer to the table of integrals in Appendix C.

$$U(x) = \left| a/(b^2 + x^2)^{1/2} \right|_{\infty}^{x}$$
$$= a/(b^2 + x^2)^{1/2}$$

Exercise 53
What is the minimum (net) energy needed to take a 1-kg object from the surface to the earth to the surface of the moon? Taken $g_M = 0.16g_E$, where $g_E = GM/R^2$ is the gravitational force per unit mass at the surface. (b) Show that this is roughly twice the work needed to put the object into an orbit close to the earth's surface.

Solution:
(a) We need the change in potential energy in going from the surface of the earth to the surface of the moon:

$$\Delta E = Gm(-M_M/R_M + M_E/R_E)$$

Since at the surface $g = GM/R^2$ and $g_M = 0.16g_E$, we have

$$\Delta E = m(-0.16g_E R_M + g_E R_E) = 60 \text{ MJ}$$

The values of R_M and R_E are on the inside front cover of the text.

(b) The object is initially at rest on the surface of the earth and finally in orbit. The initial and final energies are

$$E_i = -GmM_E/R_E; \qquad E_f = -GmM_E/2R_E$$

thus $\Delta E = E_f - E_i = mg_E R_E/2 = 31$ MJ.

Problem 5
A ball of mass m moves in a vertical circle at the end of a string. Show that the tension at the bottom is greater than at the top by 6mg.

Solution:

Suppose that the speeds at the top and the bottom are v_1 and v_2. The forces acting on the ball and the direction of the acceleration are shown in Fig. 8.6. From $\Sigma F_y = ma$, we obtain

TOP: $T_1 + mg = mv_1^2/R$

BOTTOM: $T_2 - mg = mv_2^2/R$

Thus,

$$T_2 - T_1 = 2mg + m(v_2^2 - v_1^2)/R$$

In order to relate v_1 and v_2 we use the conservation of energy. We pick $U_g = 0$ at the lowest point, so

$$E_1 = 1/2\ mv_1^2 + mg(2R)$$

$$E_2 = 1/2\ mv_2^2$$

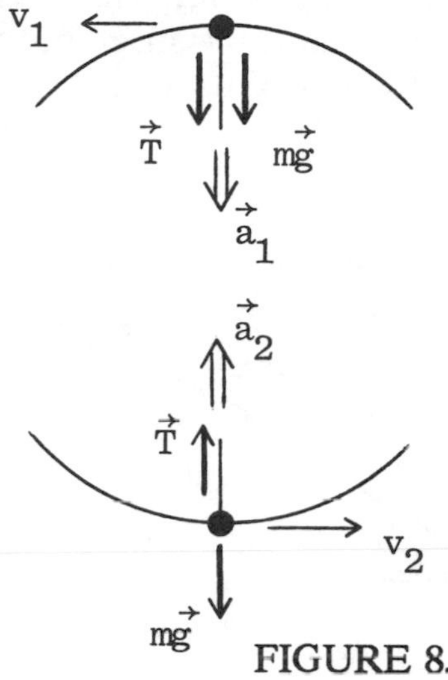

FIGURE 8.6

On setting $E_2 = E_1$ we find: $v_2^2 - v_1^2 = 4gR$. Thus $T_2 - T_1 = 6mg$.

Problem 11
An Eskimo child slides on an icy (frictionless) hemispherical igloo of radius R, as in Fig. 8.7. She starts with a negligible speed at the top. (a) At what angle to the vertical does she lose contact with the surface? (b) If there were friction, would contact be lost at a higher or a lower point?

Solution:
(a) As the child moves over the spherical surface, the radial component of Newton's second law is

$$\Sigma F_r = mg\cos\theta - N = mv^2/R$$

When the child loses contact $N = 0$, so $v^2 = gR\cos\theta$. If we set $U_g = 0$ at the top, $E_i = 0$. The vertical drop is $(R - R\cos\theta)$, so at a later time,

$$E_f = 1/2\ mv^2 - mgR(1 - \cos\theta)$$

From $E_f = E_i$ and the expression for v^2, we find $\cos\theta = 2/3$.

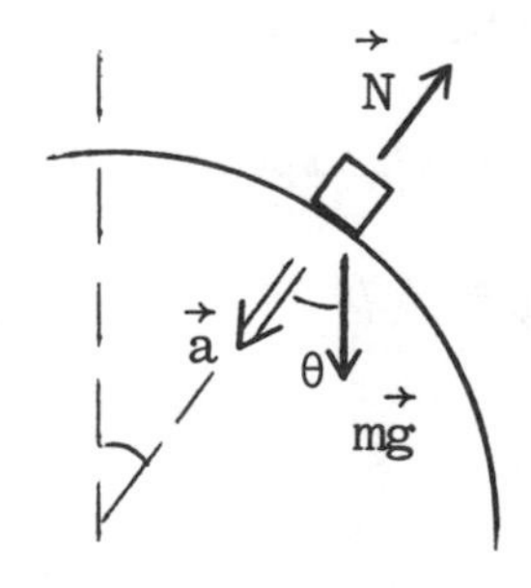

FIGURE 8.7

(b) If there were friction the speed of the child at any point would be less than before, and so the radial component of the weight could provide the centripetal force upto a lower point. The child would lose contact at a greater value of θ.

SELF-TEST

1. A block of mass 0.5 kg is attached to a spring (k = 6 N/m), as shown in Fig. 8.8. The block is on a plane inclined at 53° to the horizontal and for which the coefficient of kinetic friction is μ_k = 0.1. The block is released from rest with the spring unextended. What is the speed of the block after it has slid a distance D = 0.2 m along the incline? Use g = 10 N/kg.

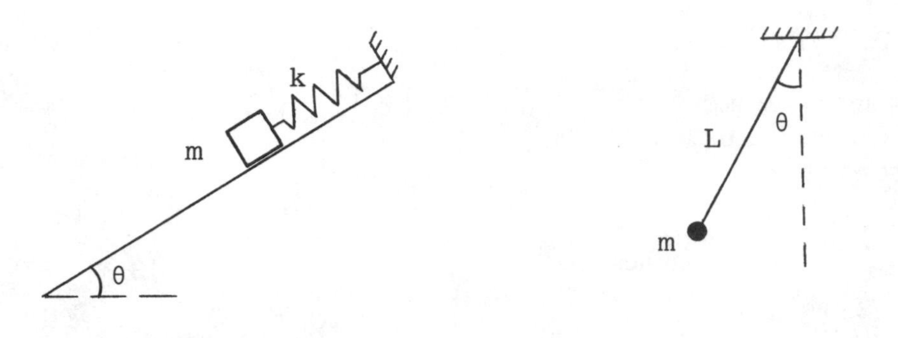

FIGURE 8.8 FIGURE 8.9

2. The bob of a simple pendulum of length L = 2 m has a mass m = 1.5 kg. It is released with the string initially at θ = 53° to the vertical, see Fig. 8.9. What is the tension in the string when it is vertical?

3. A rocket is fired with 85% of the escape speed, what is its maximum altitude in terms of the radius of the earth?

4. (a) $U = U_o \exp(-ax)$; find F_x

   (b) $F_x = -C/x$ find U(x) given U = 0 at x_o.

Chapter 9

LINEAR MOMENTUM

MAJOR POINTS

1. The conservation of linear momentum
2. (a) In an inelastic collision only momentum is conserved
   (b) In an elastic collision both momentum and kinetic energy are conserved.
3. (a) The definition of impulse
   (b) The use of impulse to find the average force acting on a body.

CHAPTER REVIEW

LINEAR MOMENTUM

The linear momentum $\mathbf{p}$ of a particle of mass m moving with velocity $\mathbf{v}$ is defined as

$$\mathbf{p} = m\,\mathbf{v}$$

The importance of linear momentum lies in the fact that it is a conserved quantity. Suppose two bodies with masses m_1 and m_2 and initial velocities $\mathbf{u}_1$ and $\mathbf{u}_2$ collide. After the collision the final velocities are $\mathbf{v}_1$ and $\mathbf{v}_2$. The principle of the conservation of linear momentum tells us that

$$m_1\,\mathbf{u}_1 + m_2\,\mathbf{u}_2 = m_1\,\mathbf{v}_1 + m_2\,\mathbf{v}_2$$

Since momentum is a vector quantity, this equation may be expressed in terms of the components. We consider only two-dimensional collisions so we write only two equations:

$$m_1\,u_{1x} + m_2\,u_{2x} = m_1\,v_{1x} + m_2\,v_{2x}$$

$$m_1\,u_{1y} + m_2\,u_{2y} = m_1\,v_{1y} + m_2\,v_{2y}$$

The biggest mistake you can make is to treat linear momentum as a scalar quantity!

It is shown in the text that the net external force acting on a system is equal to the rate of change of its momentum:

$$\mathbf{F}_{EXT} = d\mathbf{P}/dt$$

where $\mathbf{F}_{EXT} = \Sigma\,\mathbf{F}_i$ is the net external force acting on the system and $\mathbf{P} = \Sigma\,\mathbf{p}_i$ is the total linear momentum. This is a more general form of Newton's second law. It reduces to the more familiar form $\mathbf{F} = m\mathbf{a}$ if the mass of the system is constant. (The dynamics of a system of variable mass is discussed in Sec. 10.7.) From this equation we infer the condition for the applicability of the conservation of linear momentum:

$$\text{If } \mathbf{F}_{EXT} = 0, \text{ then } \quad \mathbf{P} = \Sigma\,\mathbf{p}_i = \text{constant.}$$

The linear momentum of a system is conserved only if there is no net external force acting on it.

Types of collision

Collisions are classified as being either elastic or inelastic. In both types, linear momentum is conserved. In a completely inelastic collision the bodies stick together. In an elastic collision, kinetic energy is also conserved:

(Elastic) $1/2\ m_1u_1^2 + 1/2\ m_2u_2^2 = 1/2\ m_1v_1^2 + 1/2\ m_2v_2^2$

Note that this is a scalar equation. In the special case of an elastic collision between two particles moving only in one dimension, one can combine the conservation of linear momentum and that of kinetic energy to obtain Huygen's relation:

(Elastic, One dimension.) $$v_2 - v_1 = -(u_2 - u_1)$$

The relative velocity of the particles after the collision is equal and opposite to the relative velocity before the collision.

It is usually more convenient to use this relation and the conservation of linear momentum rather than to get involved with the squared speeds in the conservation of kinetic energy equation.

EXAMPLE 1

A ball with a mass $m_1 = 2$ kg traveling east at 4 m/s and a ball with a mass $m_2 = 5$ kg traveling west at 3 m/s collide and stick together. (a) What is the final velocity of the combined mass? (b) What is the change in kinetic energy due to the collision?

Solution:

(a) We start the problem with a simple sketch and indicate the coordinate axes. We choose the x axis to point east and the y axis to point north, as in Fig. 9.1. The vector form of the conservation of linear momentum is

$$\Sigma\mathbf{p}: \quad m_1\,\mathbf{u}_1 + m_2\,\mathbf{u}_2 = (m_1 + m_2)\mathbf{V}$$

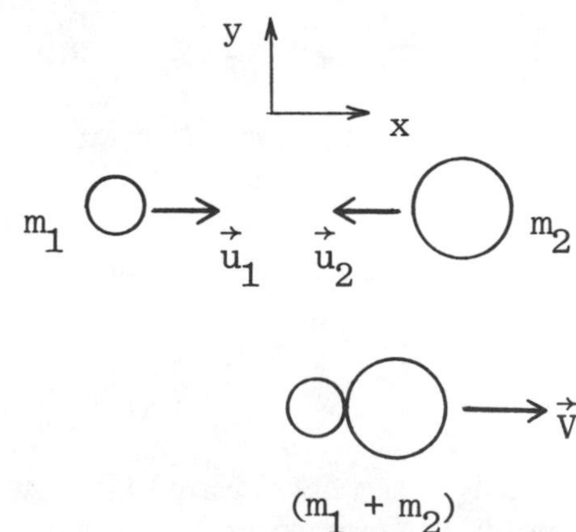

FIGURE 9.1

The x component of this equation is

$$\Sigma p_x: m_1u_1 - m_2u_2 = (m_1 + m_2)V$$

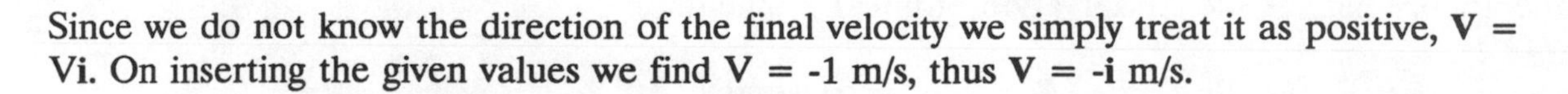

Since we do not know the direction of the final velocity we simply treat it as positive, $\mathbf{V} = V\mathbf{i}$. On inserting the given values we find $V = -1$ m/s, thus $\mathbf{V} = -\mathbf{i}$ m/s.

(b) The initial and final kinetic energies are

$$K_i = 1/2\ m_1u_1^2 + 1/2\ m_2u_2^2 = 38.5\ J$$

$$K_f = 1/2\ (m_1 + m_2)V^2 = 3.5\ J$$

The change in kinetic energy is $\Delta K = K_f - K_i = -33$ J. Since the collision is inelastic kinetic energy is not conserved.

EXAMPLE 2

A 10-kg bomb traveling horizontally at 20 m/s in the direction 30° W of N explodes into two parts. The 4-kg part moves off at 15 m/s in the direction 25° N of E. (a) What is the velocity of the other part? (b) What is the change in kinetic energy?

Solution:

We choose the x and y axes to point east and north respectively, as in Fig. 9.2. The vector form of the conservation of momentum is

$$\Sigma \mathbf{p}: \quad M\ \mathbf{u} = m_1\ \mathbf{v}_1 + m_2\ \mathbf{v}_2$$

where M = 10 kg, m_1 = 4 kg and m_2 = 6 kg. The equations for the components are

$$\Sigma p_x: \quad -\ Mu\ \sin 30° = +m_1v_1\ \cos 25° + m_2\ v_{2x}$$

$$\Sigma p_y: \quad +\ Mu\ \cos 30° = +m_1v_1\ \sin 25° + m_2\ v_{2y}$$

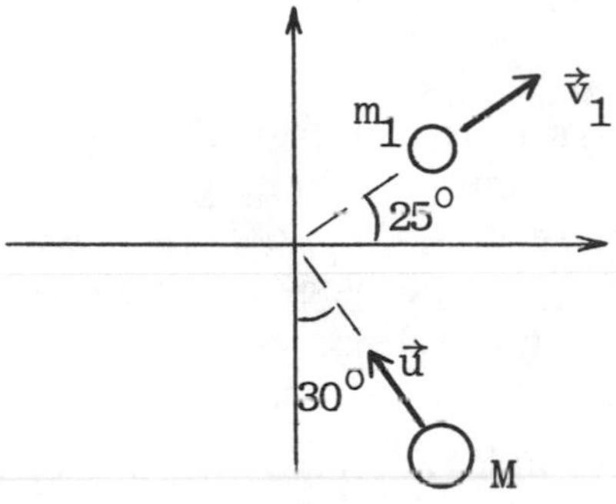

FIGURE 9.2

Notice that we express the unknown components of $\mathbf{v}_2$ simply by subscripts and not in the form $v_2\cos\theta$ and $v_2\sin\theta$, where each term involves two unknowns. With the given values we have

$$\Sigma p_x: \quad -\ (10)(20)(0.5) = (4)(15)(0.906) + 6\ v_{2x}$$

$$\Sigma p_y: \quad (10)(20)(0.866) = (4)(15)(0.423) + 6v_{2y}$$

We find $v_{2x} = -\ 25.7$ m/s and $v_{2y} = 24.6$ m/s, or $\mathbf{v}_2 = -25.7\mathbf{i} + 24.6\mathbf{j}$ m/s. The magnitude and direction of $\mathbf{v}_2$ are given by

$$v_2^2 = (25.7)^2 + (24.6)^2\ ;\ \text{so } v_2 = 35.6$$

$$\text{Tan}\theta = v_{2y}/v_{2x} = -\ 24.6/25.7;\ \text{so } \theta = -43.7°$$

Note that v_{2x}, is negative and v_{2y} is positive. Thus, the velocity is in the "second quadrant", that is, in the direction 43.7° N of W.

(b) The initial and final kinetic energies are

$$K_i = 1/2\ M\ u^2 = 2000\ J$$

$$K_f = 1/2\ m_1 v_1^2 + 1/2\ m_2 v_2^2 = 4250\ J$$

Since this is an explosion, the change in kinetic energy is positive, $\Delta K = 2250$ J.

EXAMPLE 3

A bullet of mass m = 10 g has an initial horizontal velocity u = 200 m/s. It collides with and embeds in a block of mass M = 2 kg that is at rest on a frictionless horizontal surface. After the collision, the block slides on a rough surface for which the coefficient of kinetic friction is μ = 0.3, see Fig. 9.3. (a) What is the distance travelled by the block? (b) What is the thermal energy generated in the collision?

Solution:

It is important to realize that this is a two-step problem that requires two concepts: momentum and energy. First, momentum conservation is used for the collision and second, the modified conservation of energy as the block slides on the surface. Since the collision is inelastic kinetic energy is not conserved. Nonetheless one can apply the conservation of linear momentum since the block is (initially) on a frictionless part of the surface. If V is the velocity of the combined masses just after the collision, we have

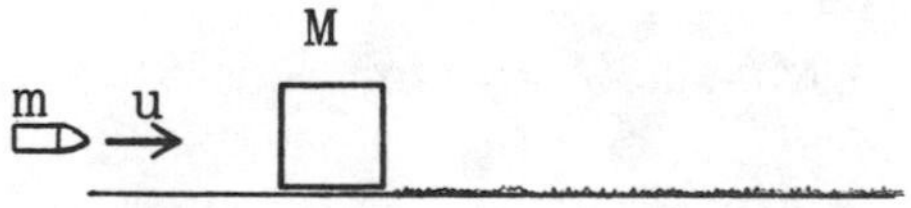

M

m u

FIGURE 9.3

$$\Sigma p: \qquad m\ u = (m + M)\ V$$

thus V = 1 m/s. For the second part, we employ the equation $\Delta E = W_{NC}$.

$$E_i = 1/2\ (m + M)V^2; \qquad E_f = 0;$$

$$W_{NC} = -\ fs = -\ \mu(m + M)g\ s$$

Thus $s = V^2/2\mu g = 0.17$ m.

(b) The initial kinetic energy of the bullet is

$$K_i = 1/2\ mu^2 = 200\ J$$

The kinetic energy just after the collision is

$$K_f = 1/2\ (m + M)V^2 = 1.0\ J$$

Thus the change in kinetic energy is $\Delta K = -199$ J. This equals the thermal energy generated during the collision. We see that most of the kinetic energy of the bullet is converted to thermal energy.

EXAMPLE 4

A block of mass $m_1 = 2$ kg is released from a height of 90 cm on an incline, see Fig. 9.4. It collides elastically with a block of mass $m_2 = 4$ kg initially at rest on the horizontal surface. To what heights to the blocks rise after the collision?

Solution:

This problem also requires the use of both linear momentum and energy. We use energy conservation to find the speed of m_1 at the bottom of the incline and then the conservation of linear momentum to find the velocities after collision. The heights are found from a second application of energy conservation. The speed of m_1 before the collision is found from energy conservation. With $U_g = 0$ at the horizontal level we have

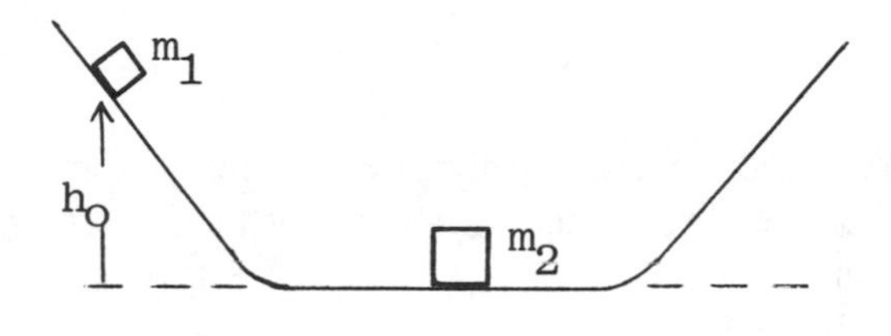

FIGURE 9.4

$$E_i = m_1 g h_o; \qquad E_f = 1/2\ m_1 u_1^2$$

thus, $u_1^2 = h_o/2g$.

To deal with the elastic collision we use the conservation of momentum and Huygen's relation for the relative velocities: $m_1u_1 = m_1v_1 + m_2v_2$, and $(v_2 - v_1) = -(u_2 - u_1)$, where $u_2 = 0$. Thus,

Σp_x: $\qquad 2\ u_1 = 2\ v_1 + 4\ v_2 \qquad$ (i)

Huygens: $\qquad u_1 = -v_1 + v_2 \qquad$ (ii)

Multiply (ii) by 2 and add (i) to find $4\ u_1 = 6\ v_2$, so $v_2 = 2u_1/3$.
Substitute this into (ii) to find $v_1 = -u_1/3$. The latter result indicates that m_1 reverses its direction of motion.

To find the final heights we again employ energy conservation: $mgh = 1/2\ mv^2$, which yields $h = v^2/2g$. Since $h_o = u_1^2/2g$, we find

$$h_1 = h_o/9 = 10 \text{ cm}; \qquad h_2 = 4h_o/9 = 40 \text{ cm}$$

EXAMPLE 5

A man of mass M = 70 kg can throw a ball of mass m = 0.2 kg at 30 m/s relative to himself. This would be the speed of the ball if he were on firm ground. What is the speed of the ball relative to the ground if he throws it horizontally when he is on a frozen (frictionless) lake?

Solution:

The notation we used in Ch. 4 for relative velocities is helpful. Thus we take

$\mathbf{v}_{BM}$	Velocity of the ball relative to the man
$\mathbf{v}_{BG}$	Velocity of the ball relative to the ground
$\mathbf{v}_{MG}$	Velocity of the man relative to the ground

These velocities are related by

$$\mathbf{v}_{BG} = \mathbf{v}_{BM} + \mathbf{v}_{MG}$$

We need consider only one component of this equation. Suppose that the ball is thrown in the negative direction, $\mathbf{v}_{BM} = -30\mathbf{i}$, and that the man's recoil velocity is $\mathbf{v}_{MG} = V\mathbf{i}$. Then,

$$v_{BG} = -30 + V$$

In order to determine v_{BG} we apply the conservation of momentum:

$$\Sigma p_x : \ 0 = M V + m(-30 + V)$$

Thus $V = 30m/(m + M) = 6/60.2 \approx 0.1$ m/s.

This example covers the essential physics of rocket propulsion. Just replace the ball with the hot exhaust gas and the man by the rocket. Of course, the rocket expels the gas continuously, so integration is required to determine the speed of the rocket (see Section 9.7).

IMPULSE

The impulse experienced by a body is defined as the change in its linear momentum

$$\mathbf{I} = \Delta\mathbf{p} = \mathbf{p}_f - \mathbf{p}_i$$

From Newton's second law in the form $d\mathbf{p} = \mathbf{F}\, dt$, we see that

$$\mathbf{I} = \int \mathbf{F}\, dt$$

An impulse may be represented by the area under the **F** versus t graph, as in Fig.9.5. The concept of impulse is useful in estimating the force experienced by a object when the details of the interaction are not available, for example when a bat hits a ball. In such cases we use the impulse approximation: We assume that the (internal) forces between the parts of a system (e.g. a bat and a ball) are very much larger than external forces (e.g. gravity or air resistance) for the short duration of the interaction. Thus we assume the ball is subject only to the force exerted by the bat.

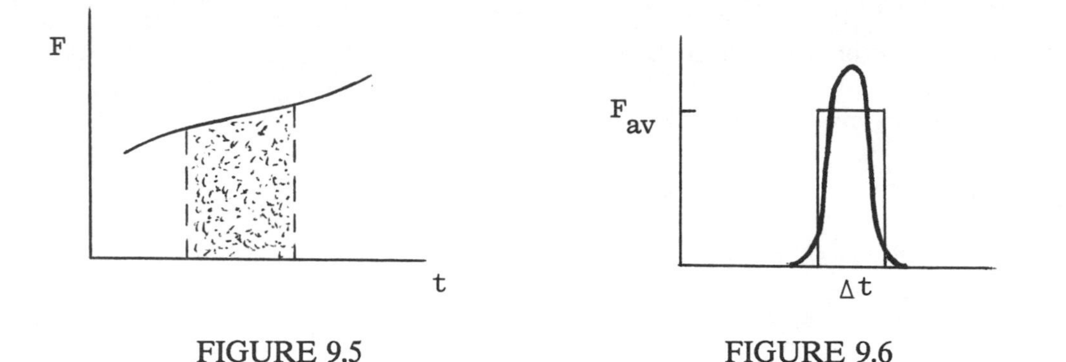

FIGURE 9.5 FIGURE 9.6

The force exerted by a bat on a ball may vary with time as shown in Fig. 9.6. Usually we do not have any information regarding this variation. However, we can measure the impulse on the ball-- which is simply its change in momentum. Then we assume the ball was subject to a constant "average" force F_{av}. The impulse due to this average force must be the same as that of the true force. That is, the area under the constant force must equal the area under the true curve.

EXAMPLE 6
A force varies with time as shown in Fig. 9.7. (a) What is the impulse? (b) If the impulse acts on a particle of mass 2 kg initially at rest, what is the final speed?

Solution:
Since $I = \Delta p = F_{av}\,\Delta t$, the impulse due to the force is given by the area under the F versus t graph.
(a) The sum of the areas of the rectangle and the triangle is

$$I = F_o\,\Delta t_1 + (F_o/2)\,\Delta t_2$$

$$= (10\text{ N})(2\text{ s}) + (5\text{ N})(2\text{ s}) = 30\text{ N.s}$$

(b) We know that $I = mv_f - mv_i = m\,v_f$. Using the value of I found above, we find that $v_f = 15$ m/s.

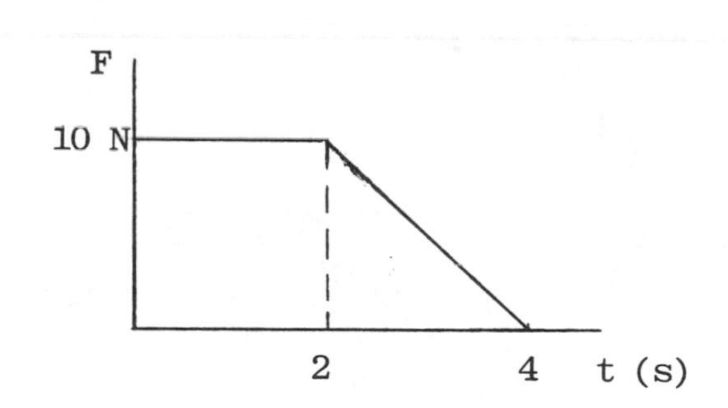

FIGURE 9.7

EXAMPLE 7
A 150-g baseball moving initially at -5**i** + 10**j** m/s is struck by a bat. After being hit, its velocity is +7**i** + 18**j** m/s. If the bat and ball were in contact for 3 ms, estimate the average force exerted by the bat on the ball.

Solution:

First we find the impulse on the ball.

$$\mathbf{I} = m\mathbf{v}_f - m\mathbf{v}_i = (0.15 \text{ kg})(12\mathbf{i} + 8\mathbf{j} \text{ m/s})$$

Using $\mathbf{I} = \mathbf{F}_{av}\, \Delta t$, we can estimate the average force

$$\mathbf{F}_{av} = \mathbf{I}/\Delta t = (1.8\mathbf{i} + 1.2\mathbf{j} \text{ N.s})/(3 \times 10^{-3} \text{ s})$$

$$= 600\mathbf{i} + 400\mathbf{j} \text{ N.}$$

SOLUTIONS TO SELECTED TEXT EXERCISE AND PROBLEMS

Exercise 3

Consider a 20-g bullet (B) and a 60-kg runner (R). (a) If they have the same momentum, what is the ratio of their kinetic energies, K_B/K_R? (b) If they have the same kinetic energy, what is the ratio of their momenta?

Solution:

The basic expressions for linear momentum and kinetic energy are $p = mv$ and $K = 1/2\ mv^2$.

(a) $K = p^2/2m$, thus

$$K_B/K_R = m_R/m_B = 3000$$

(b) $p = (2mK)^{1/2}$, so

$$p_B/p_R = (m_B/m_R)^{1/2} = 0.0183$$

Exercise 29

A 10-g bullet moving at 400 m/s strikes a ballistic pendulum of mass 2.5 kg. The bullet emerges with a speed of 100 m/s. (a) To what height does the pendulum bob rise? (b) How much work was done by the bullet in passsing through the block?

Solution:

(a) We apply the conservation of momentum to the collision. With m_1 as the mass of the bullet and m_2 as the mass of the block:

$$\Sigma p_x\text{:} \quad m_1 u_1 = m_1 v_1 + m_2 v_2$$

$$4 = 1 + 2.5\, v_2$$

Thus $v_2 = 1.2$ m/s. The maximum height is found from the equation of kinematics: $0 = v^2 - 2g\Delta y$, thus $H = v_2^2/2g = 7.3$ cm.

(b) The work done on the bullet is equal to the change in its kinetic energy. The work done by the bullet is

$$W_{Bullet} = -\Delta K = 1/2\ m(v_2^2 - v_1^2) = +750\ J$$

Exercise 35
A particle of mass m_1 makes a head-on elastic collision with a particle of mass m_2 at rest. If the initial kinetic energy is K_o, and the final kinetic energy of m_2 is K_2, evaluate the ratio K_2/K_o for a collision between a neutron of mass 1 u and each of the following: (a) a deuteron of mass 2 u; (b) a carbon nucleus of mass 12 u; (c) a lead nucleus of mass 208 u.

Solution:
We may use the conservation of momentum and Huygen's relation wit $u_2 = 0$:

Σp_x: $$m_1u_1 = m_1v_1 + m_2v_2 \qquad \text{(i)}$$

Relative velocity: $$v_2 - v_1 = u_1 \qquad \text{(ii)}$$

Substitute $v_1 = v_2 - u_1$ from (ii) into (i) to find

$$v_2 = 2m_1u_1/(m_1 + m_2) \qquad \text{(iii)}$$

The initial kinetic energy is $K_o = 1/2\ m_1u_1^2$ and $K_2 = 1/2\ m_2v_2^2$. Using (iii)

$$K_2/K_o = 4m_1m_2/(m_1 + m_2)^2$$

(a) $8/9 = 0.89$;
(b) $48/(13)^2 = 0.28$;
(c) $4(208)/(209)^2 = 0.019$.

Exercise 43
A hammer with a head of mass 0.5 kg strikes a nail at 4 m/s and is brought to rest. If the collision lasts 10^{-3} s, what is the average force on the nail? Compare this force with your weight.

Solution:
The impulse is $\Delta p = m\ \Delta v = 2$ kg.m/s. The average force is

$$F_{av} = \Delta p/\Delta t$$

$$= (2\ \text{kg.m/s})/(10^{-3}\ \text{s}) = 2000\ \text{N}$$

Exercise 51
A 60-g tennis ball strikes the ground at 25 m/s at 40° to the horizontal. It bounces off at 20 m/s at 30° to the horizontal; see Fig. 9.8. (a) Find the impulse exerted on the ball. (b) If the collision lasted for 5 ms, find the average force exerted on the ball by the court.

Solution:

The components of the change in momentum of the ball are

$$\Delta p_x = m(v_2 \cos\theta_2 - v_1 \cos\theta_1)$$

$$= -0.11 \text{ kg.m/s}$$

$$\Delta p_y = m(v_2 \sin\theta_2 + v_1 \sin\theta_1)$$

$$= +1.56 \text{ kg.m/s}$$

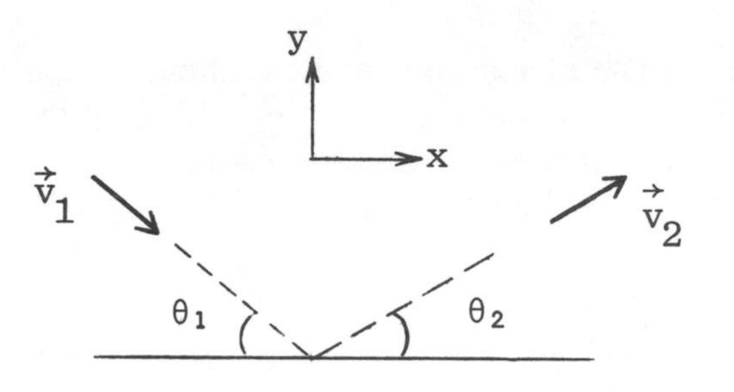

FIGURE 9.8

(a) The impulse is

$$\mathbf{I} = \Delta\mathbf{p} = -0.11\mathbf{i} + 1.56\mathbf{j} \text{ kg.m/s}$$

(b) The average force on the ball is

$$\mathbf{F}_{av} = \mathbf{I}/\Delta t = -22\mathbf{i} + 312\mathbf{j} \text{ N}$$

Problem 3
Two particles with masses m_1 and m_2 travel toward each other with speeds u_1 and u_2. They collide and stick together. Show that the loss in kinetic energy is

$$m_1m_2(u_1 + u_2)^2/2(m_1 + m_2)$$

Solution:

The conservation of linear momentum yields

$$\Sigma p_x: \quad m_1u_1 - m_2u_2 = (m_1 + m_2)V \qquad \text{(i)}$$

where V is the final velocity of the combined masses.
The change in kinetic energy is

$$\Delta K = 1/2\,(m_1 + m_2)V^2 - 1/2\,(m_1u_1^{\,2} + m_2u_2^{\,2}) \qquad \text{(ii)}$$

From (i), $V = (m_1u_1 - m_2u_2)/(m_1 + m_2)$ which we substitute into (ii).
After some algebra, which you should go through, we find

$$\Delta K = -\,m_1m_2(u_1 + u_2)^2/2(m_1 + m_2).$$

Problem 19
A vertical chain has a length L and a mass M. It is released with the bottom just touching a table; see Fig. 9.9. (a) Find the force on the table as a function of the distance fallen by the top end. (b) Show that the maximum force is 3Mg.

Solution:

(a) The force on the table has two parts. First, the weight of the length of chain that has landed. Second, the rate of momentum transfer as the chain hits the surface. Thus,

$$F = \lambda y g + v\, dm/dt$$

$$= \lambda g y + \lambda v\, dy/dt$$

$$= \lambda g y + \lambda v^2$$

Since the chain is in free-fall,
$v^2 = 2gy$, thus $F = 3\lambda g y$

(a) The maximum value of y is L, and
$M = \lambda L$, so $F_{max} = 3Mg$.

FIGURE 9.9

Problem 21
A particle of mass m_1 collides elasically with a particle of mass m_2 ($< m_1$) initially at rest. Show that the maximum angle θ_1 to the original direction of motion at which m_1 can move off is given by

$$\sin\theta_{1(max)} = m_2/m_1$$

Solution:

The x and y components of the conservation of momemtum and the conservation of kinetic energy take the form:

Σp_x: $$m_1 u_1 = m_1 v_1 \cos\theta_1 + m_2 v_2 \cos\theta_2 \quad \text{(i)}$$

Σp_y: $$0 = m_1 v_1 \sin\theta_1 - m_2 v_2 \sin\theta_2 \quad \text{(ii)}$$

ΣK: $$m_1 u_1^2 = m_1 v_1^2 + m_2 v_2^2 \quad \text{(iii)}$$

Square $m_2 v_2 \cos\theta_2$ from (i) and $m_2 v_2 \sin\theta_2$ from (ii) and add:

$$m_1^2(u_1 - v_1 \cos\theta_1)^2 + (m_1 v_1 \sin\theta_1)^2 = (m_2 v_2)^2 \quad \text{(iv)}$$

From (iii) we know

$$(m_2 v_2)^2 = m_2 m_1 (u_1^2 - v_1^2) \quad \text{(v)}$$

Substitute (v) into (iv) to obtain

$$(m_1 + m_2)v_1^2 - (2m_1u_1 \cos\theta_1)v_1 + (m_1 - m_2)u_1^2 = 0$$

For a real solution of this quadratic equation we need $b^2 = 4ac$, or $\cos^2\theta_1 = 1 - (m_2/m_1)^2$, or $\sin\theta_1 = m_2/m_1$.

SELF-TEST

1. Two pucks slide toward each other on a frictionless level surface. The initial velocity of puck 1 (m_1 = 1.2 kg) is 3 m/s due south, and that of puck 2 (m_2 = 2.4 kg) is 2 m/s due north; see Fig. 9.10. After the collision, puck 1 moves off at 2.5 m/s in the direction 30° west of south. (a) What is the velocity of puck 2 after the collision? (b) Is the collision elastic?

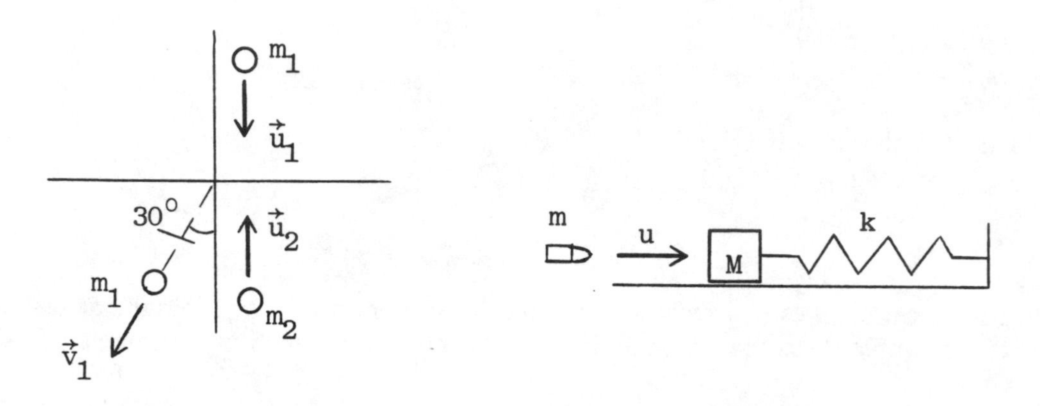

FIGURE 9.10 FIGURE 9.11

2. A bullet of mass m = 20 g moving at u = 180 m/s imbeds in a block of mass M = 0.6 kg initially at rest on a frictionless horizontal surface, see Fig. 9.11. The block is attached to a spring (k = 40 N/m). What is the maximum compression of the spring?

3. A baseball of mass 150 g is thrown at 30 m/s due east and is struck by a bat that changes its velocity to 40 m/s due north. The ball is contact with the bat for 3 ms. Find: (a) the impulse; (b) the average force on the ball.

4. A ball of mass m_1 = 2 kg moving at $\mathbf{u}_1$ = 5**i** m/s makes a one-dimensional elastic collision with an ball of mass m_2 = 3 kg moving at $\mathbf{u}_2$ = -2**i** m/s. What are the final velocities $\mathbf{v}_1$ and $\mathbf{v}_2$?

Chapter 10

SYSTEMS OF PARTICLES

MAJOR POINTS

1. The location of the center of mass (CM) of a system.
2. Newton's first law for a system: The velocity of the CM of an isolated system stays constant.
3. Newton's second law for a system: $F_{EXT} = ma_{CM}$, where a_{CM} is the acceleration of the CM
4. The kinetic energy of a system is $K_{CM} + K_{Rel}$, the energy of the CM motion plus the energy relative to the CM.

CHAPTER REVIEW

CENTER OF MASS

The position of the center of mass (CM) of a system of discrete particles is given by

$$\mathbf{r}_{CM} = \Sigma\, m_i\mathbf{r}_i/M$$

where $M = \Sigma\, m_i$ is the total mass of the system. The x and y components of this equation are

$$x_{CM} = \Sigma\, m_ix_i/M\ ; \qquad y_{CM} = \Sigma\, m_iy_i/M$$

In general there would also be a z component. For a continuous distribution of mass, the sum is replaced by an integral:

$$\mathbf{r}_{CM} = \int \mathbf{r}\, dm/M$$

where dm is the mass of an infinitesimal element of the body. The mass distribution may be specified in terms of the linear mass density λ kg/m, the areal mass density σ kg/m^2 or the volume mass density ρ kg/m^3. The elemental mass would then be one of the following

$$dm = \lambda\, d\ell;\quad dm = \sigma\, dA;\quad dm = \rho\, dV$$

where $d\ell$ is an infinitesimal length; dA an infinitesimal area and dV an infinitesimal volume.

EXAMPLE 1
Locate the center of mass of the three particles in Fig. 10.1.

Solution:

The x and y coordinates of the CM are

$x_{CM} = (m_1x_1 + m_2x_2 + m_3x_3)/M$

$= [(0.5\text{ kg})(2\text{ m}) + (1.5\text{ kg})(-2\text{ m})$

$+ (2\text{ kg})(-1\text{ m})]/4\text{ kg} = -1.0\text{ m}$

$y_{CM} = (m_1y_1 + m_2y_2 + m_3y_3)/M$

$= [(0.5\text{ kg})(2\text{ m}) + (1.5\text{ kg})(1\text{ m})$

$+ (2\text{ kg})(-1\text{ m})]/4\text{ kg} = 0.125\text{ m}$

Thus $\mathbf{r}_{CM} = -1.0\mathbf{i} + 0.125\mathbf{j}$ m

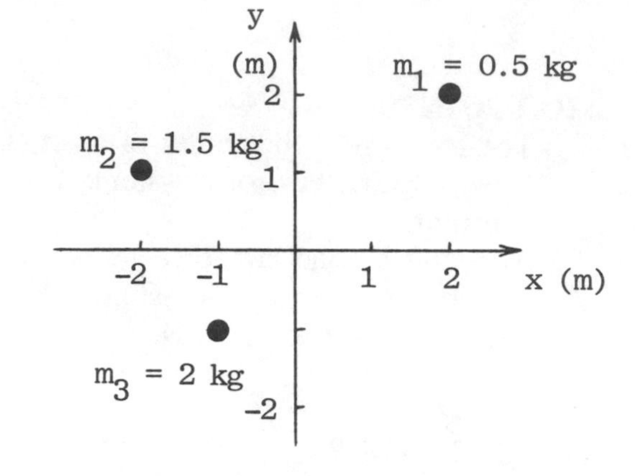

FIGURE 10.1

EXAMPLE 2
A disk of radius R = 1 m and areal mass density $\sigma = 0.2\text{ kg/m}^2$ is attached to one end of a uniform rod of length L = 4 m with a linear mass density 0.25 kg/m, see Fig. 10.2. Locate the CM relative to the free end of the rod.

Solution:

The CM of the uniform rod is at its center, $x_1 = 2$ m, and that of the disk is at its center, $x_2 = 5$ m. For the purpose of locating the CM of the composite body we may treat the rod and the disk as if all their mass were concentrated at the individiual CM's.

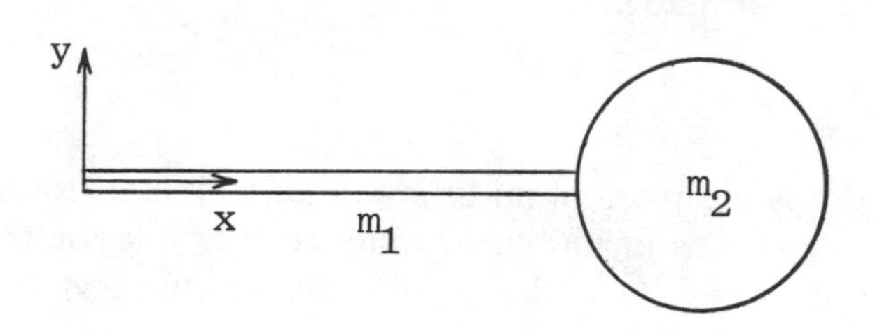

FIGURE 10.2

ROD: $m_1 = \lambda L = (0.25\text{ kg/m})(4\text{ m}) = 1\text{ kg}$

DISK: $m_2 = \sigma(\pi R^2) = 0.628\text{ kg}$

The position of the CM is given by

$x_{CM} = (m_1x_1 + m_2x_2)/M$

$= [(1\text{ kg})(2\text{ m}) + (0.628\text{ kg})(5\text{ m})]/1.628\text{ kg}$

$= 3.16\text{ m}.$

EXAMPLE 3
Locate the CM of a uniform plate in the form of a quarter of a circle of radius R.

Solution:

Since the mass of the object is continuously distributed we have to use integration. Let us say that the areal mass density is σ kg/m^2. In order to determine x_{CM} a convenient choice of element is an infinitesmal strip of width dx and height y, as shown in Fig. 10.3. Note that all parts of the element have the same value of x. The mass of the element is, $dm = \sigma\, dA = \sigma\, y\, dx$. The position of the CM is given by

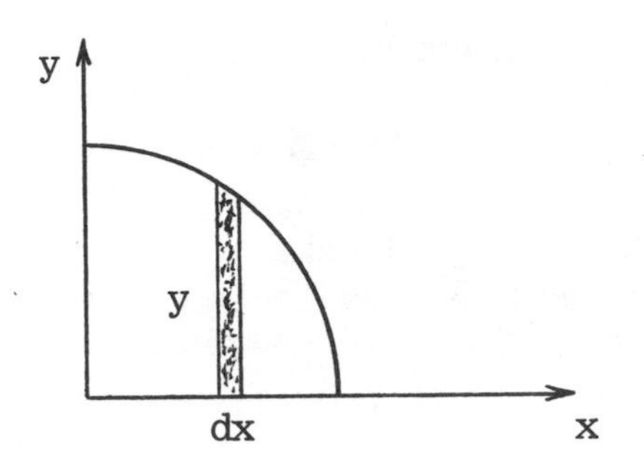

FIGURE 10.3

$$x_{CM} = \int x\, dm/M = \int \sigma\, x\, y\, dx/M$$

where $M = \sigma\pi R^2/4$. To evaluate the integral y must be expressed in terms of x. Since the plate is circular, we use the equation for a circle of radius R centered at the origin: $x^2 + y^2 = R^2$ and the integral takes the form

$$x_{CM} = \int_0^R \sigma\, x\, (R^2 - x^2)^{1/2}\, dx/M$$

If we let $u = x^2$, so $du = 2x\, dx$, the integral becomes

$$x_{CM} = \int_0^{R^2} (\sigma/2M)\, (R^2 - u)^{1/2}\, du$$

$$= (\sigma/2M) \left| -2/3\, (R^2 - u)^{3/2} \right|_0^{R^2}$$

$$= (\sigma/2M)(2R^3/3) = 4R/3\pi$$

By symmetry, $y_{CM} = 4R/3\pi$. The integral could also be evaluated by using the substitutions $x = R\cos\theta$ and $y = R\sin\theta$. Try it.

MOTION OF THE CENTER OF MASS

Since $\mathbf{v} = d\mathbf{r}/dt$, by taking the derivative of $\mathbf{r}_{CM}$ we obtain the velocity of the CM

$$\mathbf{v}_{CM} = \Sigma\, m_i\mathbf{v}_i/M$$

By rewriting this equation in the form

$$M\,\mathbf{v}_{CM} = \Sigma\, m_i\mathbf{v}_i = m_1\mathbf{v}_1 + m_2\mathbf{v}_2 + ... + m_N\mathbf{v}_N$$

we see that the right hand side is the total linear momentum of the particles in the system. The total momentum is equal to the momentum of an imaginary particle of mass M moving with the velocity of the CM.

Newton's second law for a system of particles is

$$\mathbf{F}_{EXT} = d\mathbf{P}/dt$$

where $\mathbf{P} = M\mathbf{v}_{CM} = \Sigma\, m_i\mathbf{v}_i$ is the total linear momentum and $\mathbf{F}_{EXT}$ is the net external force. We infer that if the net external force acting on a system is zero, its total linear momentum is conserved, that is

$$\text{If } \mathbf{F}_{EXT} = 0, \text{ then } \mathbf{v}_{CM} = \text{constant.}$$

This is Newton's first law as it applies to a system of particles. Internal forces between parts of a system may change its size or shape, but they cannot change the motion of the CM.

Since $\mathbf{a} = d\mathbf{v}/dt$, by taking the derivative of $\mathbf{v}_{CM}$ we find $M\mathbf{a}_{CM} = \Sigma\, m_i\,\mathbf{a}_i = \Sigma\, \mathbf{F}_i$, or

$$\mathbf{F}_{EXT} = M\,\mathbf{a}_{CM}$$

This is Newton's second law for a system of fixed mass. It says that the CM of the system accelerates as if the net external force acts on a particle of mass M at this point.

WORK-ENERGY THEOREM FOR A SYSTEM

The work-energy theorem may be extended from one particle to a system of particles. However, there are several subtle points that need to be considered and are discussed in optional sections in the text. For our purposes, we state a simple relationship:

$$W_{CM} = \mathbf{F}_{EXT} \cdot \mathbf{s}_{CM} = \Delta\, K_{CM}$$

where $\mathbf{s}_{CM}$ is the displacement of the CM. The quantity W_{CM} is called the center-of-mass work and K_{CM} is the translational kinetic energy of the CM.

The kinetic energy of a system of particles can be divided in to the translational kinetic energy of the CM, K_{CM}, and kinetic of the particles relative to the CM, K_{Rel}:

$$K = K_{CM} + K_{Rel}$$

The term K_{Rel} may involve translation, vibration, or rotation relative to the CM.

EXAMPLE 4
A ball of mass 1.5 kg initially moving east at 4 m/s and a ball of mass 2.5 kg is initially moving west at 8 m/s. (a) What is the velocity of the CM? (b) Find K_{CM} and K_{rel}

Solution:
(a) The velocity of the CM is

$$v_{CM} = [(1.5 \text{ kg})(4 \text{ m/s}) + (2.5 \text{ kg})(-8 \text{ m/s})]/(4 \text{ kg})$$

$$= -3.5 \text{ m/s}$$

(b) The kinetic energy of the CM is

$$K_{CM} = 1/2\ (m_1 + m_2)v_{CM}^2$$

$$= 1/2\ (4 \text{ kg})(-3.5 \text{ m/s})^2 = 24.5 \text{ J}$$

The velocities relative to the CM are $v_1' = 4 - (-3.5) = 7.5$ m/s; and $v_2' = -8$ m/s $-(-3.5$ m/s$) = -4.5$ m/s. Thus the kinetic energy relative to the CM is

$$K_{rel} = 1/2\ m_1 v_1'^2 + 1/2\ m_2 v_2'^2$$

$$= 1/2\ (1.5 \text{ kg})(7.5 \text{ m/s})^2 + 1/2\ (2.5 \text{ kg})(-4.5 \text{ m/s})^2 = 67.5 \text{ J}$$

The total kinetic energy is 92 J.

SOLUTIONS TO SELECTED TEXT EXERCISES AND PROBLEMS

Exercise 5
A square of side 2R has a circular hole of radius R/2 removed. Relative to the center of the square the center of the hole is located at (R/2, R/2) as shown in Fig. 10.4. Locate the CM with respect to the center of the square. (Hint: Treat the hole as an object of negative mass.)

Solution:
When you deal with an object that has some sort of hole in it, the hole may be treated as having negative mass. To avoid complications with signs, place the object in the first quadrant of the coordinate system. Let σ kg/m^2 be the areal mass density.

(Plate) $m_1 = \sigma(4R^2)$; $\quad x_1 = y_1 = R$

(Hole) $m_2 = -\sigma\pi(R/2)^2$; $\quad x_2 = y_2 = 3R/2$

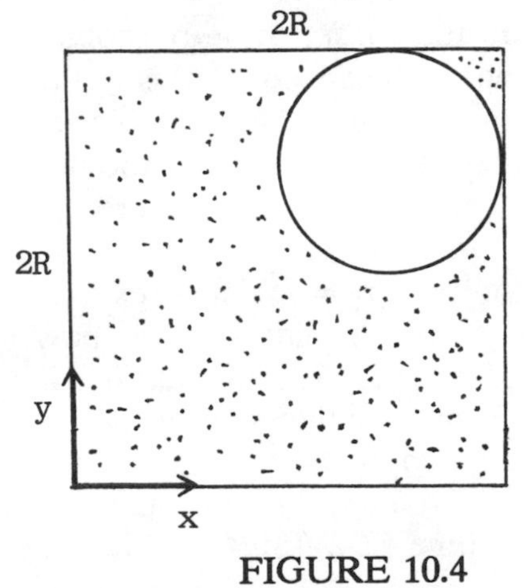

FIGURE 10.4

The position of the CM relative to the origin is

$$x_{CM} = y_{CM} = \frac{\sigma 4R^2(R) - (\sigma \pi R^2/4)(3R/2)}{\sigma(4 - \pi/4)R^2} = 0.878R$$

With respect to the center, $x_{CM} = y_{CM} = 0.122R$.

Exercise 11
Jack, of mass 75 kg, is at x = 0, whereas Jill, of mass 60 kg, is at x = 5 m on a frictionless frozen lake. When theu pull on a rope, Jill moves 1.5 m. (a) How far apart are they at this time? (b) Where do they finally meet?

Solution:
(a) The CM of an isolated system cannot move, that is, $\Delta x_{CM} = 0$. Since $Mx_{CM} = m_1x_1 + m_2x_2$, we have

$$0 = m_1 \Delta x_1 + m_2 \Delta x_2$$

If Jack moves by d, then 0 = 75d - (60 kg)(1.5 m), so d = 1.2 m. Their

separation is [5 - (1.5 + 1.2)] = 2.3 m.
(b) They will meet at the CM: x_{CM} = (60 kg)(5 m)/135 kg = 2.3 m.
(The answers in (a) and (b) are equal by mere coincidence.)

Exercise 15
Two blocks of masses m and 2m are held against a massless compressed spring within a box of mass 3 m and length 4L whose center is at x = 0 (see Fig. 10.5a). All surfaces are frictionless. After the blocks are released they are each at a distance L from the ends of the box when they lose contact with the spring. Show that the position of the center of the box shifts by L/6 after both blocks collide with and stick to it.

Solution:
From the conservation of momentum we know that if the speed of m is u, then the speed of 2m is u/2. After the block of mass m collides with the box, the box moves at v_1, thus $-mu = 4mv_1$, therefore $v_1 = -u/4$.

In the time $t_1 = L/u$ it takes the block of mass m to collide with the box, the block of mass 2m moves L/2, thus it is still L/2 from the other side of the box.

The speed of the box relative to 2m is (u/2 + u/4) = 3u/4, therefore the box moves for a time (L/2)/(3u/4) = 2L/3u before

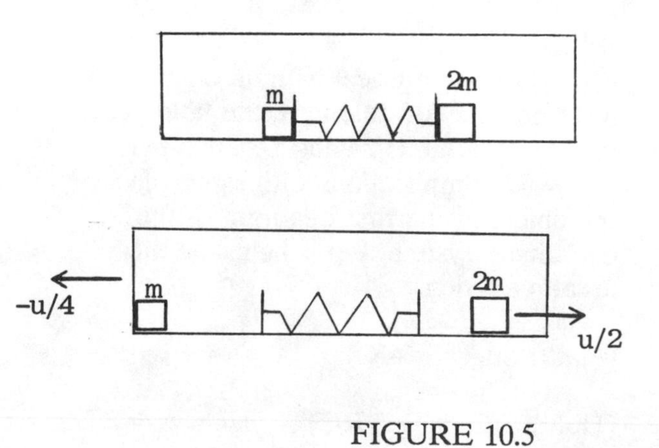

FIGURE 10.5

colliding with the block of mass 2m and coming to rest (How do we know this?). The displacement of the box in this time is (2L/3u)(-u/4) = -L/6.

Exercise 23
A 1000-kg Honda Accord heading east at 15 m/s makes a completely inelastic collision with a 1800-kg Jaguar XJ-6 moving north at 10 m/s on an icy road. (a) What is the velocity of the CM before the collision? (b) Where is the CM 3 s after the collision?

Solution:
(a) Substituting into the equation for $\mathbf{v}_{CM}$ we have:

$$\mathbf{v}_{CM} = [(1000 \text{ kg})(15\mathbf{i} \text{ m/s}) + (1800 \text{ kg})(10\mathbf{j} \text{ m/s})]/2800 \text{ kg}$$

$$= 5.36\mathbf{i} + 6.43\mathbf{j} \text{ m/s}$$

(b) $\mathbf{v}_{CM}$ does not change, so $\mathbf{r}_{CM} = \mathbf{v}_{CM}\, t = 16.1\mathbf{i} + 19.3\mathbf{j}$ m.

Exercise 31
A particle of mass m_1 = 5 kg moving at 4**j** m/s makes a one-dimensional elastic collision with a particle of mass m_2 = 3 kg at rest. Find the kinetic energy (a) relative to the CM, and (b) of the CM motion.

Solution:
We are given $\mathbf{v}_1$ = 5**j** m/s and $\mathbf{v}_2$ = 0, so the velocity of the CM is

$$\mathbf{v}_{CM} = [(5 \text{ kg})(4\mathbf{j}) + 0]/8 \text{ kg} = 2.5\mathbf{j} \text{ m/s}$$

The velocities of the particles relative to the CM are

$$\mathbf{v}_1' = \mathbf{v}_1 - \mathbf{v}_{CM} = 1.5\mathbf{j} \text{ m/s}; \quad \mathbf{v}_2' = \mathbf{v}_2 - \mathbf{v}_{CM} = -2.5\mathbf{j} \text{ m/s}$$

(a) $K_{Rel} = 1/2\,(5 \text{ kg})(1.5 \text{ m/s})^2 + 1/2\,(3 \text{ kg})(2.5 \text{ m/s})^2 = 15$ J;

(b) $K_{CM} = 1/2\, M v_{CM}^2 = 1/2\,(8 \text{ kg})(2.5 \text{ m/s})^2 = 25$ J.

Problem 3
Locate the CM of a thin wire frame that consists of a quarter of a circle and two radial lines of length R, as shown in Fig. 10.6.

Solution:
Let us first locate the CM of the arc, whose mass is $\lambda\pi R/2$. The mass of an infinitesmal element of the arc is $dm = \lambda\, R d\theta$, and its coordinates are $x = R\cos\theta$, and $y = R\sin\theta$. However, from symmetry we see that $x_{CM} = y_{CM}$, so we need to calculate just one of these.

$$x_{CM} = y_{CM} = \lambda R^2 \int_0^{\pi} \frac{\sin\theta \, d\theta}{\pi\lambda R/2} = 2R/\pi$$

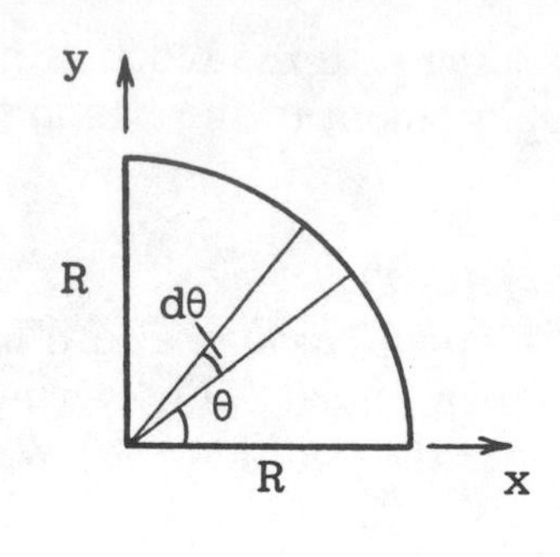

For the whole figire we take $m_1 = m_2 = \lambda R$, and $m_3 = \lambda\pi R/2$ for the arc

$$x_{CM} = y_{CM} = \frac{(\lambda R)(R/2) + (\pi R\lambda/2)(2R/\pi)}{(2 + \pi/2)\lambda R}$$

$$= 3R/(\pi + 4) = 0.420R$$

FIGURE 10.6

Problem 12

Figure 10.7 shows a spring within a tube that has a latch to hold the spring in a compressed position. The total mass is M. A particle of mass m and speed u collides with the endplate of the spring. If the energy of the spring when compressed is E, show that the minimum energy required of the particle to make an inelastic collision is $(M + m)E/M$.

Solution:

From momentum conservation we know that

$$mu = (M + m)V$$

where V is the velocity of the combined masses and hence also the velocity of the CM. The energy of the CM motion cannot change, thus only the kinetic energy relative to the CM is available to excite the system, that is, to make an inelastic collision. The kinetic energy relative to the CM is

FIGURE 10.7

$$E = K_{rel} = 1/2 \, m(u - V)^2 + 1/2 \, MV^2$$

$$= mMu^2/2(M + m)$$

Thus, the kinetic energy of the incoming particle is $1/2 \, mu^2 = (M + m)E/M$.

SELF-TEST

1. The CM of a uniform plate in the form of a right angled triangle with base b and height h is located at (b/3, h/3), as shown in Fig. 10.8a. Use this to locate the center of mass of the shape in Fig. 8b.

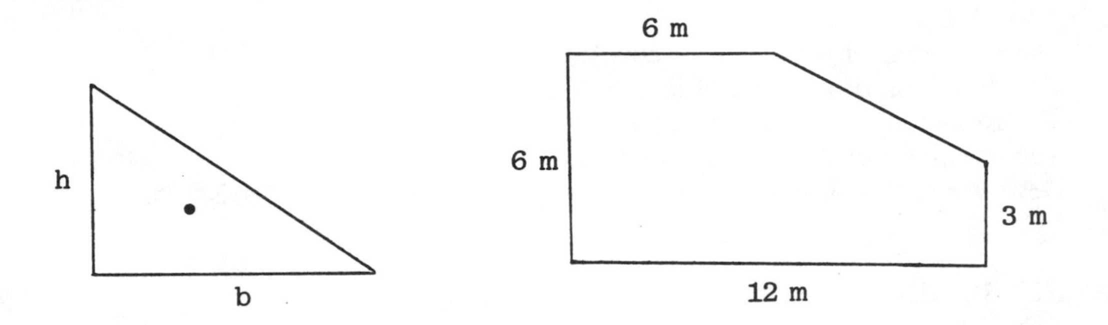

FIGURE 10.8

2. Jack (70 kg) and Jill (50 kg) are seated 3 m apart on a boat. When they exchange places, how far does the boat move relative to land? Ignore the mass of the boat.

Chapter 11

ROTATION OF A RIGID BODY ABOUT A FIXED AXIS

MAJOR POINTS

1. The equations of rotational kinematics
2. The definition of moment of inertia
3. The rotational kinetic energy of a rigid body
4. The definition of torque. The concept of a lever arm
5. Newton's second law for rotation of a rigid body about a fixed axis.

CHAPTER REVIEW

EQUATIONS OF ROTATIONAL KINEMATICS

In a rigid body, the particles maintain their relative positions. That is the size and shape of the body are fixed. The direction of a fixed axis of rotation does not change, although the position may change (for example, when a cylinder rolls down an incline).

The orientation of a rigid body is specifed by its angular position θ relative to some reference line. This angle is measured in radians:

$$\theta = s/r$$

where s is the arc length subtended by θ on a circle of radius r. One complete revolution corresponds to $s = 2\pi r$, thus

$$1 \text{ rev} = 2\pi \text{ rad}$$

The magnitude of the angular velocity of a body is defined as the rate of change of the angular position θ:

$$\omega = d\theta/dt$$

The SI unit of ω is rad/s. The unit r.p.m. (revolutions per minute) is a rotational frequency that should be converted to rad/s:

$$1 \text{ rpm} = \pi/30 \text{ rad/s}$$

The angular velocity of a rigid body is the same about any point on the body. Angular acceleration is defined as the rate of change of ω:

$$\alpha = d\omega/dt$$

The SI unit of α is rad/s^2.

The equations of rotational kinematics for constant angular acceleration have the same form as the equations of (linear) kinematics for constant acceleration:

$$v = v_o + at \qquad \omega = \omega_o + \alpha t$$

$$x = x_o + v_o t + 1/2\ \alpha t^2 \qquad \theta = \theta_o + \omega_o t + 1/2\ \alpha t^2$$

$$v^2 = v_o^2 + 2a(x - x_o) \qquad \omega^2 = \omega_o^2 + 2\alpha(\theta - \theta_o)$$

If a body rotates about an axis fixed in position, the tangential speed v, and tangential acceleration, $a_t = dv/dt$, of a point at a distance r from the axis are related to ω and α:

$$v = \omega r; \qquad a_t = \alpha r$$

Rolling

When a wheel, or sphere, of radius R rolls without slipping, the speed of the center of the wheel is related to the angular velocity by $v_{CM} = \omega R$.

EXAMPLE 1

An disk of radius 15 cm accelerates uniformly from rest to 45 rpm in 2.5 s and then maintains this rotational rate. Find: (a) the radial and the tangential accelerations of a point on the rim at 1.5 s; (b) the number of revolutions completed in 4 s. (c) When a brake is applied, the disk stops in 8 revolutions. How long does this take?

Solution:

(a) The final angular velocity is $\omega = 1.5\pi$ rad/s = 4.71 rad/s. The angular acceleration is $\alpha = \Delta\omega/\Delta t = 1.88$ rad/s^2. At 1.5 s, the angular vleocity is $\omega = \alpha t = 2.83$ rad/s. The radial and tangential acclerations at r = 0.15 m are

$$a_r = \omega^2 r = 1.20 \text{ m/s}^2; \qquad a_t = \alpha r = 0.28 \text{ m/s}^2.$$

The magnitude of the total linear acceleration is $a = (a_r^2 + a_t^2)^{1/2} = 1.23$ m/s^2.

(b) We need to find the sum of the angular displacements while the disk is accelerating and while it rotates at constant angular velocity.

$$\Delta\theta_1 = 1/2\ \alpha t_1^2 = 0.5(1.88 \text{ rad/s}^2)(2.5 \text{ s})^2 = 5.88 \text{ rad}$$

$$\Delta\theta_2 = \omega t_2 = (4.71 \text{ rad/s})(1.5 \text{ s}) = 7.07 \text{ rad}$$

The number of revolutions is $(\Delta\theta_1 + \Delta\theta_2)/2\pi = 2.06$ rev.

(c) We first find the angular acceleration from $\omega^2 = 0 = \omega_o^2 + 2\alpha\Delta\theta$,

$$0 = (4.71 \text{ rad/s})^2 + 2\alpha\ (16\pi \text{ rad})$$

thus $\alpha = -0.221$ rad/s^2. The time to stop is given by $\omega = \omega_o + \alpha t$, which leads to t = 21.3 s.

ROTATIONAL KINETIC ENERGY AND MOMENT OF INERTIA

The moment of inertia of a system of discrete particles is defined as

$$I = \Sigma\, m_i\, r_i^2$$

where r_i is the perpendicular distance of the i th particle from the axis.
For a continuous distribution, the body is divided into infinitesimal elements of mass, dm. Thus,

$$I = \int r^2\, dm$$

The moment of inertia of a rigid body about a given axis is a measure of its rotational inertia, that is, its resistance to angular acceleration.

Parallel Axis Theorem

The moment of inertia about an arbitrary axis is related to the moment of inertia about a parallel axis through the center of mass by the parallel axis theorem:

$$I_{new} = I_{CM} + Mh^2$$

where h is the perpendicular distance between the two axes.

Kinetic Energy

The rotational kinetic energy of a rigid body is given by

$$K = 1/2\, I\omega^2$$

Note that this has the same form as $K = 1/2\, mv^2$ for translational kinetic energy.

The total kinetic energy of a rotating body is the sum of the translational kinetic energy associated with the motion of the center of mass, and the rotational kinetic energy relative to the CM:

$$K = 1/2\, mv_{CM}^2 + 1/2\, I_{CM}\omega^2$$

EXAMPLE 2

Three balls are connected by light (masssless) rods. Their masses and positions are $m_1 = 2$ kg at (2 m, 4 m), $m_2 = 1$ kg at (-2 m, 2 m), and $m_3 = 3$ kg at (1 m, -2 m). Find the moment of inertia about the x, y, and z axes respectively.

Solution:

In the equation $I = \Sigma\, m_i r_i^2$, r_i is the perpendicular distance to the axis. Thus, for the moment of inertia about the x axis, we use the y coordinate, and for the moment of inertia about the y axis we use the x coordinate.

$I_x = \Sigma\, m_i\, y_i^2 = (2\text{ kg})(4\text{ m})^2 + (1\text{ kg})(2\text{ m})^2 + (3\text{ kg})(-2\text{ m})^2 = 48\text{ kg.m}^2$

$I_y = \Sigma\, m_i\, x_i^2 = (2\text{ kg})(2\text{ m})^2 + (1\text{ kg})(-2\text{ m})^2 + (3\text{ kg})(1\text{ m})^2 = 15\text{ kg.m}^2$

$I_z = \Sigma\, m_i\, r_i^2$, but $r^2 = x^2 + y^2$, so $I_z = I_x + I_y = 63\text{ kg.m}^2$

EXAMPLE 3
A right-angled triangular plate has a base b and a height h. What is its moment of inertia about the vertical side? See Fig. 11.1.

Solution:
This is a continuous body so we must use integration to determine the moment of inertia. Since we are looking for the moment of inertia about a vertical axis, an appropriate choice for an infinitesimal mass element is a vertical strip of width dx and height y, as shown in Fig. 11.1. If the areal mass density is σ kg/m², the mass of the element is

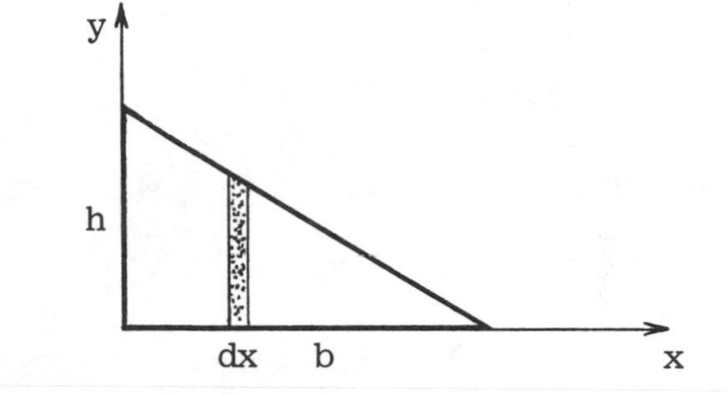

$$dm = \sigma\, dA = \sigma\, y\, dx$$

FIGURE 11.1

We need to express y in terms of x. The equation for the sloping side is $y = h - hx/b$, thus $dm = \sigma h(1 - x/b)dx$ and the moment of inertia of this element is

$$dI = x^2\, dm = \sigma h(x^2 - x^3/b)\, dx$$

The moment of inertia of the whole plate is the integral

$$I = \int dI = \sigma h \left| x^3/3 - x^4/4b \right|_0^b = Mb^2/6$$

where we have used $M = \sigma bh/2$.

EXAMPLE 4
An object has the shape of a table tennis bat. It consists of a rod of mass 1.5 kg and length 80 cm and a disk of mass 0.8 kg and radius 60 cm attached to one end of the rod, as in Fig. 11.2. What is the moment of inertia about an axis through the end of the handle? The moment of inertia of a disk about its center is $MR^2/2$ and of a rod about its center is $ML^2/12$.

Solution:

Since the given axes do not pass through the centers of either part of the composite object, we must use the parallel axis theorem for both, that is,

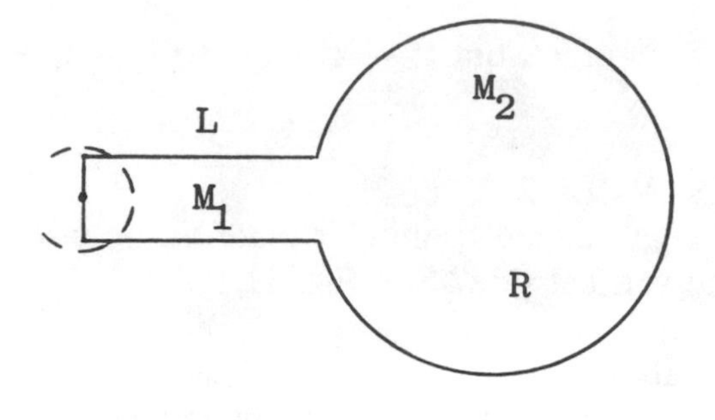

FIGURE 11.2

$$I_{new} = I_{CM} + Mh^2$$

ROD: $\quad I_1 = M_1L^2/12 + M_1(L/2)^2$

$$= M_1L^2/3 = 0.48 \text{ kg.m}^2$$

DISK: $\quad I_2 = M_2R^2/2 + M_2(L + R)^2$

$$= 1.71 \text{ kg.m}^2$$

The total moment of inertia is $I_1 + I_2 = 2.19$ kg.m^2

EXAMPLE 5

A block of mass m = 2.4 kg is attached to a vertical spring (k = 3 N.m) by a string that hangs over a pulley (disk) of mass M = 0.8 kg and radius R = 40 cm, as shown in Fig. 11.3. The system starts at rest with the spring unextended. What is the speed of the block after is falls by 50 cm?

Solution:

The presence of the spring with its nonconstant force makes the use of Newton's second law difficult. Instead we use the conservation of energy, $\Delta E = \Delta K + \Delta U = 0$. Recall that with this approach we do not need to set the $U_g = 0$ level.

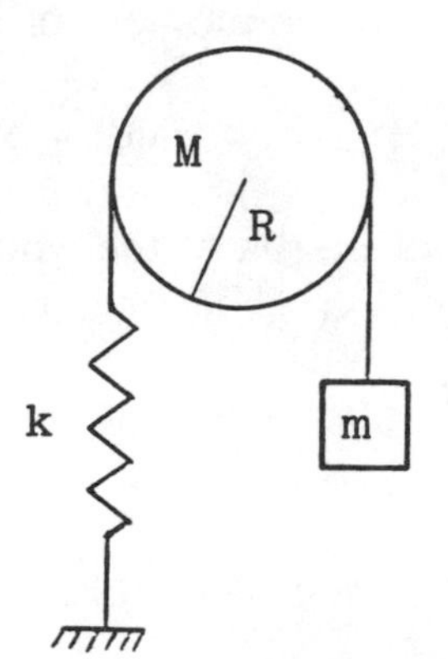

FIGURE 11.3

$$\Delta E = 1/2\ mv^2 - mgd + 1/2\ kd^2 + 1/2\ I\omega^2 = 0$$

Assuming that the string does not slip over the pulley, we have $v = \omega R$, thus

$$1/2\ I\omega^2 = 1/2\ (MR^2)(v/R)^2 = 1/2\ Mv^2$$

From $\Delta E = 0$, we find

$$1/2\ (m + M)v^2 = mgd - 1/2\ kd^2$$

Using the given values we find v = 2.67 m/s (check it).

TORQUE

Torque is a measure of the ability of a force to cause a rotation of a body about a given axis. Just as force causes linear acceleration, torque causes angular acceleration. When a force is exerted on a wrench, only the component of the force perpendicular to the arm of the wrench is effective in turning it. Equivalently, one can say that the full force acts at a shorter distance from the axis. The torque due to a force may be written in several ways:

$$\tau = r\,F_{\perp} = r_{\perp}\,F$$

$$= r\,F\,\sin\theta$$

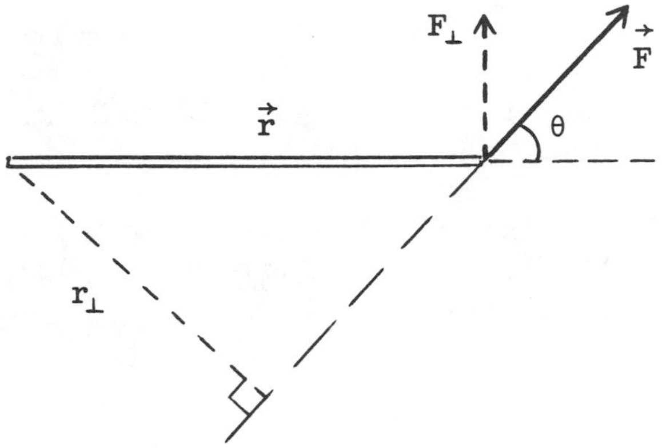

FIGURE 11.4

The quantity $r_{\perp}$, which is the shortest distance from the axis to the line of action of the force, is called the lever arm (see Fig. 11.4). The SI unit of torque is N.m (which should not be written as joules). For the moment, we will specify only the sense of the torque as being either clockwise or counterclockwise.

EXAMPLE 6

A uniform rod of length L and weight W is pivoted freely at one end. It is supported by two ropes as shown in Fig. 11.5. Find the torque due to each force about the pivot.

Solution:

To find the torque we could use either $rF_{\perp}$ or $r_{\perp}F$. Students usually find $rF_{\perp}$ to be more convenient, but we will illustrate both approaches. We will take counterclockwise torques to be positive, as shown by the circular arrow.

$$\tau_1 = r_1\,F_{1\perp} \;=\; -(L)(T_1\,\sin\theta)$$

$$\tau_2 = r_2\,F_{2\perp} \;=\; +(0.75L)(T_2\,\sin\alpha)$$

$$\tau_3 = r_{3\perp}\,F_3 \;=\; -(0.5L\,\cos\theta)W$$

If the rod is in equilibrium, the net torque is zero.

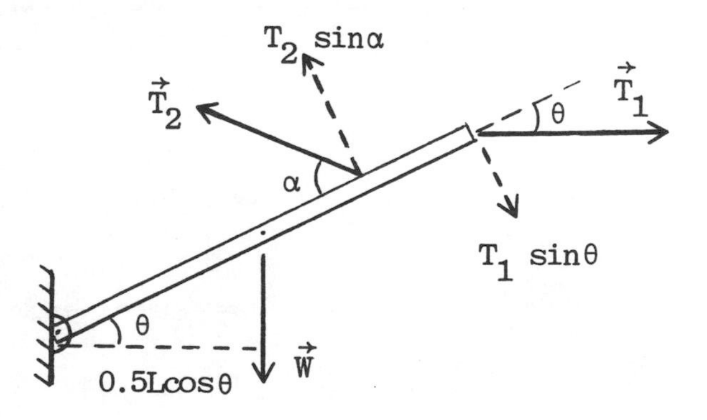

FIGURE 11.5

ROTATIONAL DYNAMICS

Newton's second law for the rotation of a rigid body about a fixed axis is

$$\tau = I\,\alpha$$

where τ is the net torque acting on the body. This equation has the same form as F = ma. It may be applied in two cases:

(1) The axis is fixed in direction and position.
(2) The axis is fixed in direction and passes through the center of mass.

ENERGY AND POWER

If a torque τ acts on a body that rotates through angle θ, the work done is

$$W = \tau\,\theta$$

The work-energy theorem for rotational motion is

$$W = 1/2\ I\omega_f^2 - 1/2\ I\omega_i^2$$

If a torque is applied to a body when its angular velocity is ω, the power supplied to the body is given by

$$P = \tau\omega$$

This has the same form as P = Fv for linear motion.

EXAMPLE 7

A 1.5 kg grinding wheel (disk) of radius 12 cm rotates initially at 3000 rpm. It is brought to rest in 10 rev by pressing against the rim. (a) What was the tangential force acting on it? (b) What was the work done on the wheel?

Solution:

(a) We use $\omega^2 = \omega_o^2 + 2\alpha\Delta\theta$ to find α, with $\omega = 0$ and $\Delta\theta = 20\pi$ rad. Since 1 rpm = $\pi/30$ rad/s, $\omega_o = 100\pi$ rad/s. Thus,

$$0 = (100\pi)^2 + 2\alpha(20\pi)$$

and $\alpha = -250\pi$ rad/s^2. If F is the tangential force, the torque acting on the wheel is

$$\tau = FR = I\alpha$$

where $I = 1/2\ MR^2$. Thus,

$$F = I\alpha/R = MR\alpha/2 = 70.7 \text{ N}$$

(b) The work done on the wheel is equal to the change in its kinetic energy

$$\Delta K = -1/2\, I\omega_o^2$$

where $I = MR^2/2 = 1.08\text{x}10^{-2}$ kg.m^2 and $\omega_o = 100\pi$ rad/s. We find $\Delta K = -533$ J

EXAMPLE 8

Two blocks with masses $m_1 = 2$ kg and $m_2 = 5$ kg hang on either side of pulley (disk) of mass $M = 2$ kg and radius $R = 20$ cm, see Fig. 11.6. Find (a) the angular acceleration, (b) the number revolutions made in 2 s if the system starts from rest; (c) the power supplied to the pulley at 1 s. Take $I = 1/2\, MR^2$.

Solution:

(a) We apply $F = ma$ to the blocks and $\tau = I\alpha$ to the pulley.

$$T_1 - m_1 g = m_1 a$$

$$m_2 g - T_2 = m_2 a$$

$$(T_2 - T_1)R = I\alpha = Ia/R$$

Divide the third equation by R and add all three equations to find

$$a = (m_2 - m_1)g/(m_1 + m_2 + I/R^2)$$

$$= 3.68 \text{ m/s}^2$$

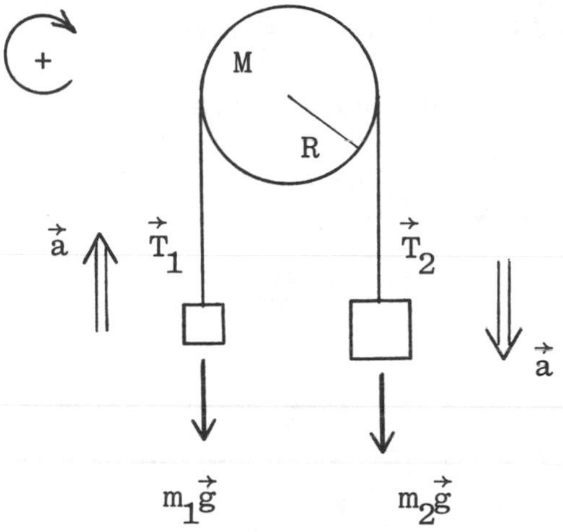

FIGURE 11.6

(b) The angular acceleration is $\alpha = a/R = 18.4$ rad/s^2. The angular displacement is

$$\Delta\theta = 1/2\, \alpha t^2 = 36.8 \text{ rad}$$

Since 1 rev = 2π rad, there are 5.86 revolutions.

(c) At 1 s, the angular velocity is $\omega = \alpha t = 18.4$ rad/s. The power delivered to the pulley at this time is

$$P = (\tau_2 - \tau_1)\omega$$

$$= I\omega/R = MR\omega/2 = 3.68 \text{ W}$$

EXAMPLE 9
A disk of mass M and radius R is pivoted freely at its rim. Initially the center is at the level of the pivot. What is the speed of the bottom of the disk when the center is below the pivot?

Solution:
The moment of inertia of the disk about a point on its rim is given by the parallel axis theorem.

$$I = MR^2/2 + MR^2 = 1.5MR^2$$

When the line joining the pivot to the center is at θ to the vertical, as in Fig. 11.7, the torque due to the weight is $\tau = -MgR \sin\theta$ (the sense of τ is CW but θ increases CCW). From $\tau = I\alpha$ we have

$$-MgR \sin\theta = (1.5MR^2)\alpha$$

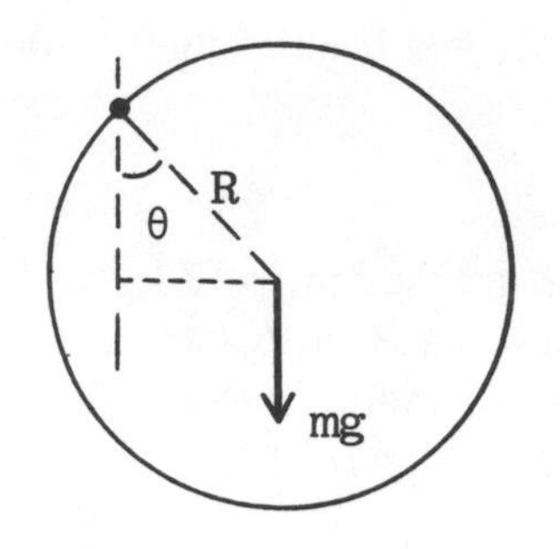

FIGURE 11.7

In order to find the final angular velocity we need to express α in terms of θ. Using the chain-rule we have

$$\alpha = d\omega/dt = (d\omega/d\theta)(d\theta/dt) = \omega\, d\omega/d\theta$$

thus,

$$\int_0^{\omega} \omega\, d\omega = \int_{\pi/2}^{0} \alpha\, d\theta = (g/1.5R) \int_{\pi/2}^{0} -\sin\theta\, d\theta$$

$$| \omega^2/2 | = (g/1.5R) | \cos\theta |$$

We find $\omega = (4g/3R)^{1/2}$. The velocity of the lowest point is $v = \omega(2R) = (16gR/3)^{1/2}$. This result can be obtained more simply by using energy conservation. See Exercise 43 in the text.

SOLUTIONS TO SELECTED TEXT EXERCISES AND PROBLEMS

Exercise 13
A wheel of a car has a radius of 20 cm. It initially rotates at 120 rpm. In the next minute it makes 90 revolutions. (a) What is the angular acceleration? (b) How much further does the can travel before coming to rest? There is no slipping.

Solution:
(a) $\Delta\theta = \omega_o t + 1/2\, \alpha t^2$, where $\Delta\theta = 180\pi$ rad and $\omega = 4\pi$ rad/s, thus

$$180\pi = 4\pi(60) + 1/2\, \alpha(60)^2$$

We find $\alpha = -\pi/30 = -0.105$ rad/s^2.

(b) $\omega = \omega_o + \alpha t = 4\pi - (\pi/30)(60) = 2\pi$ rad/s, and $v_o = \omega r = 0.4\pi$ m/s for the last leg, thus

$$0 = v_o^2 - 2a\Delta x$$

which gives $\Delta x = 37.7$ m.

Exercise 29

Two solid spheres of mass m and radius R are stuck to the ends of a thin rod of mass m and length 3R. Find the moment of inertia of the system about the axis at the midpoint of the rod and perpendicular to it, as shown in Fig. 11.8.

Solution:

The moment of inertia of a rod about its center is $mL^2/12$ and of a solid sphere about a diameter is $2mR^2/5$. The distance between the center of each sphere and the center of the rod is $h = 2.5R$. This must be used in the parallel axis theorem $I = I_{CM} + mh^2$ to find the moment of inertia of the spheres about the center of the rod,

$$I = 2m[2R^2/5 + (2.5R)^2] + m(3R)^2/12$$

$$= 14.1mR^2$$

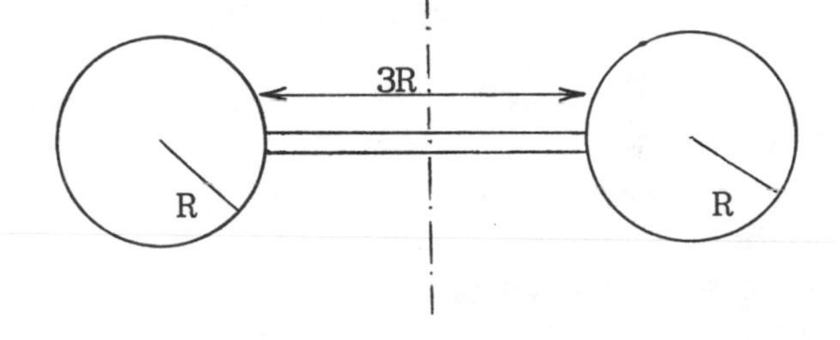

FIGURE 11.8

Exercise 41

A solid sphere and a disk of the same mass and radius roll up an incline. Find the ratio of the heights, h_S/h_D, to which they rise if , at the bottom, they have the same (a) kinetic energy; (b) speed.

Solution:

The initial and final energies are

$$E_i = 1/2\ mv^2 + 1/2\ I\omega^2; \quad E_f = mgh$$

(a) If the initial energies are equal, the final heights are also equal. Thus the ratio of heights is one.

(b) With $I = 2mR^2/5$ for the sphere we find $E_i = 7mv^2/10$. With $I = mR^2/2$ for the disk we find $E_i = 3mv^2/4$.

When we set $E_f = E_i$, we find $h_S = 7v^2/10g$ for the sphere and $h_D = 3v^2/4g$ for the disk. The ratio is $h_S/h_D = 14/15$.

Exercise 53
A wheel starts from rest and rotates through 150 rad in 5 s. The net torque due to the motor and friction is constant at 48 N.m. When the motor is switched off, the wheel stops in 12 s. Find the torque due to (a) friction, and (b) the motor.

Solution:
(a) From $\Delta\theta = 1/2\ \alpha t^2$ we find $\alpha = 12\ \text{rad/s}^2$. From $\tau = I\alpha$,

$$I = \tau/\alpha = 48/12 = 4\ \text{kg.m}^2$$

When the motor is switched off $\omega = 60$ rad/s, thus

$$\alpha = \Delta\omega/\Delta t = -60/12 = -5\ \text{rad/s}^2$$

thus, $\tau_f = I\alpha = -20$ N.m.

(b) We are given the total torque $\tau = \tau_M + \tau_f = 48$ N.m, thus $\tau_M = 68$ N.m

Exercise 65
In November 1984, astronaut Joe Allen of the shuttle Discovery attached a device to the disabled Palaba B salellite that was spinning at 2 rpm (see Fig. 11.63 in the text). The satellite was a solid cylinder of mass 7000 kg and had a radius of 80 cm. He fired a 20-N jet tangential to the cylindrical surface to halt the rotation. (a) What was the initial kinetic energy of the satellite? (b) How long did it take to stop the spinning?

Solution:
(a) The angular velocity is $\omega = 4\pi/60 = \pi/15$ rad/s. The initial kinetic energy is

$$K = 1/2\ (1/2\ MR^2)\omega^2 = 49.1\ \text{J}$$

(b) From $\tau = I\alpha$ we find

$$\alpha = \tau/I = 7.14\text{x}10^{-3}\ \text{rad/s}^2$$

Then, $t = \Delta\omega/\alpha = 29.3$ s

Problem 3
A uniform rod of length L is held vertically on a frictionless floor. A very slight nudge causes it to fall. (a) What is its angular velocity on landing? (b) Find the speeds of its ends when it lands.

Solution:
(a) The mass of the rod may be taken to be concentrated at it center. Thus, the initial potential energy is

$$E_i = mgL/2$$

The final energy of the rod consists of translational kinetic energy of the CM and rotational kinetic energy about the CM:

$$E_f = 1/2\ mv_{CM}^2 + 1/2\ I_{CM}\ \omega^2$$

On landing, the speed of the left side is $v_L = 0$, and for the right side it is $v_R = 2v_{CM}$. Thus $\omega_f = v_{CM}/(L/2)$. Substituting into $E_i = E_f$ we find $\omega = (3g/L)^{1/2}$

(b) $v_L = 0$, $v_R = \omega L = (3gL/4)^{1/2}$.

Problem 9
Find the moment of inertia of a hollow cone of mass M, height h and apex angle 2α about the central axis of symmetry. (Hint: Use the distance from the apex along the surface as the variable.)

Solution:
We divide the cone into infinitesimal rings of radius $\ell\sin\alpha$ and thickness $d\ell$, as shown in Fig. 11.9. The surface area of a ring is $dA = (2\pi\ell\sin\alpha)\ d\ell$ and its mass is

$$dm = \sigma(2\pi\ell\sin\alpha)\ d\ell$$

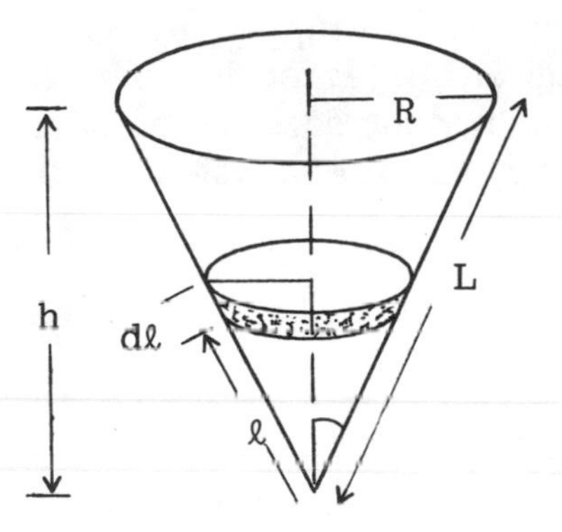

FIGURE 11.9

The total mass of the cone is

$$M = 2\pi\sigma\sin\alpha(L^2/2)$$

or $M = \sigma\pi RL$. The moment of inertia of a ring is

$$dI = dm\ r^2 = [2\pi\sigma\ell\sin\alpha\ d\ell](\ell\ \sin\alpha)^2$$

$$= 2\pi\sigma\sin^3\alpha\ \ell^3\ d\ell$$

On integrating from 0 to L and using $L = h/\cos\alpha$, we find,

$$I = 1/2\ Mh^2\ \tan^2\alpha$$

Problem 13
Use the perpendicular axis theorem (Exercise 34 in the text) and what you know about the moments of inertia of a thin rod, to find the moment of inertia of a thin rectangular plate with sides of length a and b about a cnetral axis perpendicular to its plane, as shown in Fig. 11.10. (The final result is not restricted to a thin plate. Why not?)

Solution:

According to the perpendicular-axis theorem the moment of inertia of a planar body about the z axis perpendicular to its plane is $I_z = I_x + I_y$, where I_x and I_y are the moments of inertia about the x and y axes. If one looks along the x or y axis, the plate appears to be a rod- which has a moment of inertia $ML^2/12$. For the plate $I_x = Ma^2/12$ and $I_y = Mb^2/12$, thus $I_z = M(a^2 + b^2)/12$. The result is not restricted to a thin plate because a thick plate can be treated as a series of thin plates.

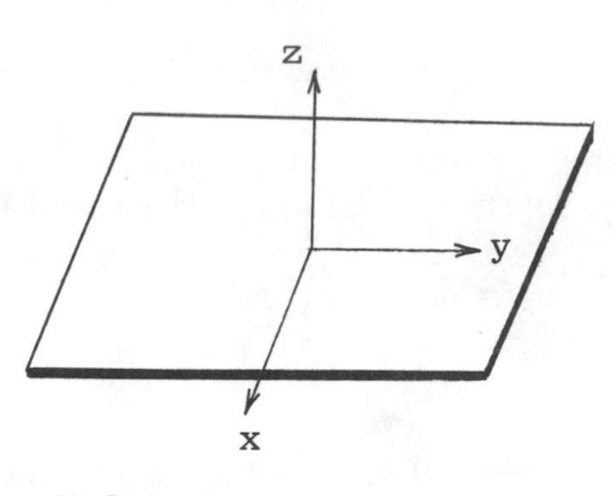

FIGURE 11.10

SELF-TEST

1. A block of mass m = 2 kg slides on a frictionless incline (θ = 30°) and is connected to a rope that is wrapped around a pulley of mass M = 4 kg and radius R = 0.5 m, as in Fig. 11.11. (Treat the pulley as a disk with $I = 1/2\ MR^2$). Find: (a) The tension in the rope and the acceleration of the block; (b) The number of revolutions made by the pulley in 3 s. The system starts from rest. Use g = 10 N/kg

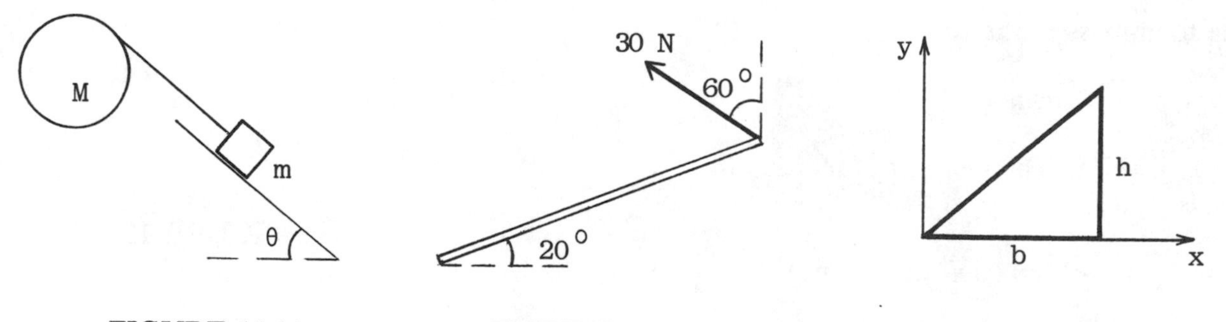

FIGURE 11.11 FIGURE 11.12 FIGURE 11.13

2. A uniform rod of length 60 cm and mass 5 kg is freely pivoted at one end, as in Fig. 11.12. It is at 20° to the horizontal. A rope is connected to the other end and is directed at 60° to the vertical. The tension in the rope is 30 N. What is the torque about the pivot due to the weight and the tension?

3. Determine the moment of inertia of the triangular plate in Fig. 11.13 about the y axis.

4. Two blocks, m_1 = 5 kg and m_2 = 2 kg hang vertically on either side of a pulley (disk) of mass M = 2 kg and radius R = 50 cm. The system starts at rest. What is the speed of m_2 after it has fallen by 60 cm? Use the energy approach.

Chapter 12

ANGULAR MOMENTUM AND STATICS

MAJOR POINTS

1. The definition of torque as a vector
2. (a) The angular momentum of a particle
   (b) The angular momentum of a rigid body about a fixed axis.
3. Rotational dynamics
4. The conservation of angular momentum
5. The conditions for static equilibrium

CHAPTER REVIEW

THE TORQUE VECTOR

The torque due to a force **F** exerted at a position **r** is defined as

$$\boldsymbol{\tau} = \mathbf{r} \times \mathbf{F}$$

The SI unit of τ is N.m. You should review the cross product in Chapter 2. The direction of the torque is given by the right-hand rule (see Fig. 12.2 in the text). The magnitude of the torque is

$$\tau = rF\sin\theta$$

$$= r_{\perp} F = rF_{\perp}$$

where θ is the smaller angle between **r** and **F**. The two other expressions employ the lever arm, $r_{\perp}$, or the component of the force $F_{\perp}$, perpendicular to **r**.

EXAMPLE 1

A force **F** = 2**i** - 3**j** N acts at **r** = 2.5**j** m. What is the torque about the origin?

Solution:

The force and position vectors are shown in Fig. 12.1. The torque is

$$\boldsymbol{\tau} = \mathbf{r} \times \mathbf{F}$$

$$= (2.5\mathbf{j}\ \text{m}) \times (2\mathbf{i} - 3\mathbf{j}\ \text{N})$$

$$= -5\mathbf{k}\ \text{N.m}$$

Use the right hand rule to confirm the direction.

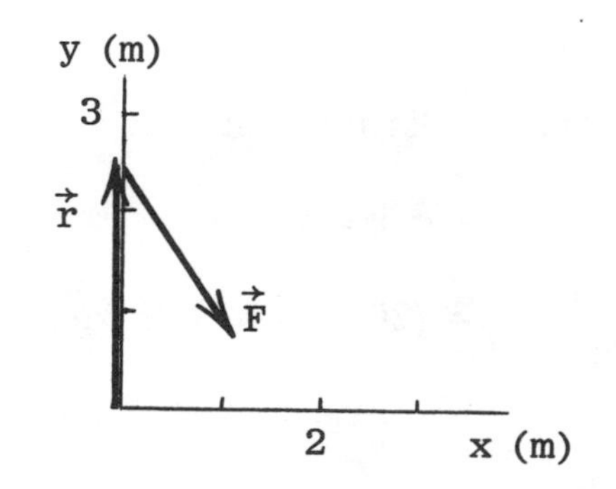

FIGURE 12.1

ANGULAR MOMENTUM

The angular momentum of a single particle that has a linear momentum **p** is defined as

$$\boldsymbol{\ell} = \mathbf{r} \times \mathbf{p}$$

where r is the vector position of the particle. Note that $\boldsymbol{\ell}$ depends on the choice of origin. The SI unit of angular momentum is kg.m^2/s or J.s.
The magnitude of $\boldsymbol{\ell}$ is

$$\ell = r\,p\,\sin\theta$$

$$= r_\perp\,p = r\,p_\perp$$

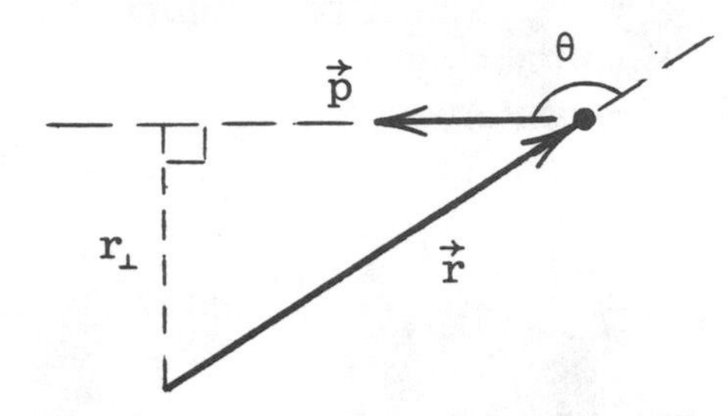

FIGURE 12.2

where **θ** is the angle between **r** and **p**, as in Fig. 12.2. Again, one can use the "moment arm", $r_\perp$, or the component of **p** perpendicular to **r**. The total angular momentum of a system is

$$\mathbf{L} = \Sigma\,\boldsymbol{\ell}_i$$

The angular momentum of a rigid body rotating about a fixed axis (the z axis)

$$L_z = I\omega$$

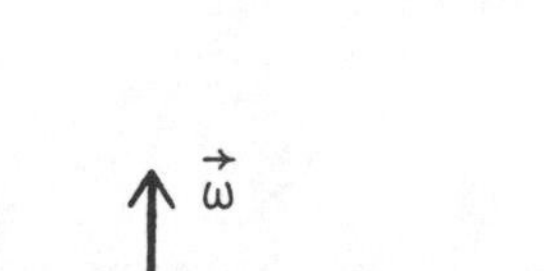

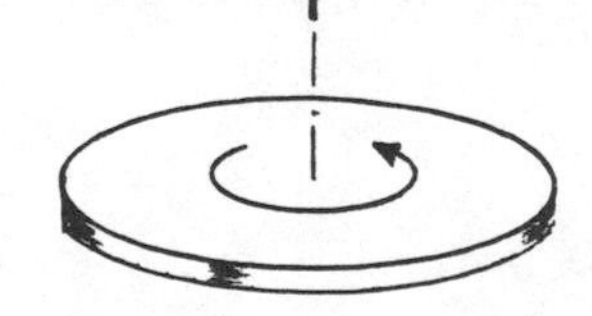

FIGURE 12.3

Note that this is not a vector relationship, it refers only to the component of **L** along the direction of **ω**, which is given by the right hand rule shown in Fig. 12.3.

EXAMPLE 2

The position and linear momentum of a particle are **r** = 2**i** + 3**k** m and **p** = 4**k** kg.m/s. What is its angular momentum?

Solution:

The angular momentum is

$$\boldsymbol{\ell} = \mathbf{r} \times \mathbf{p}$$

$$= (2\mathbf{i} + 3\mathbf{k}\text{ m})\times(4\mathbf{k}\text{ kg.m/s})$$

$$= -8\mathbf{j}\text{ kg.m}^2\text{/s}$$

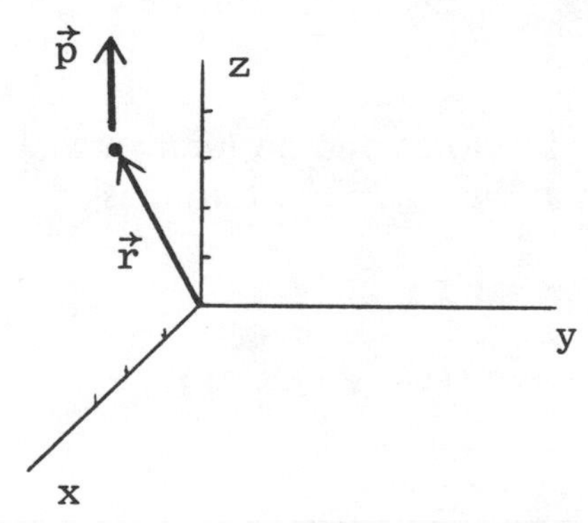

Confirm the direction with the right hand rule in Fig. 12.4.

FIGURE 12.4

EXAMPLE 3
A satellite of mass m is in circular orbit of radius r about a planet of mass M. How does its angular momentum vary with r?

Solution:

For a satellite in orbit, Newton's second law is

$$GmM/r^2 = mv^2/r$$

thus $v = (GM/r)^{1/2}$. Since the angle between the radial line and the velocity is always 90°, the angular momentum for a particle in circular motion is

$$L = mvr$$

Using the expression for the orbital speed, we find

$$L = m(GMr)^{1/2}$$

Thus the angular momentum is proportional to the square root of the radius.

ROTATIONAL DYNAMICS

The rotational analog of Newton's second law for a system, $\mathbf{F} = d\mathbf{P}/dt$, is

$$\boldsymbol{\tau} = d\mathbf{L}/dt$$

where $\boldsymbol{\tau} = \Sigma\, \boldsymbol{\tau}_i$ is the net external torque acting on the system of particles and $\mathbf{L} = \Sigma\, \boldsymbol{\ell}_i$ is the total angular momentum of the particles. Both $\boldsymbol{\tau}$ and $\mathbf{L}$ must be measured relative to either (a) or (b) below:

(a) the same origin in an inertial frame,
(b) the center of mass of the system.

In the special case of a rigid body rotating about a fixed axis, this equation reduces to the scalar form

$$\tau = I\alpha$$

which we used in Chapter 11.

EXAMPLE 4

A cylinder of radius 40 cm can rotate freely about a central axis. A rope wrapped around it has a tension of 12 N that acts for 5 s. What is the angular momentum of the cylinder at this time?

Solution:

The angular acceleration is given by $\alpha = \tau/I$ where $\tau = TR$, since the rope is tangential to the rim. The final angular velocity is

$$\omega = \alpha t = (\tau/I)t$$

The angular momentum of this rigid body is

$$L = I\omega = \tau t$$

Since $\tau = TR$, $L = TRt = (12\ N)(0.4\ m)(5\ s) = 24$ J.s. Notice that the mass of the cylinder does not appear in the final result.

CONSERVATION OF ANGULAR MOMENTUM

If the net external torque on a system is zero, then the angular momentum is constant:

$$\text{If } \boldsymbol{\tau}_{ext} = 0, \text{ then } \mathbf{L} = \text{constant.}$$

This is the principle of the conservation of angular momentum. For a rigid body rotating about a fixed axis, we have

(Rigid body, fixed axis) $$I_f\omega_f = I_i\omega_i$$

EXAMPLE 5

A child of mass m stands at the rim of a circular platform (disk) of mass M and radius R freely pivoted at its center. She walks at speed v relative to earth along the rim. What is the angular velocity of the platform?

Solution:

When the child starts walking, she has angular momentum relative to teh center of the platform:

$$L_c = m\,v\,R$$

Since there is not external torque acting on the system, the angular momentum cannot change from its initial value of zero, that is

$$L_c + L_p = 0$$

Thus, the platform must rotate with an angular momentum equal and opposite to that of the child.

$$L_p = I\,\omega = -mvR$$

Since $I = MR^2/2$, we find $\omega = -2mv/MR$. If the girl walks clockwise, the platform rotates counterclockwise. If the girl's angular momentum is directed upward, the platform's angular momentum is directed downward (both using the right hand rule).

EXAMPLE 6

A uniform rod of mass M and length d is pivoted freely about its center. A ball of putty of mass m traveling at speed u, as shown in Fig. 12.5, collides with and sticks to one end of the rod. What is the angular velocity after the collision?

Solution:

(a) We cannot use the conservation of linear momentum because of the external force exerted by the pivot. However, the pivot exerts no torque, thus we may apply the conservation of angular momentum.

Even though the putty is initially moving in a straight line it has angular momentum about the pivot:

$$L_i = r_\perp\, p = mud/2$$

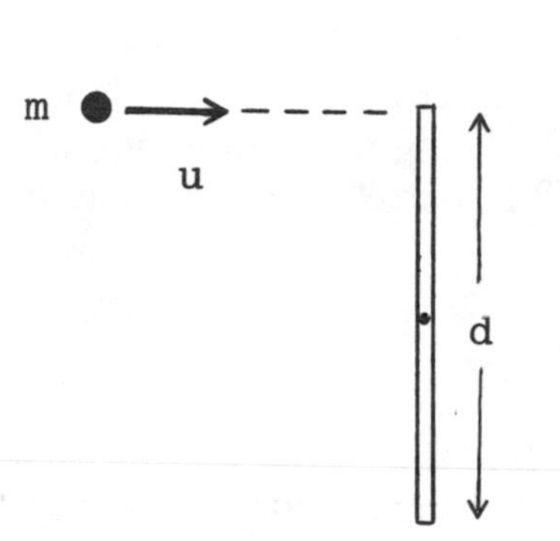

FIGURE 12.5

When the putty sticks to the rod, the moment of inertia increases. The moment of inertia of the rod about its center is $ML^2/12$ so the final angular momentum is

$$L_f = (I_p + I_r)\omega = (md^2/4 + Md^2/12)\omega$$

Notice that we did not simply add the mass of the putty to M; the putty does not spread uniformly over the rod! On setting $L_f = L_i$ we find

$$\omega = \frac{mu}{(md/4 + Md/12)}$$

CONDITIONS FOR STATIC EQUILIBRIUM

When the acceleration of a body is zero, it is said to be in translational equilibrium. When the angular acceleration is zero, it is in rotational equilibrium. The special case in which the object is at rest, is called static equilibrium. Two conditions must be satisifed for an object to be in equilibrium:

$$\Sigma\, \mathbf{F} = 0; \qquad \text{and } \Sigma\, \tau = 0$$

If the forces are confined to the xy plane and the body can rotate only about the z axis, we have

$$\Sigma F_x = 0, \qquad \Sigma F_y = 0, \qquad \Sigma\tau = 0$$

where τ is taken about the z axis.

The center of gravity of a body is the point about which the net gravitational torque is zero. In calculating the torque due to the weight, we treat a body as if all its weight acts at the CG. The CM and CG coincide only in a uniform gravitational field.

EXAMPLE 7

A rod of mass 2.5 kg and length 60 cm is freely pivoted at one end and is inclined at 30° to the horizontal. It is supported by two ropes as shown in Fig. 12.6. The rope at the end has a tension T_1 = 10 N, the other rope is attached 45 cm from the pivot and is inclined at 20° to the horizontal as shown. Determine the tension T_2 and the horizontal and vertical forces exerted on the rod at the pivot.

Solution:

By choosing the pivot as the point about which to take torques we eliminate the force at the pivot. Thus,

$$\Sigma\tau: \quad -(L/2)mg - (L)(T_1 \sin\theta)$$

$$+ (0.75L)[T_2 \sin(\alpha + \theta)] = 0$$

Therefore T_2 = 30.0 N.

To find the components of the force exerted by the pivot we may use the condition $\Sigma\mathbf{F} = 0$, that is

$$\Sigma F_x = H - T_2 \cos\alpha + T_1 = 0$$

$$\Sigma F_y = V + T_2 \sin\alpha - W = 0$$

We find H = 18.2 N and V = 21.1 N.

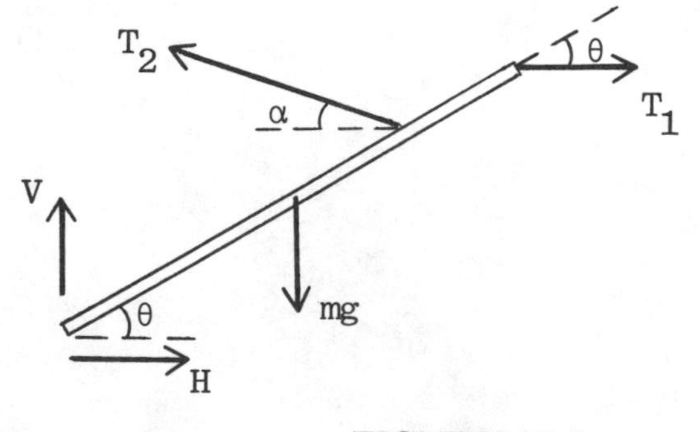

FIGURE 12.6

SOLUTIONS TO SELECTED TEXT EXERCISES AND PROBLEMS

Exercise 3

Two particles of equal mass have the same speed and travel in opposite directions along two parallel lines. Show that the total angular momentum is independent of the choice of origin.

Solution:

We choose the origin at a point that is a perpendicular distance a from the path of one particle. The angular momentum is given by $\ell = r\,p$. One must keep in mind that ℓ is a vector given by the right hand rule. From Fig. 12.7 we see that

$$L = mv(a + d) - mva = mvd$$

This depends only on d.

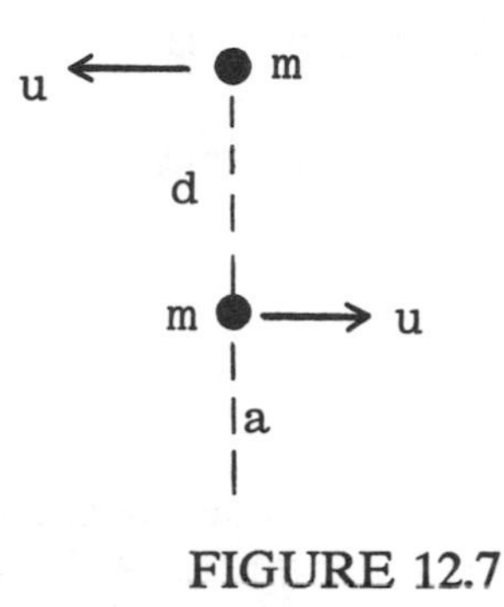

FIGURE 12.7

Exercise 11

A particle of mass M moves in the xy plane. Its coordinates as a function of time are given by $x(t) = At^3$; $y(t) = Bt^2 - Ct$, where A, B, and C are constants. (a) Find its angular momentum about the origin. (b) What force acts on it?

Solution:

(a) The position is $\mathbf{r} = At^3\mathbf{i} + (Bt^2 - Ct)\mathbf{j}$. Thus, the velocity is $\mathbf{v} = 3At^2\mathbf{i} + (2Bt - C)\mathbf{j}$ and the angular momentum is

$$\mathbf{L} = \mathbf{r} \times \mathbf{p} = M\,(At^3\mathbf{i} + (Bt^2 - Ct)\mathbf{j}) \times (3At^2\mathbf{i} + (2Bt - C)\mathbf{j})$$

$$= MAt^3(2Bt - C)\mathbf{k} - M(Bt^2 - Ct)(3At^2)\mathbf{k}$$

$$= MAt^3(2C - Bt)\mathbf{k}$$

(b) From the velocity we find the acceleration $\mathbf{a} = 6At\mathbf{i} + 2B\mathbf{j}$. Then,

$$\mathbf{F} = m\mathbf{a} = M(6At\mathbf{i} + 2B\mathbf{j})$$

Exercise 21
An 80-kg man stands at the rim of a 100-kg carousel of radius 2 m initially at rest. He starts to walk along the rim at 1 m/s relative to the carousel. What is the angular speed of the carousel? Treat the carousel as a disk.

Solution:

When the man walks along the rim, the carousel rotates in the opposite sense keeping the total angular momentum zero. If the speed of a point on the rim is v, then the speed of the man relative to the ground is (1 - v) and so his angular momentum is

$$L_M = mR(1 - v)$$

The angular vleocity of the carousel is $\omega = v/R$, thus

$$L_C = -I\omega = -(1/2\ MR^2)(v/R)$$

The negative sign is included because the direction of L_C is opposite that of L_M (both given by the right hand rule) From $L_M + L_C = 0$, we find
$160(1 - v) = 100v$, which leads to $v = 0.615$ m/s, and $\omega = 0.308$ rad/s.

Exercise 41
A uniformly packed crate of weight W = 200 N and of dimensoins a = 0.4 m and b = 1.0 m is placed on a dolley. What howizontal force applied at P is needed to hold the system in rotational (but not translational) equilibrium in the position shown in Fig. 12.8? Take h = 1.1 m and $\theta = 30°$.

Solution:

From the geometry in Fig. 12.8 we see that the distance from the corner of the box to the center is $H = (a^2 + b^2)^{1/2}/2 = 0.539$ m and that the angle α is given by $\tan\alpha = a/b = 0.4$, thus $\alpha = 21.8°$. Taking torques about the lower left corner:

$$\Sigma\tau = Fh\cos\theta - WH\cos(\alpha + \theta) = 0$$

where $(\alpha + \theta) = 51.8°$. Thus

$$F = \frac{(200\ \text{N})(0.539\ \text{m})\cos 51.8°}{(1.1\ \text{m})(\cos 30°)} = 70\ \text{N}$$

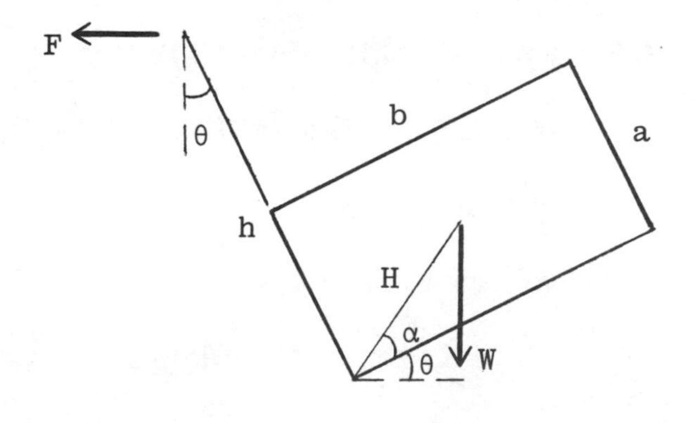

FIGURE 12.8

Problem 1

A particle of mass m = 0.5 kg moving at speed u = 4 m/s strikes a dumbbell consisting of two blocks of equal mass M = 1 kg separated by a massless rod of length 2 m (see Fig. 12.9). The dumbbell and the particle are free to slide on a horizontal surface. Find: (a) the speed of the center of mass of the system after the particle sticks to one of the blocks; (b) the angular velocity of the system about the center of mass.

Solution:

(a) Since the dumbell is not pivoted we may apply the conservation of linear momentum to the collision:

$$\Sigma p: \ mu = (2M + m)v_{CM}$$

which leads to $v_{CM} = 0.8$ m/s.

(b) The position of the CM after the collision is a distance d = MR/(M + m) = 0.8 m from the upper block. Taking the origin at the CM, the conservation of angular momentum tells us

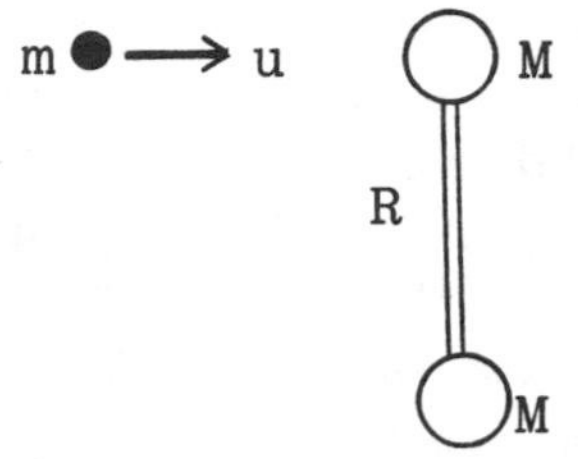

FIGURE 12.9

$$\Sigma L: \quad mud = (M + m)d^2\omega + M(R - d)^2\omega$$

From this we find $\omega = 2/3$ rad/s.

Problem 5

A cylinder of mass M and radius R is rotating at angular velocity ω_o when it is placed vertically on a horizontal surface for which the coefficient of kinetic friction is μ_k. (a) Write the linear and the rotational forms of the second law. (b) Show that it takes a time $\omega_o R/3\mu_k g$ for the cylinder to start rolling without slipping. (c) How far does it travel before it rolls without slipping?

Solution:

(a) Fig. 12.10 shows the force of friction on the cylinder, $f = \mu(Mg)$. Newton'ssecond law is f = Ma, so, $a = \mu g$. The torque acts to slow the rotation:

$$fR = I\alpha$$

so, $\alpha = 2\mu g/R$ (counterclockwise).

(b) When there is no slipping, $v = \omega R$. This may be written as

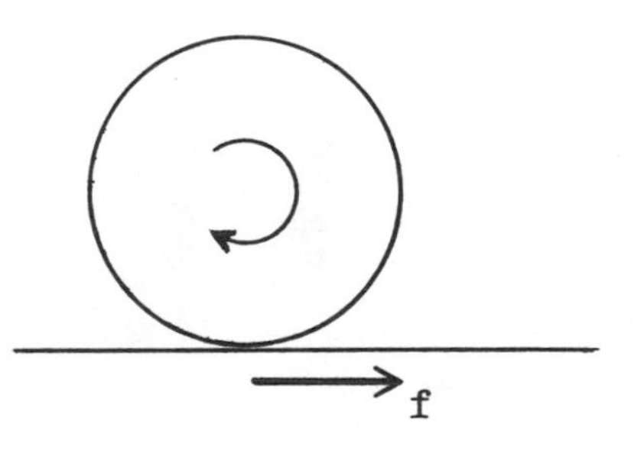

FIGURE 12.10

$$a\,t = (\omega_o - \alpha t)R$$

Thus $t = \omega_o R/3\mu_k g$.

(c) The displacement is $x = 1/2\ at^2 = (\omega_o R)^2/18\mu g$

SELF-TEST

1. A particle of mass 1.5 kg is located at (2 m, 3 m) and moving at 7 m/s at 25° above the +x axis. What is its angular momentum?

2. A circular platform of mass 50 kg and radius 2 m is rotating freely at 2.4 rad/s. A child of mass 30 kg is initially standing on the ground beside the platform. (a) When the child gets on to the platform does the angular velocity change? If so, what is the new value? (b) The child then walks half-way to the center of the platform. Does the angular velocity change? If so, what is the new value?

3. A horizontal rod of length 80 cm and weight 30 N is pivoted freely at one end and supported by a rope at the other end as shown in Fig. 12.11. A block of weight 12 N hangs from the other end. Find the tension in the rope and the horizontal and vertical components of the force exerted by the pivot on the rod.

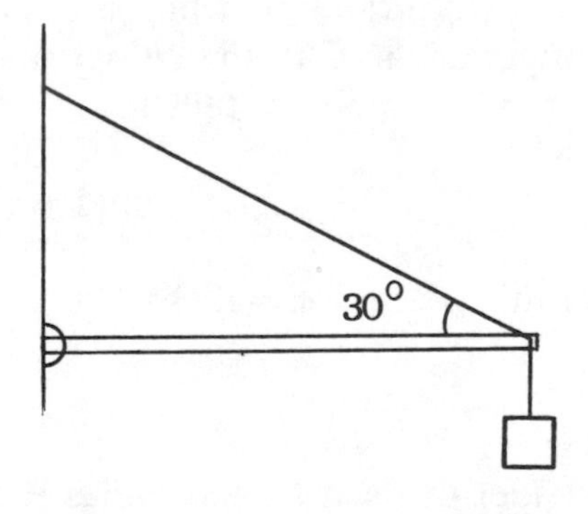

FIGURE 12.11

Chapter 13

GRAVITATION

MAJOR POINTS

1. Newton's law of gravitation
2. The distinction between inertial mass and gravitational mass
3. Kepler's laws of planetary motion
4. Continuous distributions of mass, the point mass theorem

CHAPTER REVIEW

Note that the dynamics of circular orbits, including Kepler's third law, has been discussed in Chapter 6 and the energy aspects of circular orbits are treated in Chapter 8.

NEWTON'S LAW OF GRAVITATION

Consider two point particles of mass m_1 and m_2 separated by a distance r. According to Newton's law of gravitation there is an attractive force betweeen these particles given by

$$F = Gm_1m_2/r^2$$

where the constant $G = 6.67\text{x}10^{-11}$ N.m^2/kg^2. If there are several particles present, the net force on a given particle, say m_1, is the vector sum of the individual forces due to each of the other masses:

$$\mathbf{F}_1 = \mathbf{F}_{12} + \mathbf{F}_{13} + \ldots.. + \mathbf{F}_{1N}$$

where $\mathbf{F}_{12}$ is the force exerted on m_1 by m_2. This is called the principle of linear superposition. It means that the force between any given pair of particles is not affected by other particles.

Spherical Mass Distribution

Although, in general, Newton's law of gravitation applies only to point particles, there is an exception. According to the "point mass theorem" a spherically symmetric mass distribution (such as a uniform sphere or shell) attracts a point particle as if all its mass were concentrated as the center. Thus we may take r to be the distance from the point particle to the center. (See Example 13.4.) This result is a consequence of the $1/r^2$ nature of the force law.

Another consequence of the inverse square nature of the force law is that a point particle inside a uniform spherical shell experiences no net force.

EXAMPLE 1
Three particles are located as follows: m_1 = 10 kg at the origin, m_2 = 5 kg at (0, 2 m) and m_3 = 20 kg at (3 m, 2 m). What is the net force on m_3?

Solution:
We first draw the directions of the forces on m_3, as shown in Fig. 13.1. Then we calculate the magnitudes

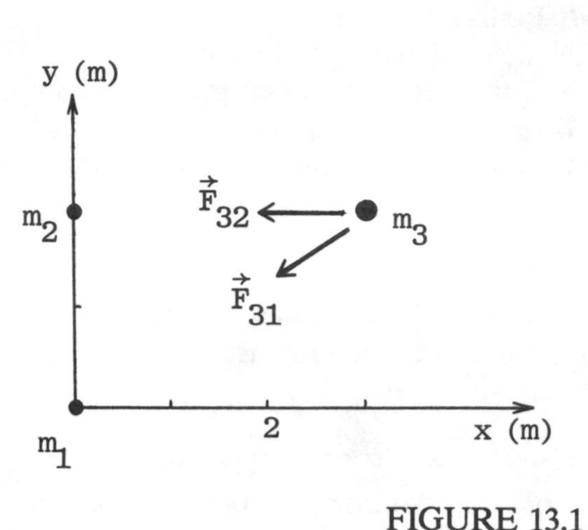

FIGURE 13.1

$F_{31} = G(200\ kg^2)/13\ m^2 = 15.4G$

$F_{32} = G(100\ kg^2)/9\ m^2 = 11.1G$

The components of the net force on m_3, $\mathbf{F}_3 = \mathbf{F}_{31} + \mathbf{F}_{32}$, are

$F_{3x} = -F_{31}\cos\theta - F_{32} = -23.9G$

$F_{3y} = -F_{31}\sin\theta = -8.5G$

Thus $\mathbf{F}_3 = (-1.59\mathbf{i} - 0.57\mathbf{j})\text{x}10^{-9}$ N

Gravitational Field Strength

In Chapter 5, we defined the weight of an object as the net gravitational force acting on it. Consider a particle of mass m at a distance r from the center of a planet of mass M and radius R. We have

$$F = GmM/r^2 = mg$$

Thus, the gravitational force per unit mass, or gravitational field strength, is

$$g = F/m = GM/r^2$$

At the surface of the planet, r = R, and $g_o = GM/R^2$. At the surface of the earth, $g_o \approx 9.81$ N/kg.

EXAMPLE 2
Show that the fractional change in the gravitational field strength dg/g is related to the fractional increase in distance from a planet, dr/r, by dg/g = -2dr/r. Estimate the value of g at the peak of Mount Everest at a height of 29,000 ft (8.84 km) given that the value at sea level is 9.807 N/kg. The mean radius of the earth is 6370 km.

Solution:

From the previous section we know that $g = GM/r^2$, thus

$$dg/dr = -2GM/r^3$$

which may be rewritten as $dg = -2g\ dr/r$, or

$$dg/g = -2\ dr/r$$

We are given that $r = R_E = 6370$ km, $dr = 8.84$ km and $g = 9.807$ N/g, thus

$$dg = -2(8.84/6370)(9.807) = 0.0272 \text{ N/kg}$$

At Mount Everest the field strength is 9.780 N/kg.

EXAMPLE 3

A block of mass 1 kg is suspended by a vertical string at the equator. What is the tension in the string given that the gravitational field strength at that point is 9.807 N/kg?

Solution:

The block at the equator is in uniform circular motion and thus it has a centripetal acceleration, $a_c = v^2/R$. The gravitational force on the block is its weight mg, where $g = 9.807$ N/kg. From Fig. 13.2, we see that Newton's second law for this motion is

$$mg - T = ma_c$$

Thus, $T = mg - ma_c$. At the equator $a_c = 4\pi^2R/T^2$ where $R = 6370$ km and $T = 1$ day. We find $a_c = 3.38\text{x}10^{-2}$ m/s², then, $T = 9.773$ N.

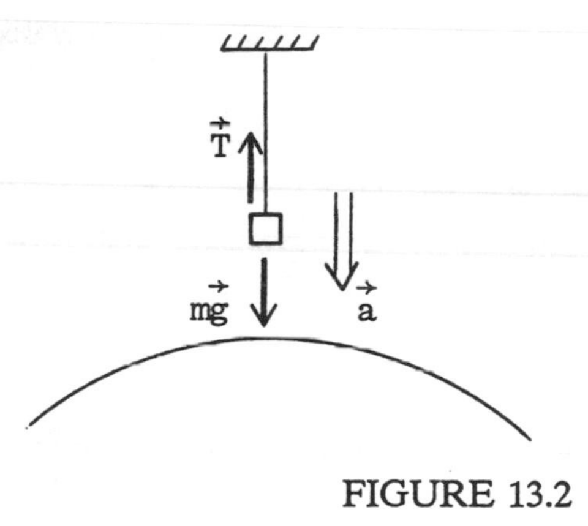

FIGURE 13.2

KEPLER'S LAWS OF PLANETARY MOTION

In the early 17th century, Kepler formulated three laws for the motion of the planets about the sun. In fact, these laws are valid for orbits about any central body, such as the earth or Jupiter.

1. The planets move in elliptical orbits with the sun at one focus.

2. The radial line from the sun to a planet sweeps out equal areas in equal time intervals.

3. The square of the period of a planet is proportional to the cube of its mean distance (semimajor axis) from the sun.

(semimajor axis) from the sun.

$$T^2 = \kappa a^3$$

The constant $\kappa = 4\pi^2/GM$, where M is the mass of the central body.

A brief discussion of the ellipse, and how one can be drawn, is given in the text. Examples of two equal areas referred to in Law 2 are shown in Fig. 13.3. Law 2 is a consequence of the conservation of angular momentum - which in turn follows from the fact that the gravitational force is central: It acts along the line joining a planet to the sun and thus exertes no torque. In particular, at the perihelion (closest) and aphelion (farthest) points in an orbit about the sun, law 2 tells us that

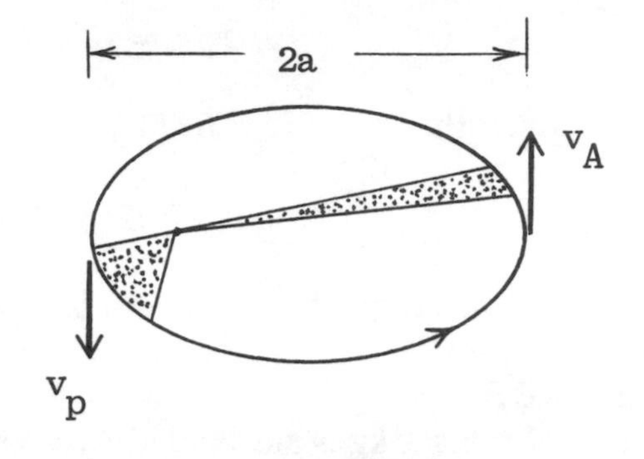

FIGURE 13.3

$$r_A v_A = r_P v_P$$

The third law is similar to that already discussed in Ch. 6, with the change from the radius r to the semimajor axis a.

ENERGY

It can be shown that the energy of a satellite or a planet of mass m in an elliptical orbit is

$$E = -GmM/2a$$

where $2a = r_A + r_P$ is the major axis.

EXAMPLE 4

The mean distance (semimajor axis) of the earth from the sun is 1.50×10^{11} m and that of Venus is 1.08×10^{11} m. What is the period of the orbit of Venus?

Solution:

From Kepler's third law $T^2 = \kappa a^3$ where κ depends only on the central body, in this case the sun.

$$T_E^2/T_V^2 = a_E^3/a_V^3$$

We find $T_V = 0.61$ y.

EXAMPLE 5
The aphelion and perihelion distances for the orbit of Mercury are $r_A = 7.0\text{x}10^{10}$ m and $r_P = 4.6\text{x}10^{10}$ m. The speed at the perihelion is $v_P = 59$ km/s. Find (a) the speed at the aphelion v_A; (b) the orbital energy. The mass of Mercury is $3.3\text{x}10^{23}$ kg.

Solution:
(a) We know that $r_A v_A = r_P v_P$, thus

$$v_A = r_P v_P / r_A = 39 \text{ km/s}$$

(b) The major axis is $2a = r_A + r_P = 11.6\text{x}10^9$ m, and the energy is

$$E = -GmM/2a$$

Using the mass of the sun as $M = 1.99\text{x}10^{30}$ kg, we find $E = -3.78\text{x}10^{33}$ J

SOLUTIONS TO SELECTED TEXT EXERCISES AND PROBLEMS

Exercise 5
In Fig. 13.4 a particle of mass $M_1 = 20$ kg is at the origin while a particle of mass $M_2 = 80$ kg is at (0, 1 m). Find the force on a third particle of mass $M_3 = 10$ kg at (2 m, 0).

Solution:
The directions of the forces on M_3 are shown in Fig. 13.4. The magnitudes of the forces on M_3 are

$$F_{31} = G(10 \text{ kg})(20 \text{ kg})/4 \text{ m}^2 = 50G$$

$$F_{32} = G(10 \text{ kg})(80 \text{ kg})/5 \text{ m}^2 = 160G$$

The components of the net force $\mathbf{F}_3$ are ($\cos\theta = 0.894$, and $\sin\theta = 0.447$)

$$F_{3x} = -F_{31} - F_{32}\cos\theta = -1.29\text{x}10^{-8} \text{ N}$$

$$F_{3y} = F_{32}\sin\theta = 4.77\text{x}10^{-9} \text{ N}$$

Thus $\mathbf{F}_3 = -1.29\text{x}10^{-8}\,\mathbf{i} + 4.77\text{x}10^{-9}\,\mathbf{j}$ N

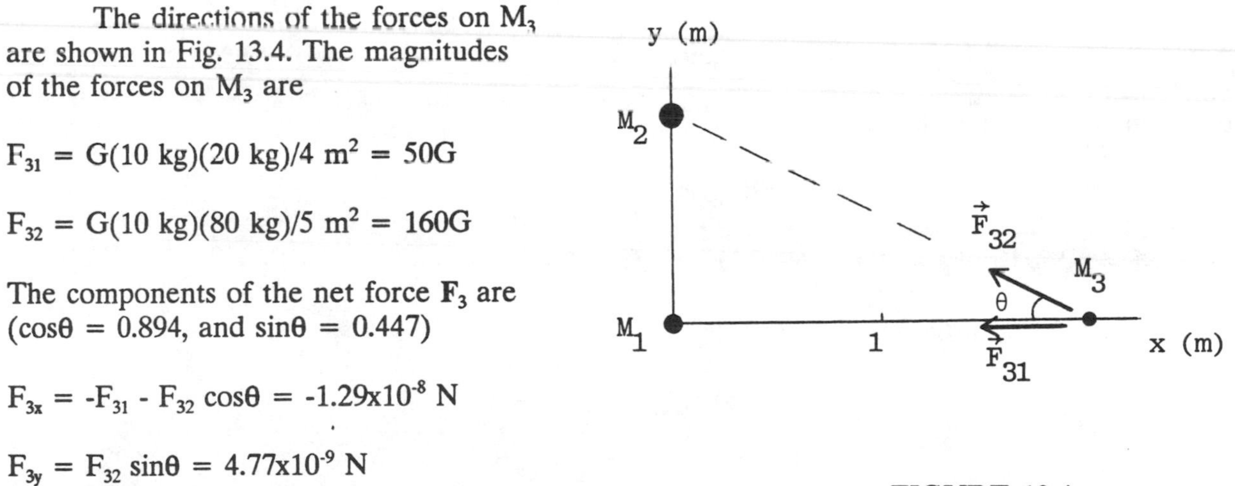

FIGURE 13.4

Exercise 17
(a) How does the gravitational field strength g at the surface of a planet depend on its radius R and its density ρ? How would g change in the following cases: (b) the mass is kept fixed and the radius is halved; (c) the density is halved and the radius is doubled; (d) the density is kept fixed but the volume is doubled?

Solution:
(a) From $GmM/R^2 = mg_o$, we know that $g_o = GM/R^2$ at the surface. The mass $M = \rho V = 4\pi\rho R^3/3$, therefore,

$$g_o = 4\pi\rho RG/3 = C\,\rho R$$

where $C = 4\pi G/3$.

(b) Since $g_o = GM/R^2$, if $R' = R/2$, then $g_o' = 4g_o$

(c) $g_o' = C\,\rho' R' = C(\rho/2)(2R) = g_o$

(d) Since $V' = 2V$ and $V \propto R^3$, we find $R' = (2)^{1/3}R = 1.26R$. Since ρ is unchanged, $g_o' = 1.26g_o$.

Exercise 25
The first Russian satellite, Sputnik I, of mass 83.5 kg, was launched on October 4, 1957, and placed in an orbit for which the distances from the center of the earth at perigee and apogee were $r_P = 6610$ km and $r_A = 7330$ km. Find: (a) the mechanical energy of the satellite; (b) its period; (c) the speed at perigee.

Solution:
(a) We have that $2a = r_A + r_P = 13940$ km, thus

$$E = -GmM/2a = -2.39\text{x}10^9 \text{ J}$$

(b) From Kepler's third law

$$T^2 = (4\pi^2/GM)a^3$$

Using $M = 5.98\text{x}10^{24}$ kg and $a = 6970$ km, we find $T = 96.5$ min.

(c) From Eq. 13.7a in the text we have

$$v_P^2 = (GM/a)(r_A/r_P)$$

thus $v_P = 7.97$ km/s.

Problem 5
A rocket is fired from the earth with 85% of the escape speed. Find its maximum distance from the earth's center if it is fired: (a) vertically, (b) horizontally. Ignore friction and the earth's rotation. (Hint: You need two conservation laws for (b). You will encounter a quadratic equation.)

Solution:

When a rocket is fired with the escape speed, its mechanical energy is zero, that is,

$$E = 1/2\ mv_{esc}^2 - GmM/R = 0$$

so, $v_{esc} = (2GM/R)^{1/2}$.

(a) If a rocket is fired with 85% of the escape speed, its initial energy is

$$E_i = 1/2\ m(0.85v_{esc})^2 - GmM/R = -0.278GmM/R$$

At the maximum distance the energy is pure potential: $E_f = -GmM/d$,
On setting $E_f = E_i$, we find $d = 3.6R$
(b) The initial energy is the same as part (a):

$$E_i = -0.278GmM/R$$

In this case, however, the rocket will go into elliptical orbit. At the maximum distance d from the center of the earth

$$E_f = 1/2\ mv^2 - GmM/d$$

The angular momentum of the rocket is also conserved. At both the initial and final positions, the velocity is perpendicular to the radial line, so

$$L = m(0.85v_{esc})R = mvd$$

Next we use $v = 0.85v_{esc}R/d$, where v_{esc} is given above, in $E_f = E_i$ to obtain a quadratic equation:

$$0.278d^2 - Rd + 0.723R^2 = 0$$

The solutions are $d = R$ and $2.60R$. The appropriate choice is $d = 2.6R$.

Problem 7
Obtain an expression for the energy needed to send an object initially at rest on the earth's surface (a) vertically to a maximum height H, (b) into orbit at the same height. For what value of H, in terms of R_E, the radius of the earth, would the value for part (b) be twice that in (a). Ignore the earth's rotation.

Solution:
(a) Both the initial and final energies are purely potential

$$E_i = -GmM/R;$$

$$E_f = -\ GmM/(R + H);$$

Thus, $\Delta E_1 = E_f - E_i = GmM[1/R - 1/(R + H)]$

(b) The final energy in circular orbit is $E = -GmM/2r$, thus

$$E_i = -GmM/R;$$

$$E_f = -GmM/2(R + H)$$

Thus, $\Delta E_2 = E_f - E_i = GmM[1/R - 1/2(R + H)]$

(c) The condition $\Delta E_2 = 2\ \Delta E_1$ leads to $H = R/2$. (Check it.)

SELF-TEST

1. At what altitude above the surface is the gravitational field strength 20% of its value at the earth's surface? Express your answer in terms of the earth's radius, R.

2. Three particles with masses $m_1 = 2$ kg, $m_2 = 2.5$ kg and $m_3 = 1.2$ kg are located as shown in Fig. 13.5. What is the net force on m_2?

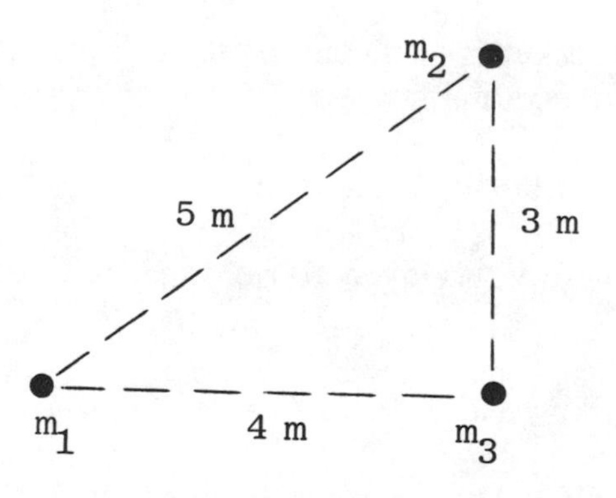

FIGURE 13.5

3. A satellite is in circular orbit about the earth (R = 6370 km) at an altitude of 600 km. Its period is 97 min. Estimate the average density of the earth.

4. The planet Mars, whose mass is 6.42×10^{23} kg, is in an elliptical orbit for which the aphelion distance is 2.49×10^{11} m and the perihelion distance is 2.07×10^{11} m. The speed at the perihelion point is 26.4 km/s. Find: (a) the period of the orbit; (b) the energy; (c) the speed at the aphelion point. The mass of the sun is 2×10^{30} kg

Chapter 14

SOLIDS AND FLUIDS

MAJOR POINTS

1. The elastic properties of a material are characterized by Young's modulus, the shear modulus and the bulk modulus
2. Pascal's principle applies to an enclosed fluid subject to external pressure
3. Archimedes' principle: The buoyant force on an object equals the weight of the displaced fluid
4. The equation of continuity is a statement of the conservation of mass
5. Bernoulli's equation applies to the laminar flow of an ideal, incompressible fluid

CHAPTER REVIEW

DENSITY

The density of a substance is defined as the mass per unit volume

$$\rho = m/V$$

The SI unit is kg/m^3, although one often encounters the cgs unit, g/cm^3, where 1 g/cm^3 = 1000 kg/m^3.

ELASTIC MODULI

Young's modulus is defined as

$$Y = \text{Stress/Strain} = (F/A)/(\Delta L/L_o)$$

where $\Delta L/L_o$ is the fractional change in length (Fig. 14.1a).

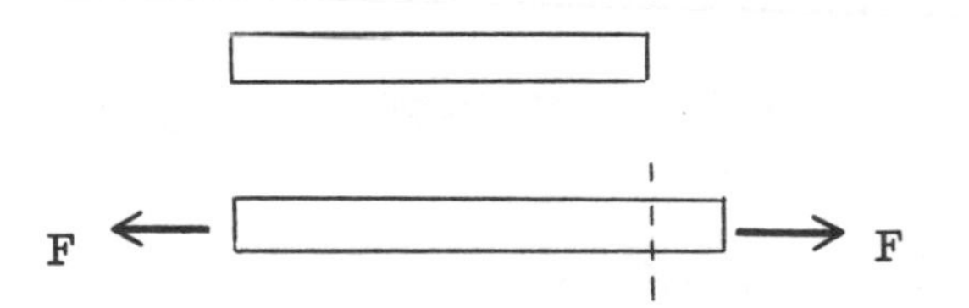

The shear modulus is defined as

$$S = \text{Shear stress/Shear strain} = (F_t/A)/(\Delta x/h)$$

where the force F_t is tangential to the surface of area A (Fig. 14.1b).

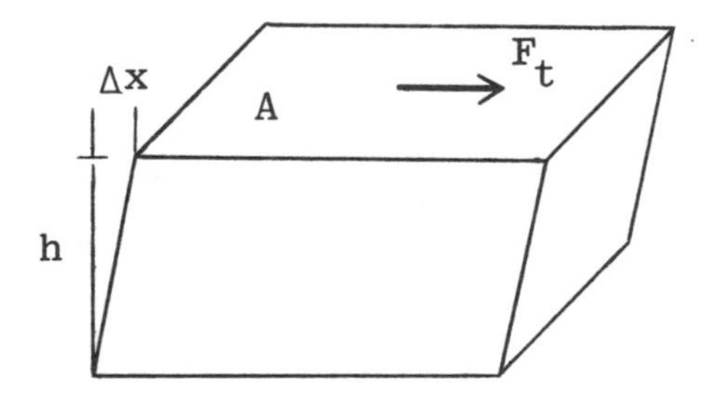

The bulk modulus is defined as

$$B = \text{Volume stress/Volume strain} = -\Delta P/(\Delta V/V)$$

The negative sign is included because an increase in pressures results in a decrease in volume.

FIGURE 14.1

EXAMPLE 1

(a) What pressure is required to increase the density of a sample of water by 0.01%? (b) A steel wire of radius 1 mm and length 1.5 m is subject to a tensile force of 2 kN. What is the change in its length?

Solution:

(a) Since $\rho = m/V$, we have $\Delta\rho/\rho = -\Delta V/V = -10^{-4}$

$$\Delta P = -B(\Delta V/V) = -(2.1\text{x}10^9\ N/m^2)(-10^{-4}) = 210\ kPa.$$

(b) Since $Y = FL/A\Delta L$, thus

$$\Delta L = FL/AY$$

$$= (2000\ N)(1.5\ m)/(\pi\text{x}10^{-6}\ m^2)(2\text{x}10^{11}\ N/m^2) = 4.77\ mm$$

PRESSURE

The pressure on a surface is the normal force per unit area:

$$P = F/A$$

The SI unit is N/m^2 or pascal (Pa). One atmosphere (1 atm) = 101.3 kPa. Note that gauge pressure is the difference between the absolute pressure and atmospheric pressure, P_o. At a depth h below the surface of a liquid, the absolute pressure is

$$P = P_o + \rho g h$$

Thus, for a fluid at rest, all points at a given depth are at the same pressure.

PASCAL'S PRINCIPLE

An external pressure applied to an enclosed fluid is transmitted undiminished to every part of the fluid and to the walls of the container.

EXAMPLE 2

A nurse uses a hypodermic needle of radius 0.05 mm and a piston of radius 4 mm. If the pressure in a vein is 24 mm Hg, what force is needed on the piston to give an injection?

Solution:

Since the pressure in the fluid is the same throughout, we have $F_1/A_1 = F_2/A_2$. The unit 1 mm Hg = 133 Pa, so $P_2 = F_2/A_2 = 3190$ Pa. The force needed is

$$F_1 = P_2 A_1 = (3190\ Pa)(16\text{x}10^{-6}\ m^2) = 0.05\ N$$

ARCHIMEDES' PRINCIPLE

When a body is wholly or partially immersed in a fluid, it experiences a buoyant force equal to the weight of the displaced fluid:

$$\text{Buoyant force} = \text{Weight of displaced fluid}$$

$$F_B = \rho_f g V$$

where ρ_f is the density of fluid and V is the volume of the dispalced fluid.

EXAMPLE 3

A cube of wood has sides of length 25 cm and a density of 675 kg/m^3. It floats in seawater (1025 kg/m^3). (a) What is the height of the top surface above the water? (b) How much mass must be added on top of the cube to just submerge the cube?

Solution:

(a) From Archimedes' principle the buoyant force equals the wieght of the displaced liquid. If L is the side of the cube and h is the submerged depth, then

$$\rho_w(L^3)g = \rho_f g(L^2 h)$$

thus, $h = \rho_w L/\rho_f = 16.5$ cm. The exposed height is 8.5 cm.

(b) The submerged volume is L^3. If m is the added mass, then

$$\rho_f g(L^3) = \rho_w g(L^3) + mg$$

Thus, $m = (\rho_f - \rho_w)L^3 = 5.47$ kg.

EQUATION OF CONTINUITY

When a fluid undergoes laminar flow, its motion can be represented by streamlines, which can be made visible by a colored dye or smoke. The velocity of a particle is tangent to a streamline. We make the following assumptions regarding fluid flow:

1. The fluid is nonviscous: There is no internal friction between adjacent layers
2. The flow is steady: The velocity and pressure at each point are constant in time
3. The flow is irrotational: There is no angular momentum about any point.

In steady flow, the path followed by a fluid element is called a streamline. The velocity of a particle is tangent to the streamline at that point. A group of streamlines form a tube flow, which particles do not enter or leave. This leads to the equation of continuity:

(Steady flow) $$\rho_1 A_1 v_1 = \rho_2 A_2 v_2$$

where 1 and 2 refer to two different points along the flow. If the fluid is incompressible, then $\rho_1 = \rho_2$, so $A_1 v_1 = A_2 v_2$.

BERNOULLI'S EQUATION

Bernoulli's equation applies to the laminar (streamline) flow of an ideal (nonviscous, incompressible) fluid:

$$P + \rho v^2/2 + \rho gy = \text{constant}$$

The second and third terms are the kinetic and potential energies per unit volume. This equation is a disguised form of the work-energy theorem. From Bernoulli's equation we can infer that for a given value of y, the pressure is low in regions where the speed is high, and the pressure is high where the speed is low.

EXAMPLE 4

The absolute water pressure at a firehydrant 350 kPa. The radius of the hose is 2 cm and the water emerges from a nozzle of radius 1.4 cm. What is the pressure at the top of a ladder of height 20 m if the exit speed of the water is 8 m/s?

Solution:

Let $y = 0$ at the basement level. Since $A_2 = 0.49A_1$ and $A_1v_1 = A_2v_2$, we see that $v_1 = 0.49v_2 =$ 3.92 m/s.

$$P_1 + \rho v_1^2/2 = P_2 + \rho v_2/2 + \rho gh$$

thus,

$$P_2 = P_1 - \rho gh + \rho(v_1^2 - v_2^2)/2$$

We find that $\rho gh = (10^3 \text{ kg/m}^3)(9.8 \text{ m/s}^2)(10 \text{ m}) = 196$ kPa, and

$$\rho(v_1^2 - v_2^2)/2 = (10^3 \text{ kg/m}^3)(15.6 - 64) = -48.6 \text{ kPa}$$

Thus, $P_2 = 350 - 244.6 = 105$ kPa.

EXAMPLE 5

The effective area of one side of an airfoil is 20 m². If the speed of air flow above the foil is 130 m/s and that underneath is 120 m/s, what is the left force?

Solution:

Since there is a negligible difference in height between the top and bottom surfaces,

$$P_1 + \rho v_1^2/2 = P_2 + \rho v_2^2/2$$

The lift force is

$$F = (P_2 - P_1)A = \rho A(v_1^2 - v_2^2)/2$$

$$= 0.5(1.29 \text{ kg/m}^3)(20 \text{ m}^2)(130^2 - 120^2)\text{m}^2/\text{s}^2 = 3.23\text{x}10^4 \text{ N}$$

SOLUTIONS TO SELECTED TEXT EXERCISES AND PROBLEMS

Exercise 5
A circular steel wire of length 1.8 m must not stretch more than 1.5 mm when a load of 400 N is applied. What is the minimum diameter required?

Solution:

Rewriting the definition of Young's modulus, we have

$$A = (Y/F)(\Delta L/L) = \pi D^2/4$$

where $Y = 2\text{x}10^{11}$ N/m^2. This leads to D = 1.75 mm

Exercise 17
A U tube of inner radius 0.4 cm contains 60 mL of mercury. WHen 25 mL of water is added to one arm, what is the difference in the levels of the liquid-air interfaces?

Solution:

Let y = 0 at the mercury-water interface in one arm. The pressure is the same at this level in the other arm containing just Hg.

$$V_w = Ah_w = 25 \text{ mL}$$

so $h_w = 25/\pi(0.4)^2 = 49.74$ cm. If h_m is the height of Hg above the y = 0 level, then

$$A\rho_m h_m = A\rho_w h_w$$

thus $h_m = (49.74)(1)/13.6 = 3.66$ cm. The difference in heights is $\Delta h = 49.74 - 3.66 = 46.1$ cm.

Exercise 33
A 400-g cubic block of wood floats with 40% of its volume submerged. How much weight should be placed on the wood to cause it to just become fully submerged in water?

Solution:

The buoyant force equals the weight, so

$$\rho_f(0.4V)g = (0.4 \text{ kg})g$$

This leads to $V = 10^{-3}$ m^3. When it is fully submerged by the addition of mass M:

$$(10^3 \text{ kg/m}^3)Vg = (0.4 \text{ kg})g + Mg$$

Thus M = 0.6 kg

Exercise 39
A 60-kg person floats vertically in a pool with just her head, of volume 2.5 L, exposed. What is her (average) density?

Solution:
If V is the volume of the person, the submerged volume is (V - 2.5 L). If the density of the person is ρ_p, then the buoyant force must equal the weight:

$$\rho_f g(V - 2.5x10^{-3} \text{ m}^3) = (60 \text{ kg})g$$

thus,

$$V = 62.3x10^{-3} \text{ m}^3$$

Since 60 = $\rho_p V$, we have ρ_p = 960 kg/m³.

Exercise 49
A 40-m/s wind blows past a roof of dimensions 10 m x 15 m. Assuming that the air under the roof is at rest, what is the net force on the roof?

Solution:
The difference in pressure arises from the difference in airspeed on either side of the roof. From Bernoulli's equation,

$$P_1 + 1/2 \ \rho v_1^2 = P_2 + 1/2 \ \rho v_2^2$$

Here $P_2 - P_1 = \Delta P = 1/2 \ \rho v^2$, so

$$F = \Delta P.A = (1/2 \ \rho \ v^2)(150 \text{ m}^2)$$

$$= 155 \text{ kN}$$

Problem 11
Water emerges from a small opening at a height h from the bottom of a large container, as in Fig. 14.2, which is filled to a constant depth H. (a) Show that the distance R from the base at which the water hits the ground is given by $R = 2[h(H - h)]^{1/2}$. (b) At what other height would a similar opening lead to the same point of impact?

Solution:
(a) From Bernoulli's equation

$$P_1 + \rho \ gy = P_2 + 1/2 \ \rho \ v^2$$

where $P_1 = P_2 = P_o$. Thus, the speed at which the water emerges is given by

$$v^2 = 2gy = 2g(H - h)$$

From $-h = -1/2 \ gt^2$, we find $t = (2h/g)^{1/2}$. The horizontal range is

$$R = vt = 2[h(H - h)]^{1/2}$$

(b) The other height is H - h.

FIGURE 14.2 FIGURE 143.

SELF-TEST

1. A log floats in water with 20% of its volume exposed in fresh water. When it moves into polluted water, 25% of its volume is exposed. What is the density of the polluted water?

2. What is the difference in height in the liquid levels h_1 - h_2 when water flows through a tube whose cross-sectional area decreases from $A_1 = 3\ cm^2$ to $A_2 = 1.5\ cm^2$ as in Fig. 14.3. Take $v_1 = 0.4$ m/s.

Chapter 15

OSCILLATIONS

MAJOR POINTS

1. The characteristics of simple harmonic oscillation (SHO)
2. The conditions for simple harmonic motion (SHM)
3. The differential equation for SHO

CHAPTER REVIEW

SIMPLE HARMONIC OSCILLATION

In a simple harmonic oscillation, the variation of the fluctuating quantity is given by

$$x = A \sin(\omega t + \phi)$$

where A is the amplitude, $\omega = 2\pi f = 2\pi/T$ is the angular frequency (measured in rad/s) and ϕ is the phase constant. The quantities A and ϕ are illustrated in Fig. 15.1. The horizontal axis is in terms of ωt (in radians). A positive ϕ means that the curve shifts to the left; that is the peak occurs at an earlier time. In simple harmonic oscillation the amplitude is constant and the frequency is independent of amplitude.

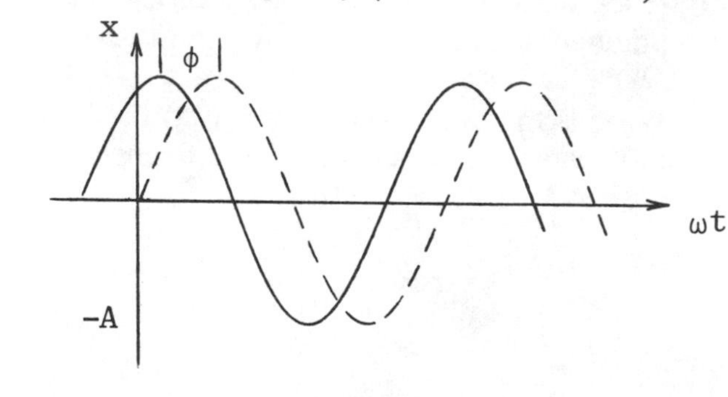

FIGURE 15.1

If x respresents the position of a particle, the velocity and acceleration are

$$v = dx/dt = \omega A \cos(\omega t + \phi)$$

$$a = d^2x/dt^2 = -\omega^2 A \sin(\omega t + \phi)$$

On comparing the expressions for x and a we see that

$$d^2x/dt^2 + \omega^2 x = 0$$

This form of differential equation applies to all simple harmonic oscillations. The term simple harmonic motion applies to mechanical examples of SHO. For SHM to occur, three conditions must be met:

1. There must be a position of stable equilibrium
2. There must be no loss in energy
3. The system must obey some form of Hooke's law, which means that

$a = -\omega^2 x$. That is, the acceleration is proportional to the displacement but is in the opposite direction.

EXAMPLE 1

The position of a particle is given by $x = 0.12 \sin(2t + \pi/3)$ m. Find: (a) the maximum speed and when it occurs for the first time; (b) the acceleration at $t = \pi/2$ s, (b) the velocity when $x = 8$ cm.

Solution:

(a) The velocity is

$$v = dx/dt = \omega A \cos(\omega t + \phi)$$

$$= 0.24 \cos(2t + \pi/3) \text{ m/s}$$

The maximum speed is $v_{max} = \omega A = 0.24$ m/s and occurs when

$$\cos(2t + \pi/3) = 1$$

thus $(2t + \pi/3) = 0, 2\pi$. The first time (> 0) is $t = (2\pi - \pi/3)/2 = 2.62$ s.

(b) The acceleration is

$$a = -\omega^2 A \sin(\omega t + \phi)$$

$$= -0.48 \sin(2t + \pi/3)$$

At $t = \pi/2$ s, $a = -0.48 \sin(4\pi/3) = +0.416 \text{ m/s}^2$.

(c) The condition $x = 0.08$ occurs when

$$0.08 = 0.12 \sin(2t + \pi/3)$$

thus $\sin(2t + \pi/3) = 0.75$, which means $\cos(2t + \pi/3) = (1 - 0.75^2)^{1/2} = +0.661$. Thus the velocity at $x = 0.08$ cm is

$$v = \omega A\cos(2t + \pi/3)$$

$$= (0.24 \text{ m/s})(+0.661) = +0.159 \text{ m/s}.$$

The two values correspond to different directions of motion.

BLOCK-SPRING SYSTEM

The force exerted by an ideal spring is given by Hooke's F_{sp} = -kx, where x is the displacement from the equilibrium position. If a block of mass m is attached to the free end of the spring, Newton's second law, F = = ma, takes the form

$$d^2x/dt^2 + (k/m)x = 0$$

This has the form of the differential equation for simple harmonic oscillation with an angular frequency

$$\omega = (k/m)^{1/2}$$

The period is $T = 2\pi/\omega = 2\pi(m/k)^{1/2}$. The energy of the system is

$$E = 1/2\ mv^2 + 1/2\ kx^2$$

$$= 1/2\ mv_{max}^2 = 1/2\ kA^2$$

EXAMPLE 2

A block of mas 0.2 kg is attached to a spring (k = 15 N/m). It is initially held at extension of 25 cm and then released. Find: (a) the period; (b) the energy. (c) Write an expression for the position x(t).

Solution:

(a) The period is

$$T = 2\pi(m/k)^{1/2}$$

$$= 2\pi(0.2/15)^{1/2} = 0.725\ s$$

(b) The amplitude is A = 0.25 m, thus the energy is

$$E = 1/2\ kA^2 = 0.469\ J$$

(c) We know A = 0.25 m and $\omega = (k/m)^{1/2}$ = 8.66 rad/s. We also know that x = A at t = 0, that is

$$A = A \sin\phi$$

which means $\sin\phi = 1$ and $\phi = \pi/2$ rad. Thus,

$$x = 0.25 \sin(8.66t + \pi/2)\ m$$

EXAMPLE 3
A block ($m = 1.25$ kg) oscillates at the end of a horizontal spring ($k = 180$ N/m). At $t = \pi/20$ s, we are given that $x = 0.15$ m and $v = -2.4$ m/s. (a) Write $x(t)$ in the form $A\sin(\omega t + \phi)$. (b) What is the total energy? (c) Where and when is the kinetic energy equal to half the potential energy for the first time?

Solution:
(a) The angular frequency is $\omega = (k/m)^{1/2} = 12$ rad/s. Using $x = A \sin(\omega t + \phi)$ and $v = \omega A \cos(\omega t + \phi)$ we find

$$0.15 = A \sin(0.6\pi + \phi) \qquad \text{(i)}$$

$$-0.2 = A \cos(0.6\pi + \phi) \qquad \text{(ii)}$$

We square and add (i) and (ii) to find $A = (0.15^2 + 0.2^2)^{1/2} = 0.25$ m. Inserting this into (i) we find $\sin(0.6\pi + \phi) = 0.6$ which means the angle is 36.9° or 143.1°. From (ii) we see that the cosine is negative, thus, using radian measure, as we must,

$$(0.6\pi + \phi) = 143\pi/180$$

and $\phi = 0.611$ rad.
(b) The total energy is

$$E = 1/2\ kA^2 = 5.63 \text{ J}$$

We could also have used the given values in $E = 1/2\ mv^2 + 1/2\ kx^2$. (This could also have been used to find A in part (a). Try it.)

(c) If $K = U/2$, then $E = K + U = 3U/2$. But $E = 1/2\ kA^2$, so

$$U = 1/2\ kx^2 = 1/3\ kA^2$$

which means $x = \pm(2/3)^{1/2}A = \pm 0.816A$. Substituting this into $x(t)$:

$$\pm 0.816 = \sin(12t + 0.611)$$

Thus $(12t + 0.611) = 0.954$ rad, 2.19 rad; 4.10 rad; 5.33 rad. The first time is $(0.954 - 0.611)/12 = 28.6$ ms

SIMPLE PENDULUM

If the rotational form of Newton's second law, $\Gamma = I\alpha$, is applied to the motion of a simple pendulum of length L, as in Fig. 15.2, we find

$$-mgL\sin\theta = (mL^2)\, d^2\theta/dt^2$$

For small angles of oscillation, $\sin\theta \approx \theta$, thus

$$d^2\theta/dt^2 + (g/L)\theta = 0$$

For small angles, a simple pendulum executes simple harmonic motion with an angular frequency and period:

$$\omega = (g/L)^{1/2}; \qquad T = 2\pi(L/g)^{1/2}$$

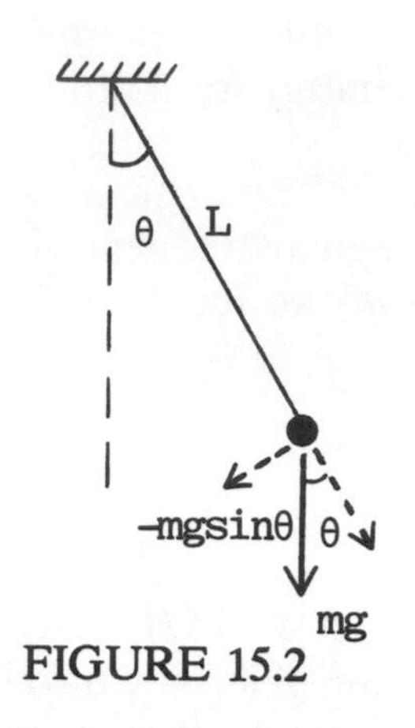

FIGURE 15.2

Note that T does not depend on the mass. The angular displacement is given by

$$\theta = \theta_o \sin(\omega t + \phi)$$

If we choose $U_g = 0$ at the lowest point, the energy of a simple pendulum is

$$E = 1/2\, mv^2 + mgL(1 - \cos\theta)$$

Note that this expression does not involve any small-angle approximation.

EXAMPLE 4

A simple pendulum has a bob of mass 50 g, a length of 1.2 m, and an angular amplitude of 0.34 rad. Find: (a) The time needed to go from -0.2 rad to +0.2 rad, (b) the energy; (c) the maximum speed of the bob.

Solution:

(a) The angular frequency is $\omega = (g/L)^{1/2} = 2.86$ rad/s. To answer the given question the phase is not relevant, so we take $\phi = 0$. Then,

$$\theta = 0.34 \sin(2.86t)$$

When $\theta = 0.2$ rad, $\sin(2.86t) = 0.588$, thus $2.86t = 0.629$ rad, so $t = 0.22$ s. The time taken to go from -0.2 rad to 0.2 rad is 0.44 s.

(b) At the maximum angle,

$$E = mgL(1 - \cos\theta_o) = mgL(1 - \cos 0.34) = 0.337 \text{ J}$$

(c) The energy at the lowest point ($\theta = 0$) is purely kinetic energy. Thus,

$$E = 1/2\ mv_{max}^2 = 0.337\ J$$

which leads to $v_{max} = 3.67$ m/s

PHYSICAL PENDULUM

Consider a solid body of mass m free pivoted about an axis at a distance d from the center of gravity, as in Fig. 15.3. From Newton's second law, $\tau = I\alpha$,

$$-mgd \sin\theta = I\ d^2\theta/dt^2$$

For small angles $\sin\theta \approx \theta$, so

$$d^2\theta/dt^2 + (mgd/I)\theta = 0$$

The body oscillates as a physical pendulum with an angular frequency and period given by

$$\omega = (mgd/I); \quad T = 2\pi(I/mgd)^{1/2}$$

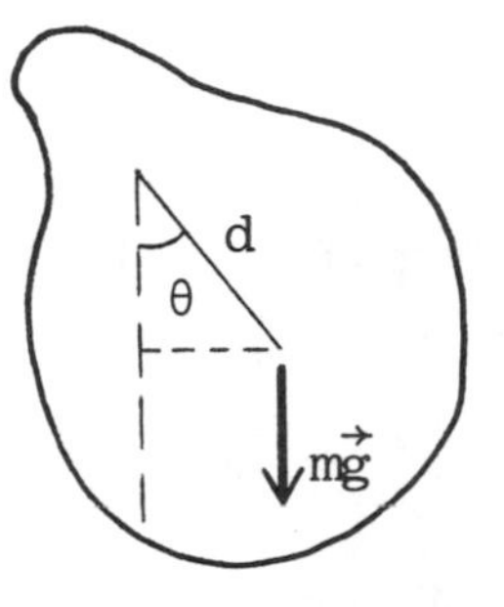

FIGURE 15.3

where I is the moment of inertia about the given axis. The variation of the angular displacement is the same as that for a simple pendulum.

EXAMPLE 5

A uniform thin rod of length 60 cm is pivoted 10 cm from the midpoint. What is its period of oscillation? $I_{CM} = ML^2/12$.

Solution:

The moment of inertia about an axis at a distance d from the CM is given by the parallel axis theorem, $I = I_{CM} + Md^2$, thus the period is

$$T = 2\pi[(I_{CM} + Md^2)/Mgd]^{1/2}$$

$$= 2\pi[(L^2/12 + d^2)/gd]^{1/2} = 1.27\ s$$

where we have used L = 0.6 m and d = 0.1 m.

TORSIONAL PENDULUM

In a torsional pendulum a body is suspended by a fiber or string. The restoring torque when the fiber is twisted through an angle θ obeys Hooke's law, $\tau = -\kappa\theta$, where κ is called the torsional constant. Newton's second law $\tau = I\alpha$ becomes

$$d^2\theta/dt^2 + (\kappa/I)\theta = 0$$

The angular frequency and period of a torsional pendulum are

$$\omega = (\kappa/I)^{1/2}; \quad T = 2\pi(I/\kappa)^{1/2}$$

EXAMPLE 6

A mass m is connected to two rubber bands of length L. Show that if it is given a small displacement perpendicular to the bands, the mass executes simple harmonic motion with an angular frequency

$$\omega = (2T/mL)^{1/2}$$

where T is the tension in each band (assumed to be essentially unchanged).

Solution:

The net force on the block is along the y direction

$$F_y = -2T\sin\theta$$

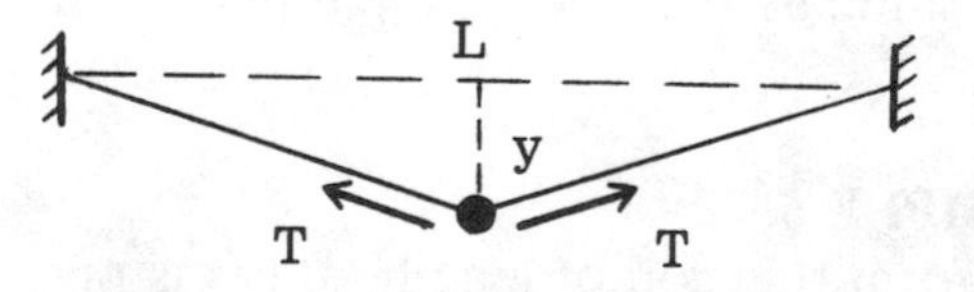

FIGURE 15.4

The negative indicates that this is a restoring force. For a small angle θ, $\sin\theta \approx \tan\theta = y/L$, so $F_y = -2Ty/L$. Newton's second law $F_y = ma_y$, takes the form

$$d^2y/dt^2 + (2T/mL)y = 0$$

which indicates SHO with an angular frequency $\omega = (2T/mL)^{1/2}$.

SOLUTIONS TO SELECTED TEXT EXERCISES AND PROBLEMS

Exercise 7

A block of mass m = 0.5 kg is attached to a horizontal spring whose spring constant is k = 50 N/m. At t = 0.1 s, the displacement x = -0.2 m and the velocity v = +0.5 m/s. Assume $x(t) = A \sin(\omega t + \phi)$. (a) Find the amplitude and the phase constant. (b) Write the equation for x(t). (c) When does the condition x = 0.2 m and v = -0.5 /s occur for the first time?

Solution:

(a) Using the standard expressions for x and v:

$$-0.2 = A \sin(10t + \phi)$$

$$0.05 = A \cos(10t + \phi)$$

On squaring and adding these equations we find A = 0.206 m. (One could also use the energy $E = 1/2\ mv^2 + 1/2\ kx^2 = 1/2\ kA^2$ to find A.) Since the sine is negative and the cosine is positive, we infer that at t = 0.1 s, $(10t + \phi)$ is in the 4th quadrant. From $\sin(1 + \phi) = -0.971$ we find $(1 + \phi) = -1.33$, and $\phi = -2.33$ rad.

(b) x(t) = 0.206 sin(10t - 2.33) m

(c) Since the sine is positive and the cosine is negative, $(10t + \phi)$ is in the 2nd quadrant: $\sin(10t + \phi) = 0.971$, thus (10t - 2.33) = 1.81, and then t = 0.414 s.

Exercise 19

A 50-g block is attached to a vertical spring whose stiffness constant is 4 N/m. The block is released at the position where the spring is unextended. (a) What is the maximum extension of the spring? (b) How long does its take the block to reach the lowest point?

Solution:

(a) Let us take $U_g = 0$ at the initial position, so $E_i = 0$. When the spring extends by x, the energy is

$$E_f = 1/2\ kx^2 - mgx$$

Setting $E_f = E_i$, we find x = 2mg/k = 0.245 m

(b) The equilibrium position of the block occurs where kx = mg, that is at $x_o = mg/k$. Since the maximum extension is $2x_o$, we infer that the block oscillates about this position with an amplitude equal to mg/k. The time required to travel between x = -A and x = A is simply one half period

$$\Delta t = T/2 = \pi(m/k)^{1/2} = 0.351 \text{ s}$$

Exercise 29

A rod suspended at its midpoint oscillates as a torsional pendulum with a period of 0.9 s. If another rod with twice the mass but half the length were used, what would be the period? Take $I = ML^2/12$.

Solution:

The period of a torsional pendulum is $T = 2\pi(I/\kappa)^{1/2}$. Since $I = ML^2/12$, we have

$$T \;\alpha\; (I)^{1/2} \;\alpha\; L(M)^{1/2}$$

Thus,

$$T_2/T_1 = (L_2/L_1)(M_2/M_1)^{1/2} = 1/(2)^{1/2}$$

so $T_2 = 0.636$ s.

Problem 3

A block of mass m is attached to a vertical spring via a string that hangs over a pulley ($I = MR^2/2$) of mass M and radius R, as shown in Fig. 15.5. The string does not slip. Show that the angular frequency of oscilllations is given by $\omega^2 = 2k/(M + 2m)$. (Hint: Use the fact that the total energy is constant in time. See Example 15.6.)

Solution:

If we set $U_g = 0$ at $x = 0$, the energy of the system is

$$E = 1/2\ mv^2 - mgx + 1/2\ I\omega^2 + 1/2\ kx^2$$

Since the string does not slip, $v = \omega R$, so $1/2\ I\omega^2 = Mv^2/4$. In Example 15.4 it is shown that for a block on a vertical spring, $F = -kx'$, where x' is the displacement from the equilibrium position $x_o = mg/k$. Thus, the potential energy of the block-spring system can be written as $U = 1/2\ kx'^2$:

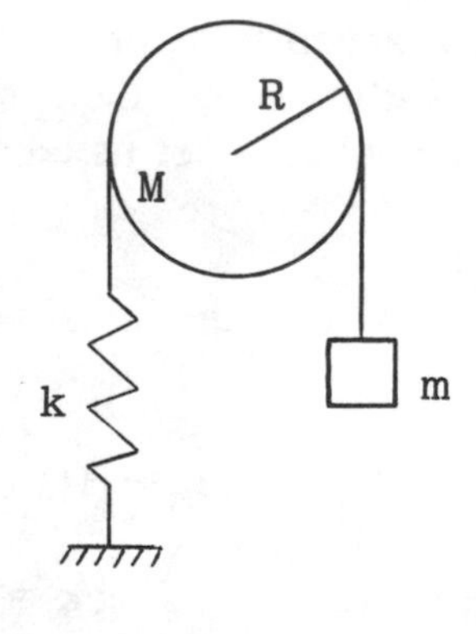

FIGURE 15.5

$$E = 1/2\ (m + M/2)v^2 + 1/2\ kx'^2$$

so,

$$dE/dt = (m + M/2)v\ dv/dt + kx'v$$

where $v = dx'/dt$. Set $dE/dt = 0$ and remove a common factor of v to find

$$d^2x'/dt^2 + [2k/(M + 2m)]x' = 0$$

which is the differential equation for SHO with an angular frequency given by $\omega^2 = 2k/(M + 2m)$.

Problem 13

Figure 15.6 shows a tunnel in a uniform planet of mass M and radius R. At a distance r from the center, the gravitational attraction is due only to the sphere of radius r (see Example 13.5). Thus,

$$F = GmM(r)/r^2 = mgr/R$$

where $M(r) = Mr^3/R^3$ and $g = GM/R^2$. Show that Newton's second law for the motion along the tunnel leads to the differential equation for simple harmonic motion:

$$d^2x/dt^2 + gx/R = 0$$

Estimate the period of oscillation for the earth.

Solution:

From the figure we see that the component of the gravitational force along the tunnel is

$$F_x = -(mgr/R)\sin\theta = -mgx/R$$

where $\sin\theta = x/r$. Applying Newton's second law $F_x = ma$, we find

$$d^2x/dt^2 + (g/R)x = 0$$

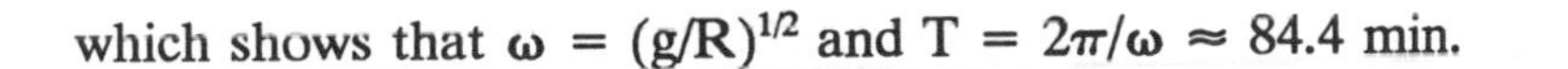

which shows that $\omega = (g/R)^{1/2}$ and $T = 2\pi/\omega \approx 84.4$ min.

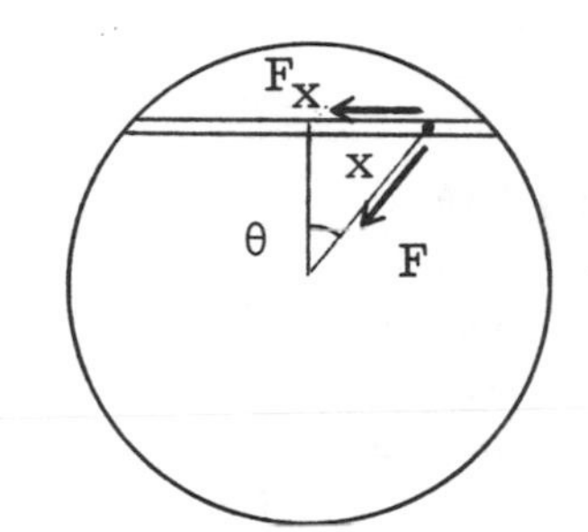

FIGURE 15.6

SELF-TEST

1. A block of mass m = 0.5 kg is attached to a horizontal spring (k = 18 N/m). At $t = \pi/24$ s, we are given that x = -2 cm and v = +12 cm/s. Use $x = A\sin(\omega t + \phi)$. Find (a) A and ϕ; (b) The total energy

2. A plate of mass 1.8 kg is pivoted 0.2 m from its center of gravity and oscillates with a frequency of 2 Hz. What is its moment of inertia about the given axis?

3. Show that at the lowest point in its swing the tension in a simple pendulum of length L oscillating with angular frequency ω is

$$T = m(g + L\omega^2\theta_o^2)$$

where θ_o, the angular amplitude, is assumed to be small.

Chapter 16

MECHANICAL WAVES

MAJOR POINTS

1. Properties of waves: Reflection, transmission across boundaries, linear superposition of waves
2. The speed of a pulse along a string
3. Wavefunction for a traveling wave; wavefunction for a harmonic traveling wave
4. Distinguish between wave velocity and particle velocity
5. Power transmission
6. Standing waves on a string
7. The wave equation

CHAPTER REVIEW

PROPERTIES OF WAVES

In a transverse wave the particle displacements are perpendicular to the direction of propagation of the wave. In a longitudinal wave, the particle displacements are along the direction of propagation. In either case, the particles of the medium do not travel with the wave; they merely oscillate about their equilibrium positions.

When a wave pulse encounters a boundary between two different media, such as two ropes of different thickness, it is partly reflected and partly transmitted. If the second rope is heavier (Fig. 16.1a) the reflected pulse is inverted. If the second rope is lighter, the reflected pulse is not inverted (Fig. 16.1b). The examples of a fixed end and a free end are special cases of the above.

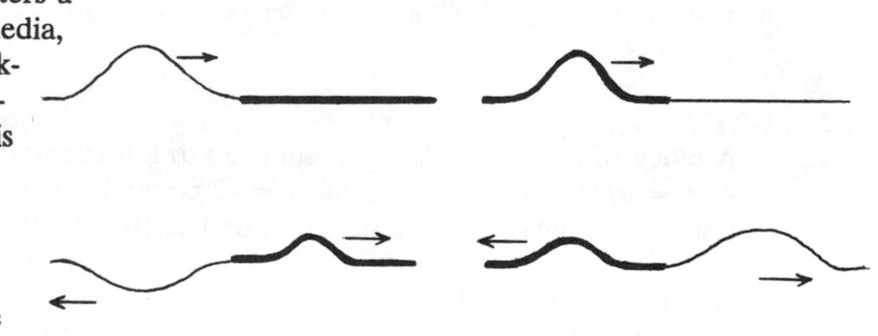

FIGURE 16.1

Speed of pulse on a string

Consider a string with a linear mass density μ (kg/m) under a tension F. The speed of a pulse on the string is

$$v = (F/\mu)^{1/2}$$

EXAMPLE 1
A steel wire of diameter 0.2 mm is under a tension of 120 N. What is the speed of a pulse? The density of steel is 7800 kg/m^3.

Solution:
For a length L, the mass is

$$m = \mu L = \rho AL$$

where $A = \pi d^2/4$ is the corss-sectional area and ρ is the volume density. Inserting $\mu = \rho A = \rho\pi d^2/4$ into the equation for v:

$$v = (4T/\pi\rho d^2)^{1/2}$$

$$= 700 \text{ m/s}$$

EXAMPLE 2
Two strings are attached end-to-end and one is connected to a source. A pulse traveling in the lighter string has a length of $d_1 = 8$ cm. The tranmitted pulse has a length of $d_2 = 5$ cm. Given that the linear mass density of the lighter string is $\mu_1 = 4$ g/m, find μ_2.

Solution:
The tensions in the strings are equal. Thus, $v_1/v_2 = (\mu_2/\mu_1)^{1/2}$. The ratio of the pulse lengths is equal to the ratio of the speeds, so

$$\mu_2/\mu_1 = (d_1/d_2)^2$$

$$= (8/5)^2 = 2.56$$

Therefore $\mu_2 = 10.2$ g/m.

TRAVELING WAVES
The wavefunction for a wave traveling in the +x direction is

$$y = f(x - vt)$$

where y is the displacement from equilibrium and v is the speed of the wave. A wave traveling in the -x direction is given by $y = f(x + vt)$.

When the source of the waves is a simple harmonic oscillator, the function f is sinusoidal, as in Fig. 16.2. For a harmonic wave traveling in the +x direction,

$$y(x, t) = A \sin(kx - \omega t + \phi)$$

where

$$k = 2\pi/\lambda$$

is called the wavenumber and $\omega = 2\pi f$ is the angular frequency. The phase constant ϕ depends on the initial conditions. As Fig. 16.2 shows, $y(0, 0) = A \sin\phi$. In a harmonic traveling wave, each particle of the medium executes simple harmonic motion about its equilibrium position.

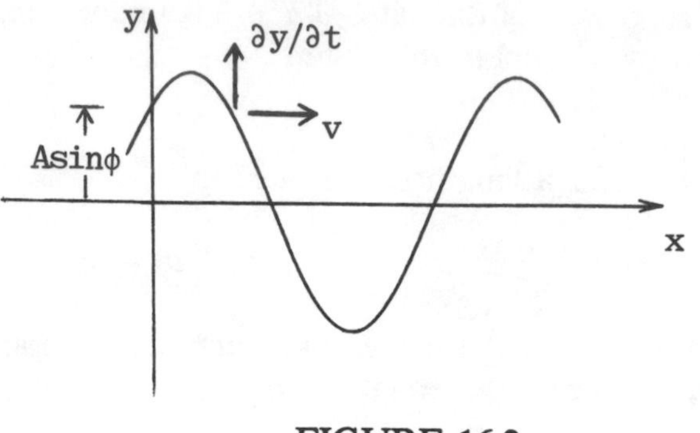

FIGURE 16.2

Since the wave travels a distance λ in one period T, so the wavespeed is $v = \lambda/T$:

$$v = f\lambda = \omega/k$$

The velocity of a particle in a harmonic traveling wave is

(Particle velocity) $$\partial y/\partial t = -\omega A \cos(kx - \omega t + \phi)$$

For a transverse wave, the particle velocity is perpendicular to the wave velocity (see Fig. 16.2).

EXAMPLE 3
A tranverse wave on a string is given by

$$y = 0.05\sin(12x + 60t - 0.2)$$

where x and y are in meters and t is in seconds. Find: (a) the wavelength and period, (b) the particle velocity at $x = 2$ m and $t = 0.1$ s.

Solution:
(a) The wavelength is

$$\lambda = 2\pi/k = 2\pi/(12\ \text{m}^{-1}) = 0.524\ \text{m}$$

The period is $T = 2\pi/\omega = 2\pi/(60\ \text{rad/s}) = 0.105$ s.
(b) The particle velocity at $x = 2$ m and $t = 0.1$ s is

$$\partial y/\partial t = 3 \cos(12x + 60t - 0.2)$$

$$= 3 \cos(28.8\ \text{rad}) = -2.59\ \text{m/s}$$

EXAMPLE 4

A wave with a frequency of 36 Hz travels along a string at 12 m/s.
(a) How long does it take for the phase at a given point to change by 45°? (b) What is the phase difference between two points 0.15 m apart?

Solution:

(a) In one period the phase changes by 2π rad. We are given $\Delta\phi = \pi/4$ rad in an interval Δt. The period is $T = 1/f = 27.8$ ms.

$$\Delta\phi/2\pi = \Delta t/T$$

Thus $\Delta t = T\ \Delta\phi/2\pi = (27.8 \text{ ms})/8 = 3.47$ ms
(b) In one wavelength the phase changes by 2π. The wavelength is $\lambda = v/f = 0.333$ m. The phase change in Δx is given by the ratio

$$\Delta\phi/2\pi = \Delta x/\lambda$$

thus $\Delta\phi = 2\pi\ \Delta x/\lambda = 2.83$ rad.

STANDING WAVES

When two sinusoidal waves of equal frequency and amplitude traveling in opposite directions are superposed we find:

$$y = A\sin(kx - \omega t) + A\sin(kx + \omega t)$$

$$= 2A\cos(\omega t)\sin(kx)$$

This represents a stationary harmonic wave, $A(t)\sin(kx)$, whose amplitude fluctuates in time $A(t) = 2A\cos(\omega t)$. The resultant wave has a permanent displacement of zero where $kx = 0, \pi, 2\pi$, etc. These points are called nodes. The positions of maximum displacement occur where $kx = \pi/2, 3\pi/2$, etc. These points are called antiodes. In an infinite medium, one can obtain standing waves of any frequency.

When one or both ends of a string are fixed, standing waves occur only at certain frequencies or wavelengths. For a string fixed at both ends, the allowed frequencies and wavelengths are

$$\lambda_n = 2L/n; \qquad f_n = nv/2L$$

EXAMPLE 5

One end of a string is attached to a prong of a tuning fork. A block of mass M hangs over a peg at the other end. The linear mass density is 4 g/m and the distance L = 80 cm. The frequency is 60 Hz. For what value of M will there be three loops in the standing waave? Treat the end at the tuning fork as a node.

Solution:

For the third harmonic $\lambda = 2L/3$ and $f = 3v/2L$ where $v = (F/\mu)^{1/2}$. The tension in the string is equal to the hanging weight, $F = Mg$, thus

$$f = (3/2L)(Mg/\mu)^{1/2}$$

From this we find,

$$M = (\mu/g)(2fL/3)^2$$

$$= 0.836 \text{ kg}$$

THE WAVE EQUATION

A traveling wave of the form $y = f(x + vt)$ satisfies the linear wave equation

$$\partial^2 y/\partial x^2 = (1/v^2)\, \partial^2 y/\partial t^2$$

When this differential equation is satisfied, the principle of linear superposition is valid. Since a standing wave is a superposition of two traveling waves, its wavefunction satisfies the wave equation.

ENERGY TRANSPORT ON A STRING

In a harmonic traveling wave each particle or element executes simple harmonic motion. The energy is equal to the maximum kinetic energy, that is, $E = 1/2\ mv_{max}^2$. The energy of an infinitesimal element of length Δx is
$\Delta E = 1/2\ (\mu\Delta x)(\omega A)^2$. If this element takes a time Δt to pass a given point on the string, the power transmitted is $P = \Delta E/\Delta t$:

$$P_{av} = 1/2\ \mu(\omega A)^2 v$$

Since the instantaneous energy of a given element of the string varies in time, this power is actually the time-averaged power transmitted. (The derivation on p. 331 of the text makes this aspect clearer.)

EXAMPLE 6

A string with a linear mass density of 6 g/m carries a wave described by

$$y = 0.02\cos(0.4x - 24t) \text{ m}$$

What is the average power transmitted along the string?

Solution:

The wavespeed is $v = \omega/k = (24 \text{ rad/s})/0.4 \text{ rad/m}) = 60$ m/s and the amplitude is $A = 0.02$ m. The average power transmitted is

$$P_{av} = 1/2\ \mu\ (\omega A)^2\ v$$

$$= 1/2\ (6\text{x}10^{-3} \text{ kg/m})(0.48 \text{ m/s})^2\ (60 \text{ m/s}) = 41.5 \text{ mW}$$

SOLUTIONS TO SELECTED TEXT EXERCISES AND PROBLEMS

Exercise 7

A stretched cord is 7.5 m long and under a tension of 30 N. If the wave speed is 20 m/s, what is the mass of the cord?

Solution:

We know that $v^2 = F/\mu = FL/m$, so

$$m = FL/v^2 = 0.563 \text{ kg}$$

Exercise 23

A transverse wave traveling in the negtive x direction has a wavelength of 2.5 cm, a period of 0.01 s, and an amplitude of 0.03 m. The displacement at x = 0 and t = 0 is y = -0.02 m and the particle velocity is positive. Write the wave function y(x, t) in the form of Eq. 16.8.

Solution:

The wave number is $k = 2\pi/\lambda = 2\pi/0.025 \text{ m} = 80\pi$ rad/m and the angular frequency is $\omega = 2\pi/T = 200\pi$ rad/s, so

$$y = 0.03 \sin(80\pi x + 200\pi t + \phi)$$

We are given that (see Fig. 16.3)

$$y(0, 0) = -0.02 \text{ m} = 0.03 \sin\phi$$

so, $\phi = -0.730$ rad, -2.301 rad. At $x = 0$ and $t = 0$, the particle velocity is $\partial y/\partial t = 60\pi \cos\phi$. Since $\partial y/\partial t > 0$, it means $\cos\phi > 0$, so $\phi = -0.730$ rad, and

$$y = 0.03 \sin[80\pi x + 200\pi t - 0.730] \text{ m}$$

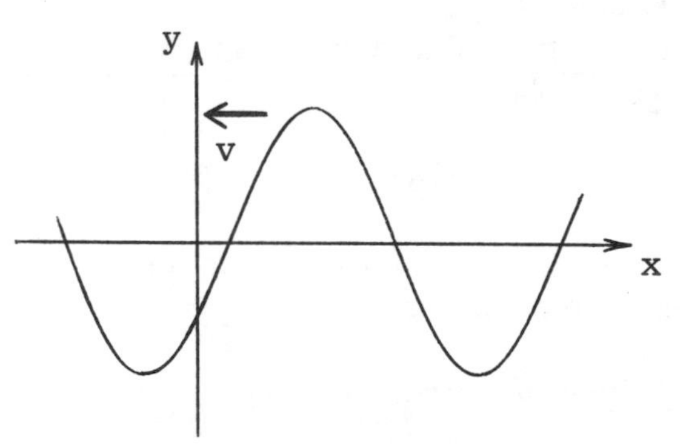

FIGURE 16.3

Exercise 29
A string fixed at both ends has consecutive standing wave modes for whcih the distnaces between adjacent nodes are 18 cm and 16 cm respectively. (a) What is the minimum possible length of the string? (b) If the tension is 10 N and the linear mass density is 4 g/m, what is the fundamental frequency?

Solution:
(a) $\lambda_n = 2L/n$ thus $2L = 36n = 32(n + 1)$, so $n = 8$.

$$L = n\lambda_n/2$$

$$= 8(36)/2 = 144 \text{ cm}$$

(b) $f_1 = (1/2L)(F/\mu)^{1/2} = 1/(2.88 \text{ m})\ (10^4/4)^{1/2} = 17.4 \text{ Hz}$

Problem 5
Consider a transverse wave of small amplitude on a string. Under the action of the wave, the string is moved upward, as shown in Fig. 16.4. (a) SHow that the power supplied by the left side to the right side is

$$P = -F\,(\partial y/\partial x)(\partial y/\partial t)$$

(b) Use the expression $y(x, t) = A\sin(kx - \omega t)$ to find the average power transmitted along the rope. Compare your result with Eq. 16.16.

Solution:
(a) The power is $P = F_y\,(\partial y/\partial t)$, where $F_y = F\sin\theta$. For small angles $\sin\theta \approx \tan\theta = \partial y/\partial x$ (which is negative at the point indicated in the figure). Thus,

$$P = -F\,(\partial y/\partial x)(\partial y/\partial t)$$

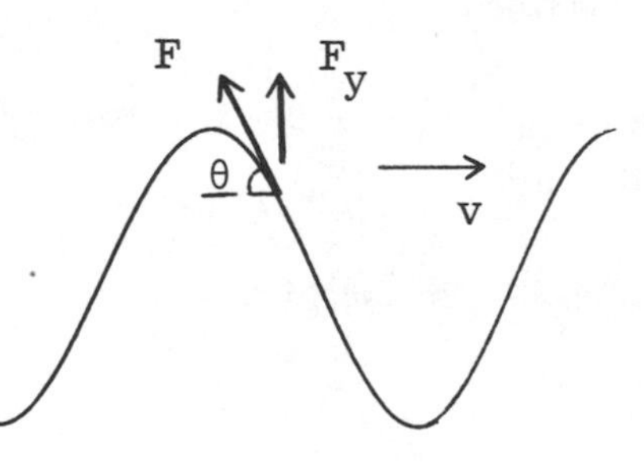

FIGURE 16.4

(b) Using the standard expression for y given earlier, we find (ignoring ϕ)

$$\partial y/\partial x = kA\cos(kx - \omega t); \qquad \partial y/\partial t = -\omega A\cos(kx - \omega t)$$

Thus,

$$P = FkA^2\omega\cos^2(kx - \omega t)$$

The average value of $\cos^2(\omega t)$ over one complete cycle at any x value is 1/2. Since $F = \mu v^2$ and $k = \omega/v$, we find the average power transmitted is

$$P_{av} = 1/2\ \mu(\omega A)^2 v$$

Problem 9
A wire is stretched from L to L + ΔL. Show that the wavespeed of transverse waves is

$$v = (Y\Delta L/\rho L)^{1/2}$$

where Y is Young's modulus (Eq. 14.3) and ρ is the density of the wire.

Solution:

$$v = (F/\mu)^{1/2} = (F/A\rho)^{1/2}$$

From Eq. 14.5 in the text, $F/A = Y\,\Delta L/L$,thus

$$v = (Y\Delta L/\rho L)^{1/2}$$

SELF-TEST

1. A vibrator connected to a string has an amplitude of 5 mm. The tension is 10 N and the mass density is 25 g/m. The waves travel along the +x axis with a wavelength of 20 cm. (a) Write an expression for the displacement y(x, t) given that $y = 0$ and $\partial y/\partial t > 0$ at $x = 0$ and $t = 0$. (b) What is the average power transmitted?

2. Two strings are of the same material and thickness. One sounds the note D while the other sounds the note A. Both are fundamental modes. If the ratio of the frequencies is $f_D/f_A = 4/3$ and $L_D/L_A = 5/6$, $M_D/M_A = 4/5$. Find F_D/F_A, where L is the length and F is the tension.

Chapter 17

SOUND

MAJOR POINTS
1. Nature of sound waves
2. Relation between displacement amplitude and pressure amplitude
3. Resonant standing sound waves in open and closed pipes
4. The Doppler effect
5. (a) The intensity of a sound wave
   (b) The decibel scale

CHAPTER REVIEW

NATURE OF SOUND WAVES

A sound wave is a longitudinal wave characterized by density or pressure fluctuations. The pressure and displacement fluctuations are one-quarter cycle out of phase, see Fig. 17.2 in the text. It is shown in Sec. 17.5, that the pressure amplitude p_o and the displacement amplitude s_o are related according to

$$p_o = \rho \omega v s_o$$

where ρ is the density of the fluid, and v is the speed of sound in the fluid, and ω is the angular frequency.

EXAMPLE 1

The pressure amplitude in a 300 Hz sound wave is 0.04 N/m². Given that the speed of sound in air is 340 m/s and the density of air is 1.29 kg/m³, find the displacement amplitude.

Solution:

The displacement amplitude is given by

$$s_o = p_o/\rho\omega v$$

$$= \frac{0.04\ \text{N/m}^2}{(1.29\ \text{kg/m}^2)(600\pi\ \text{rad/s})(340\ \text{m/s})} = 4.8\text{x}10^{-8}\ \text{m}$$

For comparison, note that the size of an atom is of the order of 10^{-10} m.

RESONANT STANDING SOUND WAVES

Resonant standing waves can be generated in the air column inside a pipe. In an open pipe, both ends are open; whereas in a closed pipe, one end is closed. Standing waves modes are usually depicted in terms of the displacement. A closed end is a displacement node whereas an open end is a displacement antinode. The resonance frequencies are

(Closed pipe) $f_n = nv/4L$ $n = 1, 3, 5,...$

(Open pipe) $f_n = nv/2L$ $n = 1, 2, 3,...$

Note that only the odd harmonics appear in the closed pipe.

EXAMPLE 2

A 440-Hz tuning fork is held over a column of water in a vertical glass tube. As the water level is lowered, the first and second resonances are
heard at the following depths $L_1 = 0.19$ m, $L_2 = 0.58$ m. Find: (a) the next depth at which resonance will be heard; (b) the speed of sound.

Solution:

(a) The first two resonance modes are depicted in Fig. 17.1. The difference between two consecutive levels is
equal to one-half wavelength:

$$\Delta L = L_2 - L_1 = \lambda/2$$

Thus, $\lambda = 0.78$ m. The next resonance will occur at

$$L_3 = L_2 + \lambda/2 = 0.97 \text{ m}$$

(b) Since f = 440 Hz, the speed of sound is $v = f\lambda = 343$ m/s.

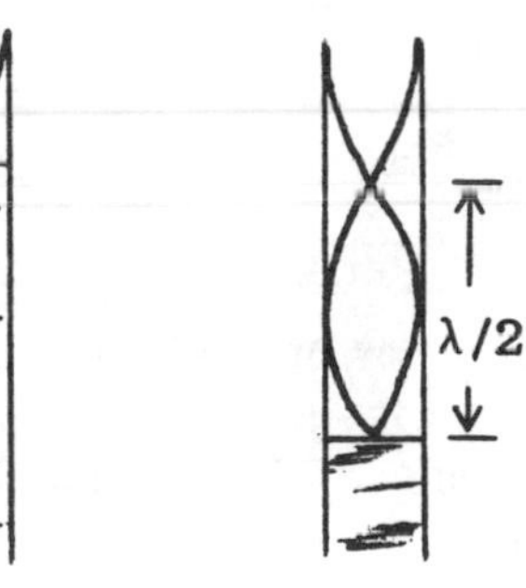

FIGURE 17.1

THE DOPPLER EFFECT

If a source has a natural frequency f_o, the apparent frequency f' heard by an observer moving relative to the source is given by

$$f' = f_o(v \pm v_O)/(v \pm v_S)$$

where v is the speed of sound, v_O is the speed of the observer and v_S is the speed of the source. The signs are chosen by noting that any motion of the source or observer toward the other tends to increase f', and any relative motion of one away from the other tends to decrease f'.

EXAMPLE 3
A police car moving at 50 m/s chases a truck moving at 25 m/s. The siren has a natural frequency of 1200 Hz. The speed of sound is 350 m/s. Find: (a) The frequency heard by the truck driver; (b) The frequency of the sound reflected off the truck heard by the policeman.

Solution:
(a) The motion of the source tends to increase the apparent frequency and the motion of the observer tends to decrease it, see Fig. 17.2a. Thus,

$$f' = f_o\,(v - v_O)/(v - v_S)$$

$$= 1300 \text{ Hz}$$

This is the frequency heard by the truck driver.

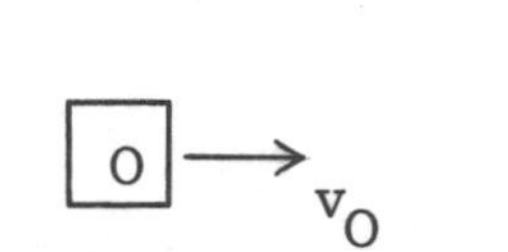

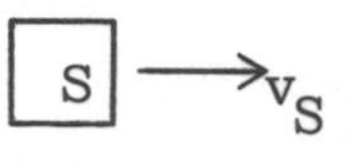

FIGURE 17.2

(b) The back of the truck now acts as the source and the policeman is the observer. The motion of the source tends to decrease the frequency whereas the motion of the observer tends to increase it, see Fig. 17.2b.

$$f'' = f'\,(v + v_O)/(v + v_S)$$

$$= 1387 \text{ Hz}$$

EXAMPLE 4
A stationary observer on a platform records the apparent frequency of a train whistle as it approaches and as it recedes. The two measured frequencies are $f_A = 570$ Hz and $f_R = 490$ Hz. What is the speed of the train. The speed of sound is 340 m/s

Solution:
The observer is at rest, so $v_O = 0$ and the motion of the source is such as to increase the apparent frequency. The apparent frequency when the train approaches is given by

$$f_A = [v/(v - v_S)]\,f_o \qquad \text{(i)}$$

When the train recedes, the apparent frequency is

$$f_R = [v/(v + v_S)]\,f_o \qquad \text{(ii)}$$

From (i) and (ii) we find $vf_o = f_A(v - v_S) = f_R(v + v_S)$, which can be solved to yield

$$v_S = v(f_A - f_R)/(f_A + f_R)$$

Substituting the given values we find $v_S = 25.7$ m/s.

SOUND INTENSITY

The average intensity of a sound wave is the power incident per unit area. In terms of the pressure amplitude, it is

$$I_{av} = P_{av}/\text{Area} = p_o^2/2\rho v$$

The intensity level β of a sound wave is defined as

$$\beta = 10 \log(I/I_o)$$

where $I_o = 10^{-12}$ W/m^2. The unit of β is the decibel (dB).

EXAMPLE 5

Two sounds with intensity levels of 65 dB and 70 dB are superposed. What is the resultant intensity level?

Solution:

First we must find the intensities of these sounds.

$$I_1 = I_o 10^{6.5}; \qquad I_2 = I_o 10^7$$

The resultant intensity level is

$$\beta = 10 \log[(I_1 + I_2)/I_o]$$

$$= 10 \log(10^{6.5} + 10^7) = 71.2 \text{ dB}$$

EXAMPLE 6

What is the pressure amplitude for a sound wave with an intensity level of 90 dB? Take ρ = 1.29 kg/m^3 and v = 330 m/s.

Solution:

The intensity of the waves is

$$I = 10^9 I_o = 10^{-3} \text{ W/m}^2$$

The pressure amplitude is given by

$$p_o^2 = 2\rho v I$$

$$= 2(1.29 \text{ kg/m}^3)(330 \text{ m/s})(10^{-3} \text{ W/m}^2)$$

Thus p_o = 0.923 N/m^2.

EXAMPLE 7
The intensity level is 100 dB at 2 m from a point source that radiates uniformly. Where is the level 85 dB?

Solution:

The energy radiated by a point source spreads unformly over ever-increasing spheres, hence the intensity varies with distance r according to

$$I = P/4\pi r^2$$

thus,

$$r_2/r_1 = (I_1/I_2)^{1/2}$$

The difference in the intensity levels is

$$\beta_2 - \beta_1 = 10\log(I_2/I_o) - 10\log(I_1/I_o)$$

$$-15 = 10\log(I_2/I_1)$$

Thus $I_2/I_1 = 0.0316$. Finally $r_2 = \ r_1 = 11.8$ m

SOLUTIONS TO SELECTED TEXT EXERCISES AND PROBLEMS

Exercise 11
Show that for a (longitudinal) wave in a fluid the pressure fluctuation is related to the particle displacement s by $p = (B/v)(ds/dt)$, where B is the bulk modulus and v is the wave speed.

Solution:

From Eq. 17.13, $s = s_o \sin(kx - \omega t)$, we obtain

$$\partial s/\partial t = -\omega s_o \cos(kx - \omega t)$$

Using $v = \omega/k$ and $p_o = Bks_o$ from Eq.17.14, we see that $\omega s_o = vp_o/B$, so

$$p = (B/v)\partial s/\partial t$$

Exercise 15
The 2nd. harmonic of a string of length 60 cm and linear mass density 1.2 g/m has the same frequency as the 3rd harmonic of a closed pipe of length 1 m. Find the tension in the string.

Solution:

Pipe: $\lambda = 4L/5 = 0.8$ m $= 340/f$, thus $f = 425$ Hz.

String: $\lambda = L = 0.6$ m $= v/f$; thus $v = 255$ m/s.

The tension in the string is $F = \mu v^2 = 78$ N

Exercise 31
A source of sound emits a signal at 600 Hz. This is observed as 640 Hz by a stationary observer as the source approaches. What is the observed frequency as the source recedes at the same speed?

Solution:

When the source approaches, the apparent frequency is higher:

$$f' = f_o v/(v - v_S)$$

$$640 = 600(340)/(340 - v_S)$$

thus $v_S = 21.25$ m/s. When the source recedes,

$$f' = f_o v/(v + v_S)$$

$$= (600 \text{ Hz})(340 \text{ m/s})/361.25 \text{ m/s}) = 565 \text{ Hz}$$

Exercise 43
The pressure variation in a cound wave is given by

$$p = 12 \sin(8.18x - 2700t + \pi/4) \text{ N/m}^2$$

Find: (a) the displacement amplitude; (b) the average intensity.

Solution:
(a) $v = \omega/k = (2700 \text{ rad/s})/(8.18 \text{ rad/m}) = 330$ m/s. The displacement amplitude is

$$s_o = p_o/\rho\omega v$$

$$= (12 \text{ Pa})/(1.29 \text{ kg/m}^3)(2700 \text{ rad/s})(330 \text{ m/s}) = 1.04\text{x}10^{-5} \text{ m}$$

(b) The intensity is

$$I = p_o^2/2\rho v = 0.169 \text{ W/m}^2$$

Problem 3
The speed of sound is given approximately by $v \approx 20T^{1/2}$, where T is the temperature in kelvins (K). (a) Show that the fractional change in the speed of sound ($\Delta v/v$) caused by a fractional change in temperature ($\Delta T/T$) is $\Delta v/v = 0.5\Delta T/T$
(b) An organ pipe has a fundamental frequency of 400 Hz at 285 K. What is the fundamental frequency at 305 k? Assume the pipe length is unchanged. (c) Compare the percentage change in frequency to that of a semitone, which is about 6%.

Solution:
(a) $dv/dT = 1/2\ (20)T^{-1/2}$. Multiply both sides by dT/v to find

$$dv/v = dT/2T$$

(b) Since $f \propto v$, we use the result in part (a):

$$\Delta f/f = \Delta T/2T$$

thus $\Delta f = (10\ K)(400\ Hz)/(285\ K) = 14\ Hz$, and $f = 414\ Hz$.
(c) $\Delta f/f = 14\ Hz/400\ Hz = 3.5\%$

Problem 7
Show that the wavefucntion of a wave emitted by a point source has the form
$y = (A/r) \sin(kr - \omega t)$. (Hint: Consider the variation of intensity with radial distance.)

Solution:
Since the energy emitted by a point source spread uniformly over ever increasing spheres, the intensity varies with radial distance according to

$$I = P/4\pi r^2 \propto 1/r^2$$

We also know that $I \propto s_o^2$, where s_o is the displacement amplitude.
Therefore, $s_o \propto 1/r$.

SELF-TEST

1. A pipe resonates at the following consecutive frequencies: 1375 Hz, and 1925 Hz. (a) Is it "open" or "closed"? (b) What is the fundamental frequency? The speed of sound is not known.

2. A panel absorbs 95% of the sound energy incident on it and transmits the rest. What is the change in intensity level?

3. A stationary observer on a platform notes that when a train approaches the apparent frequency of the whistle is f_A, and when it recedes he records f_R. Show that the natural frequency of the whistle is

$$f_o = 2f_Af_R/(f_A + f_R)$$

4. The pressure variation in a sound wave in air is given by

$$p = 0.02 \sin(1.8x - 600t)\ Pa$$

Find: (a) the displacement amplitude; (b) the intensity. Take the density of air to be $1.29\ kg/m^3$.

Chapter 18

TEMPERATURE, THERMAL EXPANSION AND THE IDEAL GAS LAW

MAJOR POINTS

1. The Celsius, Fahrenheit and Kelvin temperature scales
2. (a) The concept of thermal equilibrium
   (b) The zeroth law of thermodynamics
3. The equation of state of an ideal gas
4. Thermal expansion.

CHAPTER REVIEW

The concept of temperature is based on our sense of hot and cold. It can be measured in terms of the variation in some property of a material, for example, thermal expansion or a change in electrical resistance.

When all the state variables, such as mass, pressure and volume, of a system are constant in time, the system is said to be in thermal equilibrium. Two systems are in thermal equilibrium with each other if they are at the same temperature.

According to the zeroth law of thermodynamics, two systems in thermal equilibrium with a third body (such as a thermometer) are also in thermal equilibrium with each other.

TEMPERATURE SCALES

In the Celsius temperature scale the freezing point and boiling point of water are assigned the values 0 °C and 100 °C. In the Fahrenheit scale, these two points are assigned the values 32 °F and 212 °F. One may convert from one scale to the other with the following equations:

$$t_F = 9t_C/5 + 32; \qquad t_C = 5(t_F - 32)/9$$

The Kelvin temperature scale is related to the Celsius scale according to

$$T = t_C + 273.15$$

THE MOLE AND AVOGADRO'S CONSTANT

One mole (mol) of any substance contains as many elementary entitites (atoms or molecules) as 12 grams of the isotope C-12. This number is called Avogadro's number N_A, and its value is

$$N_A = 6.022 \times 10^{23} \text{ mol}^{-1}$$

If N is the number of molecules in a sample, then the number of moles is

$$n = N/N_A$$

The mass of one mole is called the molecular mass M, and is usually expressed in grams. Thus, for example, M = 32 g/mol for oxygen. This can be a bit tricky because in SI units we must use kg. Thus, when we substitute into an equation, we would use 32×10^{-3} kg/mol. The mass of a single molecule is M/N_A.

IDEAL GAS LAW

The pressure, volume and temperature of an ideal gas in an equilibrium state satisfy the following equation

$$PV = NkT$$

where N is the number of molecules and k is Boltzmann's constant. This equation is also expressed in terms of the number of moles, n, in the sample

$$PV = nRT$$

where $R = kN_A = 8.314$ J/mol.K is the universal gas constant.

EXAMPLE 1

Two moles of oxygen (M = 32 g/mol) are at 0 °C and 1 atm. What is the mass density of the gas?

Solution:

We need to find ρ = Mass/Volume. From the ideal gas law, the volume of 2 mole (n = 2) is

$$V = nRT/P$$

$$= 2(8.314 \text{ J/mol.K})(273 \text{ K})/(101 \text{ kPa}) = 4.48 \times 10^{-2} \text{ m}^3$$

The mass of n moles is nM, thus

$$\rho = nM/V$$

$$= 2(32 \times 10^{-3} \text{ kg/mol})/(4.48 \times 10^{-2} \text{ m}^3) = 1.42 \text{ kg/m}^3$$

EXAMPLE 2

Two moles of oxygen are at a pressure of 1 atm and a temperature of 20 °C.
Determine the final temperature given that (a) the gas is heated at constant volume until the pressure is 2.5 atm; (b) the pressure is increased by 40% and the volume is dereased by 20%.

Solution:
(a) Since the volume is fixed $P \alpha T$, so $P_2/P_1 = T_2/T_1$, thus

$$T_2 = (P_2/P_1)T_1$$

$$= (2.5)(293\ \text{K}) = 733\ \text{K}$$

(b) Since $PV \alpha T$, $P_2V_2/P_1V_1 = T_2/T_1$, thus

$$T_2 = (P_2/P_1)(V_2/V_1)T_1$$

$$= (1.4)(0.8)(293\ \text{K}) = 328\ \text{K}$$

EXAMPLE 3
A room has a volume of 75 m^3. Initially the air is at 20 °C at 1 atm. If the temperature rises to 30 °C, what mass of air leaves the room? Take $M = 29$ g/mol.

Solution:
If n_1 and n_2 are the initial and final number of moles in the room, then

$$P_1V_1 = n_1RT_1 = n_2RT_2$$

Thus,

$$n_1 - n_2 = (1/T_1 - 1/T_2)(P_1V_1/R)$$

$$= (1/293\ \text{K} - 1/303\ \text{K})(1.01\text{x}10^5\ \text{Pa})(75\ \text{m}^3)/(8.314\ \text{J/mol.K})$$

$$= 102.6\ \text{moles}$$

The mass of air is $\Delta m = \Delta n.M = 2.98$ kg.

THERMAL EXPANSION

When the temperature of an object of length L changes by ΔT, the change in length is

$$\Delta L = \alpha\ L\ \Delta T$$

where α is the coefficient of linear expansion, measured in $°C^{-1}$.

The change in volume of a body when its temperature changes is given by

$$\Delta V = \beta\ V\ \Delta T$$

where β is the coefficient of volume expansion. For an isotropic solid $\beta = 3\alpha$.

Thermal stress occurs when an object is not allowed to expand or contract normally when its temperature changes. To compute its value, one finds, in effect, the expansion, or contraction, that would have occured, and then the force needed to compress it, or stretch it, to the given length.

EXAMPLE 4
A flat plate is 80 cm x 60 cm at 20 °C. What is the change in area when the temperture rises to 45 °C? Take $\alpha = 2 \times 10^{-5}$ °C^{-1}.

Solution:

$$A(T) = A_o(1 + \alpha \Delta T)^2 = A_o(1 + 2\alpha \Delta T + \alpha^2 \Delta T^2)$$

where $A_o = L_o w_o$. Since $(\alpha \Delta T)^2 << (\alpha \Delta T)$, we find

$$A = A_o(1 + 2\alpha \Delta T)$$

Thus 2α is the coefficient of areal expansion. The change in area is

$$\Delta A = A - A_o = 2A_o \alpha \Delta T$$

$$= 4.8 \times 10^{-4} \text{ m}^2.$$

SOLUTIONS TO SELECTED TEXT EXERCISES AND PROBLEMS

Exercise 9
Write the ideal gas law in terms of the mass density (measured in kg/m^3) of the gas. At 0 °C and 1 atm., find the density of the following: (a) nitrogen; (b) oxygen; (c) hydrogen.

Solution:
The mass of a sample of n moles is m = nM, thus

$$PV = nRT = (m/M)RT$$

so, $P = \rho RT/M$, where M is in kg/mol. Thus,

$$\rho = (P/RT)M = 44.5M \text{ kg/m}^3$$

(a) Using $M = 28 \times 10^{-3}$ kg/mol, $\rho = 1.25$ kg/m^3

(b) Using $M = 32 \times 10^{-3}$ kg/mol, $\rho = 1.42$ kg/m^3

(c) Using $M = 2 \times 10^{-3}$ kg/mol, $\rho = 0.089$ kg/m^3

Exercise 17
Two moles of nitrogen are at 3 atm and 300 K. (a) What is the volume of the gas? (b) It expands at constant temperature until the pressure drops to 1 atm. What is the new volume?

Solution:
(a) From the ideal gas law,

$$V = nRT/P$$

$$= (2\text{ mol})(8.314\text{ J/mol.K})(300\text{ K})/(303\text{ kPa})$$

$$= 1.65\text{x}10^{-3}\text{ m}^3 = 16.5\text{ L}$$

(b) $P_2V_2 = P_1V_1$, so $V_2 = (P_1/P_2)V_1 = 3V_1 = 49.5$ L

Exercise 23
A copper sphere of radius 2.000 cm is placed over a hole of radius 1.990 cm in an aluminum plate at 20 °C. At what common temperature will the sphere pass through the hole? (The coefficients of linear expansion are on p. 364.)

Solution:
The radii of the sphere and of the hole are given by

$$r_s = 2.00(1 + 17\text{x}10^{-6}\Delta T)$$

$$r_h = 1.99(1 + 24\text{x}10^{-6}\Delta T)$$

We set $r_s = r_h$ to find $\Delta T = 726.7$ °C, thus $T \approx 747$ °C

Problem 7
A steel ball of radius 1.2 cm is in a cylindrical glass beaker of radius 1.5 cm that contains 20 mL of water at 5 °C. What is the change in the water level when the temperature rises to 90 °C?

Solution:
The volume of (ball + water) at 5 °C is

$$V_1 = V_B + V_W = 4\pi R^3/3 + 20\text{ mL} = 27.24\text{ mL}$$

The change in the water level depends on the "overflow" volume

$$\delta V = \Delta V_B + \Delta V_W - \Delta V_G$$

where ΔV_G is the change in volume of the glass beaker. Using the values in Table 18.2 we find (note that the volume coefficient coefficent $\beta = 3\alpha$)

$$\Delta V_B = (7.24\text{ mL})(35.1\text{x}10^{-6}\text{ K}^{-1})(85\text{ °C})$$

$$\Delta V_W = (20\text{ mL})(2.1\text{x}10^{-4}\text{ K}^{-1})(85\text{ °C})$$

$$\Delta V_G = (20\text{ mL})(27\text{x}10^{-6})](85\ °\text{C})$$

Thus $\delta V = 0.379$ mL, and the new volume of (ball + water) is $V_2 = V_1 + \delta V = 27.62$ mL. To find height of water in beaker note that the volume of water is $V = \pi r^2 h$, and that $r_2 = r_1(1 + \alpha\ \Delta T)$, thus

$$V_2/V_1 = (r_2/r_1)^2(h_2/h_1) \approx (1 + 2\alpha\Delta T)h_2/h_1$$

$$V_2/V_1 = 27.62\text{ mL}/27.24\text{ mL} = (1 + 18\text{x}10^{-6}\text{x}85)h_2/h_1$$

thus, $h_2 = (1.014)h_1/(1.0015)$. But $h_1 = 20/\pi(1.5)^2 = 2.829$ cm, so $h_2 = 2.865$ cm, and $\Delta h = 0.36$ mm

SELF-TEST

1. A glass container ($\alpha = 9\text{x}10^{-6}\ °\text{C}^{-1}$) is filled to the brim with 600 mL of gasoline ($\beta = 9.5\text{x}10^{-4}\ °\text{C}^{-1}$) at 20 °C. How much liquid spills out when the temperature is raised to 50 °C?

2. (a) If 40 L of air at 20 °C and a pressure of 150 kPa is compressed to 25 L and heated to 35 °C, what is the new pressure?
   (b) The pressure in a vacuum system is 10^{-8} Pa. What is the (number) density of molecules at 0 °C?

3. A simple pendulum consists of a small sphere suspended by a thin wire. It has a period T_1 at 20 °C. What is the percentage change in the period if the temperature rises to 40 °C? Take $\alpha = 10^{-5}\ °\text{C}^{-1}$. The period of a simple pendulum of length L is $t = 2\pi(L/g)^{1/2}$. Use the binomial exapnsion $(1 + z)^n \approx 1 + nz$ for $z << 1.0$

Chapter 19

THE FIRST LAW OF THERMODYNAMICS

MAJOR POINTS

1. The distinction between heat and internal energy
2. The definition of the specific heat of a substance
3. The first law of thermodynamics relates the work done by a system and the heat it exchanges with the environment to the change in its internal energy.
4. The work done by a system and the heat it exchanges with the environment depend on the (thermodynamic) path taken.
5. Changes in state associated with quasistatic, isobaric, isothermal and adiabatic processes
6. Heat is transferred by conduction, convection, and radiation

CHAPTER REVIEW

SPECIFIC HEAT

The heat capacity of an object is the heat required to raise its temperature by 1 C°. If the object has a mass m, then the heat may be written in the form

$$\Delta Q = m\, c\, \Delta T$$

where c, the specific heat, is the heat capacity per unit mass. The SI unit is J/kg.K. The calorie (cal) is defined as the heat required to raise the temperature of 1 g of water from 14.5 °F to 15.5 °F. The conversion between the calorie and the joule is 1 cal = 4.186 J.

It is sometimes convenient to deal with the number of moles in a sample:

$$\Delta Q = n\, C\, \Delta T$$

where n is the number of moles and C is the molar specific heat of a substance (J/mol.K). Since m = nM, where M is the molecular mass (kg/mol), we have

$$C = M\, c$$

<u>Latent Heat</u>

During a change of phase, from solid to liquid or liquid to gas, a substance absorbs heat without a change in temperature. The latent heat, L, measured in J/kg, is defined by

$$\Delta Q = mL$$

where m is the mass of the sample. When the change is in the opposite direction, from gas to liquid or liquid to solid, the latent heat flows from the sample to its surroundings.

The specific heats of substances can be measured by the method of mixtures. When bodies at different temperatures are placed in thermal contact in an isolated container, the net transfer of heat is zero. For two bodies,

$$\Delta Q_1 + \Delta Q_2 = 0$$

If heat enters a body, ΔQ is positive, and if heat leaves a body, ΔQ is negative. (Note that heat is not contained in either body. See the discussion on p. 378 of the text.)

EXAMPLE 1

A sphere of mass 200 g at 120 °C is placed into a 80 g copper cylinder containing 150 g of water at 20 °C. The final temperature is 27 °C. What is the specific heat of the sphere?

Solution:

Since the net transfer of heat is zero, we may write (we actually know that T_f = 27 °C)

$$m_1c_1(T_f - 120°C) + (m_2c_2 + m_3c_3)(T_f - 20\ °C) = 0$$

$$(0.2\ kg)c_1(-93\ K) + [(0.15\ kg)(4190\ J/kg.K) + (0.08\ kg)(385\ J/kg.K)](7\ K) = 0$$

We find c_1 = 248 J/kg.K.

EXAMPLE 2

Two kilograms of ice at -20 °C are mixed with 1.5 kg of water at 25 °C. Find final temperature. Ignore the container.

Solution:

The heat needed to raise all of the ice to 0 °C, is

$$\Delta Q_1 = (2\ kg)(2100\ J/kg)(20\ °C) = 8.4x10^4\ J$$

The heat needed to melt all the ice is

$$\Delta Q_2 = mL = (2\ kg)(3.34x10^5\ J/K) = 6.68x10^5\ J$$

The heat transfer if the water cooled to 0 °C is

$$\Delta Q_3 = (1.5\ kg)(4190\ J/kg)(-25\ °C) = -1.57x10^5\ J$$

Since $|\Delta Q_3| > |\Delta Q_1|$, all of the ice will be raised to O °C. The rest of the available heat will melt only some of the ice. So the final state is a mixture of ice and water at 0 °C. Let us find out how much ice melts. The condition for no net heat transfer is

$$\Delta Q_1 + mL + \Delta Q_3 = 0$$

$$84 \text{ kJ} + m(334 \text{ kJ/kg}) - 157 \text{ kJ} = 0$$

We find m = 0.2 kg.

WORK

The work done by a gas during a quasistatic process, in which the system is always close to equilibrium states, is given by

(Quasistatic) $$W = \int P\,dV$$

In order to evaluate this integral we need to know how P varies with V. This work, which depends on the path taken, is equal to the area under the PV curve. If $V_f < V_i$, then $W < 0$, which means that work is done on the system.
If a process is not quasistatic, the state variables, such as P, V, and T, do not have unique values and so the process cannot be depicted on a PV diagram.

Isobaric process

In an isobaric process, the pressure is constant, so the work done by the gas is

(Isobaric) $$W = P(V_f - V_i)$$

FIRST LAW OF THERMODYNAMICS

The first law of thermodynamics states that the change in internal energy of a system, ΔU, between two equilibrium states is given by

$$\Delta U = Q - W$$

where W is the work done by the system, and the heat Q is positive if heat enters the system, and negative if heat leaves the system. The first law is a generalization of the conservation of energy to include heat and internal energy. The terms work and heat involve the transfer of energy during some process; it is not correct to refer to the work or heat "in a system".
A system posseses internal energy, not heat. (See the discussion on p 378 of the text.)

In an isolated system, $Q = W = 0$, so $\Delta U = 0$.

In a cyclic process, the system returns to its initial state, so $\Delta U = 0$ and $Q = W$.

In an adiabatic process, a system is thermally isolated, which means $Q = 0$, so $\Delta U = -W$.

In an adiabatic free expansion, $W = 0$ and so $\Delta U = 0$. That is, the internal energy of any gas does not change. (In the particular case of an ideal gas the temperature is also unchanged.)

EXAMPLE 3
A system undergoes the triangular cycle shown in Fig. 19.1. (a) Determine the work done by the system in each segment. (b) The net work done during the complete cycle. (c) The internal energy of the system at B is 100 J and the heat absorbed in the constant-volume process B to C is 70 J. What is the heat transfer for the rest of the cycle, CA + AB?

Solution:
(a) There is no work done by the system in the constant-volume segment BC, that is, $W_{BC} = 0$. For the segment AB,

$$W_{AB} = P(V_f - V_i)$$

$$= (120 \text{ kPa})(-2\text{x}10^{-3} \text{ m}^3) = -240 \text{ J}$$

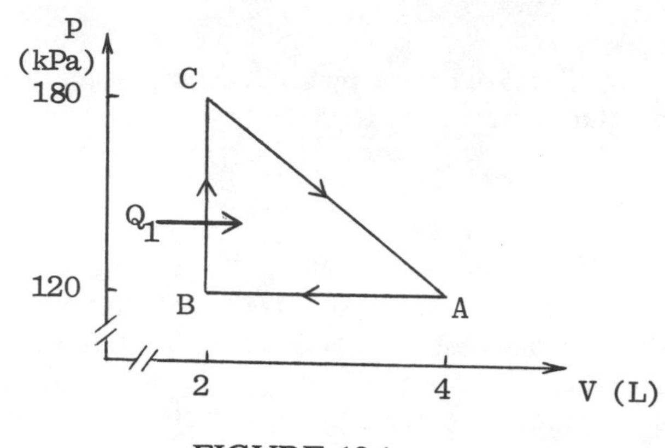

FIGURE 19.1

The work done in the process C to A is given by the area under the function. Since the function is a straight line, one can simply use the average value:

$$W_{CA} = P_{av}(V_f - V_i) = (150 \text{ kPa})(+2\text{x}10^{-3} \text{ m}^3) = +300 \text{ J}$$

(b) The net work done by the system in the complete cycle is

$$W_{net} = 300 \text{ J} - 240 \text{ J} = +60 \text{ J}$$

The net work is also simply the area enclosed by the cycle:

$$W_{net} = \text{Area} = 0.5(2\text{x}10^{-3} \text{ m}^3)(60 \text{ kPa}) = 60 \text{ J}$$

Note that the net work is positive for a clockwise cycle and negative for a counterclockwise cycle.

(c) Since no work is done in the process BC, the change internal energy in the process BC is $\Delta U = Q_1 = 70$ J, thus $U_C = U_B + 70 \text{ J} = 170$ J. The work done in the process CAB is $W = W_{CA} + W_{AB} = +60$ J, and the change in internal energy is $\Delta U = -70$ J. Thus, the heat transfer in the two processes CA + AB is:

$$Q_2 = \Delta U + W = -70 \text{ J} + 60 \text{ J} = -10 \text{ J}$$

Note that for the complete cycle: $\Delta U = (Q_1 + Q_2) - W_{net} = 0$, as it should be.

IDEAL GASES

In an isothermal process, the changes take place at constant temperature. For an ideal gas the equation $P = nRT/V$, leads to

(Isothermal) $$W = nRT \ln(V_f/V_i)$$

For an ideal gas, the difference in the molar specific heats at constant pressure and constant volume is

$$C_p - C_v = R$$

EXAMPLE 4

Three moles of an ideal gas expand from 20 L to 45 L. Find the work done by the gas given that the expansion is (a) isobaric at 160 kPa; (b) isothermal at 40 °C. (c) What is the heat transfer in (b)?

Solution:

(a) Note that $1\ L = 10^{-3}\ m^3$, thus

$$W = P(V_f - V_i) = (1.6x10^5\ N/m^2)(25x10^{-3}\ m^3) = 4\ kJ$$

(b) Converting the temperature to kelvins, we find

$$W = nRT \ln(V_f/V_i)$$

$$= (3\ mol)(8.314\ J/mol.K)(313\ K) \ln(2.25) = 6.33\ kJ$$

(c) Since the internal energy of an ideal gas depends only on temperature, and here T is constant, it follows that $\Delta U = 0$. From the first law of thermodynamics we have $Q = W = 6.33$ kJ

EXAMPLE 5

An ideal gas at 20 °C expands at a constant pressure of 120 kPa from 5 L to 8 L and in the process it absorbs 1.4 kJ of heat. Find: (a) the final temperature; (b) the change in internal energy.

Solution:

(a) Since P is fixed, $V_1/T_1 = V_2/T_2$, thus

$$T_2 = (V_2/V_1)T_1 = (8/5)(293\ K) = 469\ K$$

(b) The work done by the gas in this isobaric process is

$$W = P(V_f - V_i) = (1.2x10^5\ Pa)(3x10^{-3}\ m^3) = 360\ J$$

We are given $Q = 1.4$ kJ (positive since heat enters the system). From the first law,

$$\Delta U = Q - W = 1400 - 360 = 1040 \text{ J}$$

ADIABATIC PROCESS

When an ideal gas undergoes a quasistatic adiabatic process, the pressure and volume obey the relation

$$PV^{\gamma} = \text{constant}$$

The ideal gas law, $PV = nRT$, is valid at any point during the process.
It is shown in Example 19.6 of the text that the work done by an ideal gas in an adiabatic process may be expressed in two ways (that are equivalent):

(Adiabatic) $\quad W = (P_1V_1 - P_2V_2)/(\gamma - 1)$

(Adiabatic) $\quad W = -nC_v(T_2 - T_1)$

EXAMPLE 6
Two moles of an ideal gas ($\gamma = 1.4$) expand from 60 L to 100 L. If the initial pressure is 100 kPa, find (a) the final pressure; (b) the initial and final temperatures.

Solution:
(a) Since $P_1V_1^{\gamma} = P_2V_2^{\gamma}$

$$P_2 = P_1(V_1/V_2)^{\gamma} = (100 \text{ kPa})(0.6)^{1.4} = 48.9 \text{ kPa}$$

(b) From the ideal gas law,

$$T_1 = PV/nR = (10^5 \text{ Pa})(0.06 \text{ m}^3)/2(8.314 \text{ J/mol.K}) = 361 \text{ K}$$

$$T_2 = PV/nR = (4.89\text{x}10^4 \text{ Pa})(0.1 \text{ m}^3)/2(8.314 \text{ J/mol.K}) = 294 \text{ K}$$

EXAMPLE 7
Three moles of an ideal monatomic gas ($\gamma = 5/3$) is taken around a cycle of operations shown in Fig. 19.2. It begins at 100 kPa and a volume of 0.2 m^3. (1) Adiabatic expansion to 0.4 m^3; (2) isothermal compression to 0.2 m^3; (3) increase in pressure back to 100 kPa. Determine the work done in each segment.

Solution:

Since $P_B V_B^\gamma = P_A V_A^\gamma$, we have

$P_B = P_A(V_A/V_B)^\gamma$

$= (10^5 \text{ Pa})(1/2)^{1.67} = 31.5 \text{ kPa}$

Since $PV = nRT$ at point A,

$T_A = PV/nR$

$= (10^5 \text{ Pa})(0.2 \text{ m}^3)/3(8.314 \text{ J/mol.K})$

$= 802 \text{ K}$

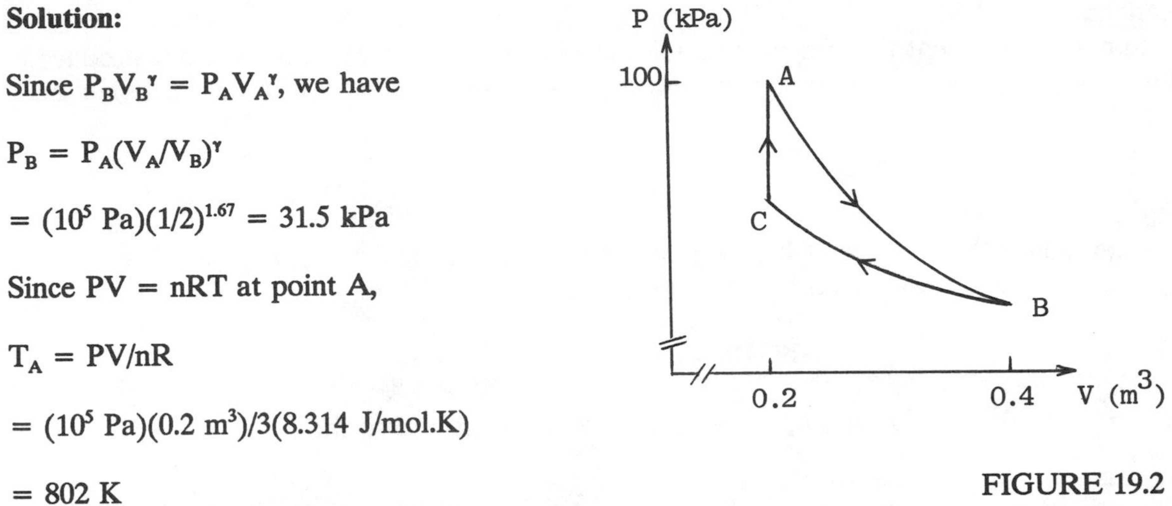

FIGURE 19.2

We can find the temperature at B by expressing the relation PV^γ = constant for an adiabatic process in the form $TV^{\gamma-1}$ = constant. Thus,

$$T_B = T_A(V_A/V_B)^{\gamma-1} = (802 \text{ K})(1/2)^{0.67} = 505 \text{ K}$$

We also know $T_C = T_B = 505$ K. Since $V_A = V_C$, from $PV = nRT$ we find

$$P_C = P_A(T_C/T_A) = (10^5 \text{ Pa})(505 \text{ K}/802 \text{ K})$$

The work done in the constant-volume segment is zero, $W_{CA} = 0$.

$$W_{AB} = (P_A V_A - P_B V_B)/(\gamma - 1)$$

$$= [(10^5 \text{ Pa})(0.2 \text{ m}^3) - (3.15\text{x}10^4 \text{ Pa})(0.4 \text{ m}^3)]/(2/3)$$

$$= 11.1 \text{ kJ}$$

$$W_{BC} = nRT \ln(V_C/V_B)$$

$$= 3(8.314 \text{ J/mol.K})(505 \text{ K}) \ln(1/2) = -8.73 \text{ kJ}$$

HEAT TRANSPORT

Conduction

The rate at which heat is conducted across a cross-sectional area A is

$$dQ/dt = -\kappa A \, dT/dx$$

where κ is the thermal conductivity of the material.

Convection

In convection, heat transfer accompanies mass transport. Forced convection is produced by a fan or a pump, whereas free convection occurs because the density of a fluid depends on the temperature.

Radiation

Radiation is the transfer of heat without an intervening medium. The power radiated by a body with a surface area A at a temperature T is

$$dQ/dt = e\sigma AT^4$$

where $\sigma = 5.67\text{x}10^{-8}$ W/m^2.K and e, the emissivity, depends on the surface.

EXAMPLE 8

A sheet of wood (κ = 0.15 W/m.K) of thickness 2 cm is placed beside a concrete wall (κ = 0.9 W/m.K) that is 25 cm thick. If the temperature difference between the inside and the outside is 30 K, what is the rate of heat transfer through 1 m^2 of surface area?

Solution:

Since we do not know the temperature difference across each layer we first obtain an expression for the total temperature difference. Omitting the negative sign, we have

$$dT/dx = (1/\kappa A)\, dQ/dt$$

Since dT/dx is constant within each layer and in the steady state the rate of heat transfer will be the same through both materials,

$$\Delta T = \Delta T_1 + \Delta T_2 = (L_1/A\kappa_1 + L_2/A\kappa_2)\, dQ/dt$$

With $A = 1$ m^2, $L_1 = 0.02$ m and $L_2 = 0.25$ m, we find dQ/dt = 73 W.

EXAMPLE 9

What is the power radiated from a sphere of coal of radius 3 cm at 600 °C. Take the emissivity to be 0.9.

Solution:

The temperature must be expressed in kelvins, thus T = 873 K. The radiated power is

$$dQ/dt = (0.9)(5.67\text{x}10^{-8}\ \text{W/m}^2.\text{K})(\pi\text{x}9\text{x}10^{-4}\ \text{m}^2)(873\ \text{K})^4$$

$$= 335\ \text{W}$$

SOLUTIONS TO SELECTED TEXT EXERCISES AND PROBLEMS

Exercise 7
How much heat is needed to convert 80 g of ice initially at -10 °C to 60 g of water and 20 g of steam at 100 °C?

Solution:
The heat required involves four terms:

$$\Delta Q = mc_i\Delta T_1 + mL_F + mc_w\Delta T_2 + m'L_V$$

where L_F and L_V are the latent heats of fusion and vaporization, m = 0.08 kg and m' = 0.02 kg.

$$(0.08\text{ kg})[(2100\text{ J/kg.K})(10\text{ K}) + 3.34\text{x}10^5\text{ J/kg} + (4190\text{ J/kg.K})(100\text{ K})] +$$

$$(0.02\text{ kg})(2.26\text{x}10^6\text{ J/kg}) = 107\text{ kJ}$$

Exercise 15
In Joule's paddle wheel experiment the two blocks had a total mass of 3.6 kg and were allowed to fall 16 times through a height of 11 m. If the barrel contained 3.5 kg of water, what would be the expected rise in temperature of the water?

Solution:
We assume that all the potential energy is converted to thermal energy in the water.

$$\Delta T = \Delta U/mc$$

$$= 16mgh/(3.5\text{ kg})(4190\text{ J/kg.K}) = 0.423\text{ C}^\circ$$

Exercise 25
When a gas undergoes a process depicted as the straight line from a to c in Fig. 19.3 the heat flow into the system is 180 J. (a) Find the work done from a to c. (b) If U_a = 100 J, find U_c. (c) What is the work done by the gas when it returns to a via b? (d) What is the heat transfer in the process cba?

Solution:
(a) The work done is equal to the area under the line, which is

$$W = P_{av}\Delta V = (150\text{ kPa})(10^{-3}\text{ m}^3) = 150\text{ J}$$

(b) From the first law

$$\Delta U = U_c - U_a = Q - W = 180\text{ J} - 150\text{ J} = 30\text{ J}$$

Thus $U_c = 130$ J

(c) There is no work done in the constant -volume segment c to b. From b to a, we have

$$W = P\Delta V = (100 \text{ kPa})(-10^{-3} \text{ m}^3) = -100 \text{ J}$$

(d) Since $\Delta U = 0$ for the complete cycle, and we know $\Delta U = +30$ J from a to c, it follows that $\Delta U = -30$ J in the process cba. The heat transfer in the process cba is

$$Q = \Delta U + W = -30 - 100 = -130 \text{ J}$$

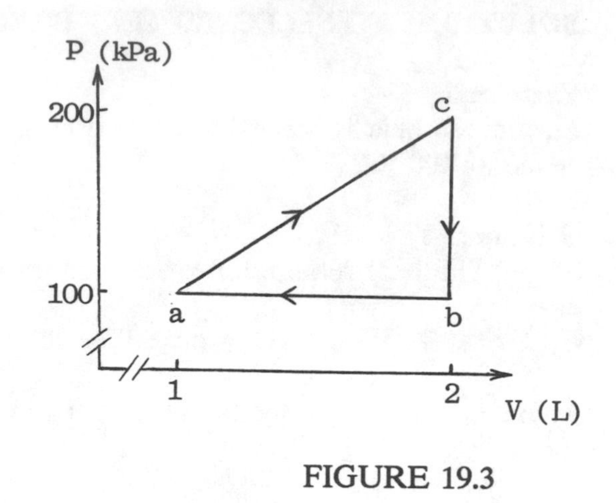

FIGURE 19.3

Exercise 31
Two moles of air ($M = 29\text{x}10^{-3}$ kg/mol) at a fixed pressure of 1 atm are heated from 0 °C to 100 °C. Find: (a) the heat input; (b) the work done by the gas; (c) the change in internal energy. Take $c_p = 1$ kJ/kg.K.

Solution:
(a) The molar specific heat is $C_p = Mc_p = 29$ J/mol.K, thus

$$Q = nC_p\Delta T = 5.8 \text{ kJ}$$

(b) At constant pressure the work done by the (ideal) gas is

$$W = P\Delta V = nR\Delta T = 1.66 \text{ kJ}$$

(c) From the first law, the change in internal energy is

$$\Delta U = Q - W = 4.14 \text{ kJ}$$

Exercise 39
One mole of air ($\gamma = 1.4$) is initially at a pressure of 100 kPa and a temperature of 300 K. It expands adiabatically to five times the initial volume. What is (a) the final pressure, and (b) the final temperature?

Solution:
(a) The equation for a quasistatic adiabatic process is PV^γ = constant.

$$P_1V_1^{1.4} = P_2(5V_1)^{1.4}$$

so $P_2 = 10.5$ kPa.

(b) From the ideal gas law $PV = nRT$, we have

$$T_2/T_1 = (P_2V_2)/P_1V_1$$

so, $T_2 = 158$ K.

Problem 3
Hot fluid passes through a cylindrical pipe of length L with inner radius a and outer radius b; see Fig. 19.4. Show that the rate of heat conduction through the wall of the pipe is

$$dQ/dt = 2\pi\kappa L(T_a - T_b)/\ln(b/a)$$

where T_a and T_b are the temperatures at the inner and outer surfaces respectively. (Hint: Note that the temperature gradient may be written as dT/dr and that the heat flows through an area $2\pi rL$.)

Solution:
The area through which the heat flows is cylindrical, $A = 2\pi rL$, so

$$dQ/dt = \kappa(2\pi rL)(dT/dr)$$

This may be rewritten an integrated

$$(dQ/dt)\int_a^b dr/r = (2\pi\kappa L)\int_{T_a}^{T_b} dT$$

thus

$$dQ/dt = 2\pi\kappa L(T_b - T_a)/\ln(b/a)$$

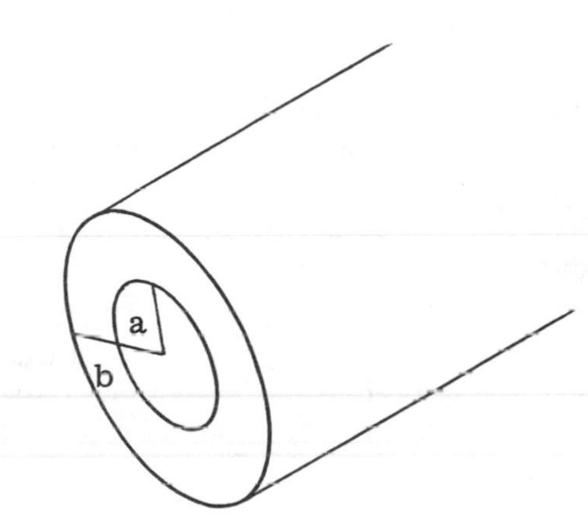

FIGURE 19.4

SELF-TEST

1. A metal sphere of mass 800 g is heated to 100 °C and then immersed into 500 g of water at 20 °C. Given that the specific heat of the metal is 450 J/kg.K, what is the final temperature? Ignore the calorimeter. The specific heat of water is 4190 J/kg.K.

2. Two moles of an ideal gas expand from 0.05 m^3 to 0.07 m^3 at a constant pressure of 100 kPa. What is the change in internal energy? Take $C_p = 29$ J/mol.K.

3. Two moles of an ideal gas initially at 301 °C expand adiabatically from 0.05 m^3 to 0.07 m^3. What is the change in internal energy? Take $\gamma = C_p/C_v = 1.4$, and $C_v = 20.7$ J/mol.K.

Chapter 20

KINETIC THEORY

MAJOR POINTS

1. The molecular model of an ideal gas
2. The kinetic theory of gases relates the macroscopically observed quantities, such as temperature and pressure, to the behavior of molecules
3. The equipartition of energy specifies how the total energy of a molecule is divided between translational, rotational and vibrational degrees of freedom.
4. Kinetic theory and the equipartition theorem can be used to predict the specific heats of an ideal gas

CHAPTER REVIEW

THE MODEL OF AN IDEAL GAS

We make the following assumptions regarding a gas:

1. The gas consists of a very large number of identical molecules moving with random velocities.
2. The molecules have no internal structure so all their energy is purely translational.
3. The molecules do not interact, except during brief collisions with each other and the walls of the container.
4. The average distance between molecules is much larger than their diameters.

KINETIC INTERPRETATION OF PRESSURE AND TEMPERATURE

According to kinetic theory, the pressure exerted by a gas arises from the constant bombardment of the walls of the container by the molecules. The pressure is related to the root mean square speed:

$$v_{rms} = [(v^2)_{av}]^{1/2}$$

which is the square root of the mean (average) value of the square of the speeds of the molecules. If there are N molecules each of mass m contained in a volume V, then the pressure is

$$P = 1/3\ \rho\ v_{rms}^2$$

where $\rho = Nm/V$ is the density of the gas.

The average kinetic of a molecule at an absolute temperature T is

$$K_{av} = 1/2\ m\ v_{rms}^2 = 3/2\ kT$$

Thus, according to kinetic theory the temperature of an ideal gas is a measure of the average kinetic energy of the molecules. The rms speed may be expressed in terms of the temperature:

$$v_{rms} = (3kT/m)^{1/2} = (3RT/M)^{1/2}$$

where M is the molecular mass (in kg/mol). If there are n moles in a sample, then its mass is m = nM and the number of molecules is $N = nN_A$, where N_A is Avogadro's number.

EXAMPLE 1

Two moles of a monatomic ideal gas have a volume of 25 L and are at a pressure of 120 kPa. Find: (a) the average kinetic energy per molecule; (b) the temperature.

Solution:

(a) We need to express the average kinetic in terms of the pressure.

$$v_{rms}^2 = 3P/\rho = 2K_{av}/m$$

Since $\rho = Nm/V$, and $N = nN_A = 2N_A$, we have

$$K_{av} = 3PV/2N$$

$$= 3(1.2\text{x}10^5 \text{ Pa})(25\text{x}10^{-3} \text{ m}^3)/4(6.022\text{x}10^{23}) = 3.74\text{x}10^{-21} \text{ J}$$

(b) Since $K_{av} = 3kT/2$, the temperature is

$$T = 2K_{av}/3k$$

$$= 2(3.74\text{x}10^{-21} \text{ J})/3(1.38\text{x}10^{-23} \text{ J/K}) = 181 \text{ K}$$

SPECIFIC HEATS

The internal energy of a monatomic ideal gas with N molecules, or n moles, is

$$U = 3/2\ NkT = 3/2\ nRT$$

For a constant volume process, W = 0, so the first law of thermodynamics, $\Delta U = Q - W$, becomes

$$\Delta U = Q = n\, C_v\, \Delta T$$

where C_v is the molar specific heat at constant volume. In the special case of an ideal gas, U depends only on T, so the expression $\Delta U = nC_v\Delta T$ is valid for <u>any</u> process, even when the volume is not fixed (in which case Q will have a different value).

For a process at constant pressure, $Q = nC_p\Delta T$ and the work done is $W = P\Delta V = nR\Delta T$. We have just seen that for an ideal gas $\Delta U = nC_v\Delta T$ for any process. Thus the first law written as $Q - \Delta U = W$, becomes (after removal of the common factor $n\Delta T$)

$$C_p - C_v = R$$

This difference in the molar specific heats of an ideal gas was also obtained in Eq. 19.15.

On comparing $U = 3/2\ nRT$ and $\Delta U = nC_v\Delta T$, we find $C_v = 3RT/2$. Thus the molar specific heats of a monatomic ideal gas are

(Monatomic) $\quad C_v = 3R/2; \quad C_p = 5R/2; \quad \gamma = C_p/C_v = 5/3$

EXAMPLE 2

An ideal monatomic gas is initially at 300 K and a pressure of 100 kPa.
The volume is held fixed at 12 L as 800 J of heat are absorbed. Find the final (a) temperature; (b) pressure.

Solution:

(a) At constant volume the heat transfer is given by

$$Q = nC_v\ \Delta T = (3nR/2)\ \Delta T$$

To find the number of moles we use the ideal gas law,

$$n = PV/RT$$

$$= (10^5\ \text{pa})(12\text{x}10^{-3}\ \text{m}^3)/(8.314\ \text{J/mol.K})(300\ \text{K} = 0.48\ \text{moles}$$

Thus,

$$\Delta T = 2Q/3nR$$

$$= 2(800\ \text{J})/3(0.48\ \text{mol})(8.314\ \text{J/mol.K}) = 134\ \text{K}$$

The final temperature is 434 K.

(b) From the ideal gas law, we have

$$P = nRT/V = (0.48\ \text{mol})R(434\ \text{K})/(0.012\ \text{m}^3) = 144\ \text{kPa}.$$

EXAMPLE 3

Two moles of an ideal monatomic gas are initially at 0 °C and 1 atm. The volume is halved adiabatically. (a) What is the change in internal energy? (b) What is the work done by the gas?

Solution:

For a quasistatic adiabatic process we know that PV^γ = constant. Using the ideal gas law $PV = nRT$, this can be reexpressed as $TV^{\gamma-1}$ = constant. Thus,

$$T_2 = T_1(V_1/V_2)^{\gamma-1} = (273\ \text{K})(2)^{2/3} = 433\ \text{K}$$

The change in internal energy of an ideal gas depends only on the temperature and is given by

$$\Delta U = nC_v\Delta T$$

$$= (2\text{ mol})(3R/2)(160\text{ K}) = 3.99\text{ kJ}.$$

(b) Since $Q = 0$ for an adiabatic process, the first law of thermodynamics, $\Delta U = Q - W$, tells us $W = -\Delta U = -3.99$ kJ. The negative sign means that work was done <u>on</u> the gas (which is why its internal energy increased).

EQUIPARTITION OF ENERGY

A degree of freedom of a molecule is a way it can have either kinetic or potential energy. According to the equipartition theorem, for a system in thermal equilibrium:

Each degree of freedom has an average energy $kT/2$.

A structureless sphere (monatomic gas) has three translational degrees of freedom associated with the three possible directions of motion.

Rigid Rotator

A molecule that can rotate, has two degrees of freedom associated with rotation about axes perpendicular to the line joining them. (Classically, one may say that the moment of inertia about the axis joining the atoms is too small to make a contribution to the energy.) Thus the total energy is

$$U = U_{trans} + U_{rot} = 3kT/2 + kT = 5kT/2$$

The molar specific heats are:

(Rigid rotator) $\quad C_v = 5R/2; \quad C_p = 7R/2; \quad \gamma = C_p/C_v = 7/5$

Nonrigid rotator

If the molecule vibrates, it has two more degrees of freedom, one kinetic, the other potential. Thus,

(Nonrigid rotator) $\quad C_v = 7R/2; \quad C_p = 9R/2; \quad \gamma = 9/7$

SOLUTIONS TO SELECTED TEXT EXERCISES AND PROBLEMS

Exercise 15
Make an estimate of the average distance between molecules of oxygen (O_2) at a temperature of 0 °C and a pressure of 1 atm. (Hint: Consider one mole.)

Solution:
The volume of 1 mole is

$$V = RT/P = 2.25\text{x}10^{-2}\ \text{m}^3$$

In 1 mole there are N_A molecules, thus the volume "occupied" by 1 molecule

$$V/N_A = 2.25\text{x}10^{-2}\ \text{m}^3/N_A = 3.73\text{x}10^{-26}\ \text{m}^3$$

If we treat this as a cube of side d, then the volume is d^3 and $d = 3.34\text{x}10^{-9}$ m.

Exercise 23
One mole of an ideal monatomic gas is heated from 0 °C to 100 °C. Find the change in internal energy and the heat absorbed given that the process takes place under the following condition: (a) constant volume; (b) constant pressure.

Solution:
(a) The heat transfer in a constant-volume process is (for n = 1 mole)

$$Q_v = C_v\Delta T = 3R\Delta T/2 = 1250\ \text{J}$$

The change in internal energy of an ideal gas depends only on the temperature, and is given by

$$\Delta U = nC_v\Delta T = 3R\Delta T/2 = 1250\ \text{J}$$

(b) At constant pressure, we have

$$Q_p = C_p\Delta T = 5R\Delta T/2 = 2080\ \text{J}$$

The change in internal energy is still $\Delta U = 3R\Delta T/2 = 1250$ J. (So why does Q_p differ from Q_v?)

Exercise 29
The mean free path of O_2 molecules in a box is $9\text{x}10^{-8}$ m. Given that the number density is $2.7\text{x}10^{19}$ molecules/cm^3, estimate the diamter of an oxygen molecule.

Solution:
The number density of molecules is

$$n_v = 2.7\text{x}10^{25} \text{ molecules/m}^3$$

From Eq. 20.18 in the text we have,

$$d = (1.41 n_v \pi \lambda)^{-1/2} = 0.304 \text{ nm}$$

Problem 3
Use the Maxwell speed distribution function to show explicitly that the total number of particles is N, that is

$$\int_0^\infty f(v)\, dv = N$$

Note that $\int_0^\infty x^2 \exp(-ax^2)\, dx = (\pi/16a^3)^{1/2}$.

Solution:
If we let α = m/2kT, then $A = 4\pi N(\alpha/\pi)^{3/2}$ and the integral is

$$\int_0^\infty f(v)\, dv = A \int_0^\infty v^2 \exp(-\alpha v^2)\, dv$$

$$= A\, (\pi/16\alpha^3)^{1/2} = N$$

SELF-TEST
1. Two moles of nitrogen (M = 28 g/mol) are at 0 °C and 100 kPa in a container. Find: (a) the average kinetic energy per molecule; (b) the rms speed; (c) the mass density; (d) the volume of the container.

Chapter 21

ENTROPY AND THE SECOND LAW OF THERMODYNAMICS

MAJOR POINTS

1. The basic principle underlying the operation of heat engines and refrigerators
2. According to the second law of thermodynamics perfect heat engines or refigerators are not possible
3. The distinction between reversible and irreversible processes
4. The Carnot cycle of operations serves as a standard by which other heat engines can be judged
5. (a) The entropy of a system is a measure of its disorder
   (b) According to the second law of thermodynamics, the entropy of an isolated system cannot decrease.

CHAPTER REVIEW

HEAT ENGINES

A heat engine is a device that uses part of the heat flowing from a hot reservoir to a cold reservoir to do work. The net work done by a heat engine is

$$W = |Q_H| - |Q_C|$$

where $|Q_H|$ is the heat input from the hot reservoir and Q_C is the heat output into the cold reservoir. The thermal efficiency of the engine is

(Heat engine) $$\epsilon = W/|Q_H| = 1 - |Q_C|/|Q_H|$$

A refrigerator, or a heat pump, requires the input of work to make heat flow from a cold reservoir to a hot reservoir. The coefficients of performance are defined as

(Refrigerator) $$COP = |Q_C|/W$$

(Heat pump) $$COP = |Q_H|/W$$

SECOND LAW OF THERMODYNAMICS

Kelvin-Planck statement:

It is impossible for a heat engine that operates in a cycle to convert its heat input completely into work

Clausius statement:

It is impossible for a cyclic device to transfer heat continuously from a cold body to a hot body without the input of work or other effect on the environment.

Reversible and Irrerversible Processes

A process is reversible if the system and its surroundings can be made to retrace its thermodynamic path back to its initial state. The system must pass through a succession of equilibrium states. For a process to be reversible three conditions must be met: (1) It must be quasistatic; (2) There must be no friction; (3) Any heat transfer must occur at constant temperature or be associated with an infinitesimal temperature difference
Any process that does not satisfy these conditions is irreversible and cannot be respesented on a PV diagram.

CARNOT CYCLE

A Carnot cycle is the most efficient cyclic process. It consists of two isothermal and two adiabatic processes. It can be shown that

$$|Q_C|/|Q_H| = T_C/T_H$$

so, the efficiency of a Carnot engine depends only on T_H and T_C:

$$\epsilon_C = 1 - T_C/T_H$$

No engine can have an efficiency greater than that of a Carnot engine operating between the same two temperatures.

EXAMPLE 1

A heat engine with a thermal efficiency of 30% deposits 100 J per cycle into the cold reservoir. Find: (a) the heat absorbed from the hot reservoir; (b) the work done. (c) If T_H = 180 °C, what is the maximum possible temperature of the cold reservoir?

Solution:

(a) The thermal efficiency of a heat engine is

$$\epsilon = 1 - |Q_C|/|Q_H|$$

With the given values, 100 J/$|Q_H|$ = 0.7, thus $|Q_H|$ = 143 J.

(b) The thermal efficency of a Carnot engine may be written in terms of the kelvin temperatures of the two reservoirs:

$$\epsilon_C = 1 - T_C/T_H$$

Since this tells us the maximum possible efficiency of any heat engine operating between two given temperatures, the maximum possible temperature of the cold reservoir is

$$T_C = (1 - \epsilon_C)T_H = (0.7)(453\text{ K}) = 317\text{ K}$$

ENTROPY

If a system absorbs or rejects an infinitesimal quantity of heat dQ_R in a reversible process from one equilibrium state to another at temperature T, the change in entropy of the system is defined as the ratio

$$dS = dQ_R/T$$

The finite change in entropy in a reversible process depends only on the initial and final equilibrium states:

$$\Delta S = S_f - S_i = \int dQ_R/T$$

To calculate the change in entropy for an irreversible process, one must find a suitable reversible path between the same initial and final equilibrium states.

<u>Reversible process (Ideal gas)</u>

For an ideal gas the change in entropy of n moles is given by

$$\Delta S = nC_v \ln(T_f/T_i) + nR \ln(V_f/V_i)$$

<u>Free Expansion</u>

In an (irreversible, adiabatic) free expansion, the entropy change of the gas is

$$\Delta S_g = nR \ln(V_f/V_i)$$

ENTROPY AND THE SECOND LAW OF THERMODYNAMICS

According to the second law of thermodynamics,

$$\Delta S > 0$$

In a reversible process the entropy of an isolated system stays constant; in an irreversible process the entropy increases.

The entropy of a system is a measure of the disorder of its constituent parts. An orderly state has low entropy. The second law of thermodynamics stated in terms of entropy tells us that natural processes tend to evolve from states of higher order to states of lower order.

EXAMPLE 2

A block of ice of mass 80 g is initially at 0 °C is placed in 200 g of water at 40 °C. What is the change in entropy of this system?

Solution:

First we must determine the final temperature of the (isolated) system. Since there is no net heat transfer,

$$m_i c_i \Delta T_1 + m_i L + m_w c_w \Delta T_2 = 0$$

$$(0.08 \text{ kg})(2100 \text{ J/kg.K})(T_f - 0) + (0.08 \text{ kg})(334 \text{ kJ/kg})$$

$$+ (0.2 \text{ kg})(4190 \text{ J/kg.K})(T_f - 40) = 0$$

$$(168 \text{ J/K})T_f + 2.672\text{x}10^4 \text{ J} + (838 \text{ J/K})(T_f - 40) = 0$$

We find $T_f = 6.76$ °C = 280 K.

Since $dQ = mcdT$, the change in entropy of a mass m may be found from

$$\Delta S = \int mc \, dT/T = mc \ln(T_f/T_i)$$

Although the process is irreversible, one can imagine that the ice is placed in contact with a series of reservoirs whose temperatures differ slightly. For the ice one must take into account the latent heat of fusion:

$$\Delta S_i = mL/T + mc\ln(T_f/T_1)$$

$$= (0.08 \text{ kg})[(334 \text{ kJ/kg})/(273 \text{ K}) + (2.1 \text{ kJ/kg.K}) \ln(280 \text{ K}/273 \text{ K})]$$

$$= 102 \text{ J/K}$$

$$\Delta S_w = mc \ln(T_f/T_2)$$

$$= (0.2 \text{ kg})(4190 \text{ J/kg.K}) \ln(280 \text{ K}/313 \text{ K}) = -93.4 \text{ J/K}$$

The total change in entropy is $\Delta S_i + \Delta S_w = +8.6$ J/K.

EXAMPLE 3

A copper rod of cross-sectional area $3\text{x}10^{-4}$ m^2 and length 15 cm conducts heat between two reservoirs at 320 K and 280 K. What is the rate of change of entropy of the system? The thermal conductivity of the rod is 400 W/m.K.

Solution:

The rate of heat conduction is given by

$$dQ/dt = -\kappa A \, dT/dx$$

where κ is the thermal conductivity and A is the cross-sectional area.
Since $dS = dQ/T$, the rate of change of entropy is $dS/dt = \pm(1/T)dQ/dt$. The net rate of increase of entropy is therefore

$$dS/dt = (1/T_C - 1/T_H)dQ/dt$$

$$= -\kappa A(1/T_C - 1/T_H)dT/dx$$

where $dT/dx = -\Delta T/L$. Inserting the given values,

$$dS/dt = (400\ \mathrm{W/m.K})(3\mathrm{x}10^{-4}\ \mathrm{m}^2)(1/280\ \mathrm{K} - 1/320\ \mathrm{K})(40\ \mathrm{K}/0.15\ \mathrm{m})$$

$$= 1.43\mathrm{x}10^{-2}\ \mathrm{J/K.s}$$

EXAMPLE 4
A heat engine that employs an ideal monatomic gas undergoes the cycle shown in Fig. 21.1. Show that the thermal efficiency is

$$\epsilon = W/Q_{in} = 1 - (\gamma - 1)\ln r/(r^{\gamma-1} - 1)$$

where $r = V_2/V_1$.

Solution:
Segment AB:
There is no work done in this constant-volume process. The heat input is

$$Q_{in} = nC_v(T_B - T_A)$$

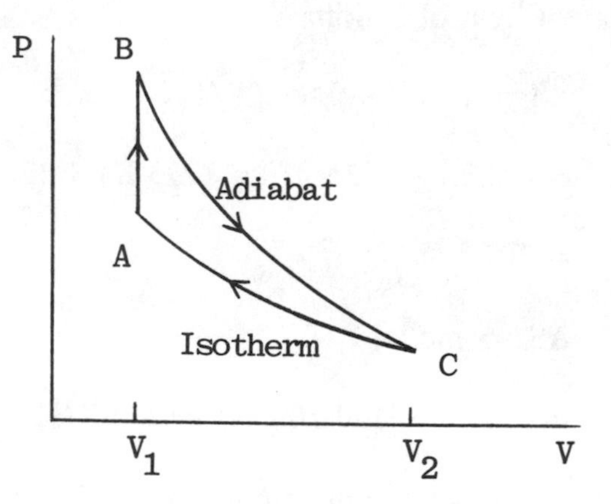

FIGURE 21.1

Segment BC:
There is no heat transfer in an adiabatic process. The work done by the gas is

$$W_{BC} = (P_BV_1 - P_CV_2)/(\gamma - 1)$$

$$= nR(T_B - T_C)/(\gamma - 1)$$

Segment CA:
The work done in this isothermal process is

$$W_{CA} = nRT_C \ln(V_1/V_2) = -nRT_C \ln(r)$$

where $r = V_2/V_1$. Since $\Delta U = 0$ for an ideal gas in an isothermal process, $Q_{out} = W_{CA}$, but this does not concern us here. The thermal efficiency is

$$\epsilon = (W_{BC} + W_{CA})/Q_{in}$$

Before we substitute we note that for a monatomic ideal gas $\gamma = 5/3$ and $C_v = 3R/2$, thus $R/C_v(\gamma - 1) = 1$. With this simplification, we find

$$\epsilon = 1 - (\gamma - 1)T_C \ln r/(T_B - T_C)$$

Next, we note that for an adiabatic process $TV^{\gamma-1}$ = constant, thus $(T_B/T_C) = r^{\gamma-1}$. Finally,

$$\epsilon = 1 - (\gamma - 1)\ln(r)/(r^{\gamma-1} - 1)$$

SOLUTIONS TO SELECTED TEXT EXERCISES AND PROBLEMS

Exercise 5
A residential heat pump with a ceofficient of performance of 4 requires 10 kWh of electrical energy during a certain period. How many kWh would be needed if the heating were accomplished with electrical resistance heaters?

Solution:
The coefficient of a heat pump is defined as $COP = Q_H/W$, thus

$$Q_H = W(COP) = 4W = 40 \text{ kWh}$$

This would be the requirement of electrical heaters.

Exercise 17
A Carnot engine operates between 200 °C and 20 °C and has a mechanical output of 360 W. If each cycle takes 0.2 s, find the heat (a) absorbed, and (b) expelled in each cycle.

Solution:
The Carnot efficiency is

$$\epsilon_C = 1 - 293 \text{ K}/472 \text{ K} = 0.381$$

Since $|Q_H| = W/\epsilon$, we have

$$dQ_H/dt = (dW/dt)/\epsilon_C = 945 \text{ W}$$

and $dQ_C/dt = 945 - 360 = 585$ W. For a 0.2 s cycle, $Q_H = 189$ J, $Q_C = 117$ J

Exercise 25
A 100-g lead ball at 100 °C is placed in 300 g of water at 20 °C in an insulated container. (a) What is the final equilibrium temperature? (b) What is the change in entropy (b) of the ball, (c) of the water, (d) of the universe? Ignore the container.

Solution:
(a) Since the net heat transfer is zero, we have

$$m_1c_1(T_f - T_1) + m_2c_2(T_f - T_2) = 0$$

$$(0.1\ \text{kg})(130\ \text{J/kg.K})(T_f - 100) + (0.3\ \text{kg})(4190\ \text{J/kg.K})(T_f - 20) = 0$$

This leads to $T_f = 20.8$ °C.

(b) The change in entropy of the ball is

$$\Delta S_1 = m_1c_1\ln(T_f/T_1) = -3.09\ \text{J/K}$$

(c) The change in entropy of the water is

$$\Delta S_2 = m_2c_2\ln(T_f/T_2) = +3.4\ \text{J/K}$$

(d) The change in entropy of the universe is

$$\Delta S_u = 3.4\ \text{J/K} - 3.09\ \text{J/K} = 0.31\ \text{J/K}$$

Problem 3
Two moles of an ideal diatomic gas ($\gamma = 7/5$) operate in the cycle of Fig. 21.2, where $T_a = 400$ K, $T_c = 250$ K and $P_c = 100$ kPa. Find: (a) the work done per cycle; (b) the efficiency of the engine.

Solution:
(a) $V_a = V_c = nRT_c/P_c = 0.04157\ \text{m}^3$.

$$P_a/T_a = P_c/T_c$$

thus $P_a = 160$ kPa. Since $P_c = P_b$,

$$P_aV_a^\gamma = P_cV_b^\gamma$$

which gives $V_b = 1.6^{5/7}\ V_b = 1.4V_a$.
The work in the segments are

$$W_{bc} = P\Delta V = P_c(V_c - V_b)$$

$$= P_c(V_a - 1.4V_a) = -1663\ \text{J};$$

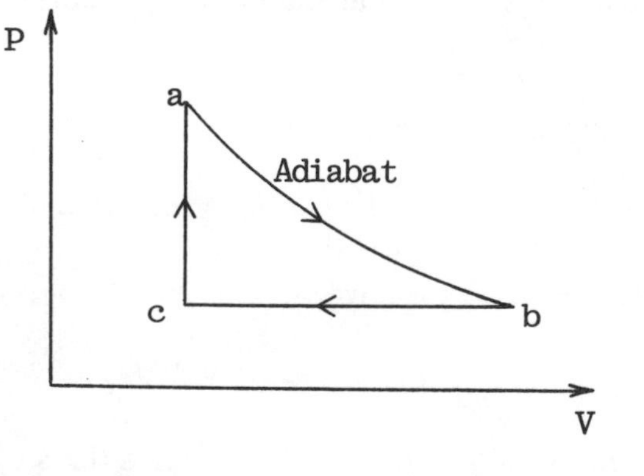

FIGURE 21.2

$W_{ca} = 0$

$W_{ab} = \int PdV = \int K\, dV/V^{\gamma} = K(V_b^{1-\gamma} - V_a^{1-\gamma})/(1 - \gamma)$

where $K = P_aV_a^{\gamma}$. We find that

$$W_{ab} = P_aV_a(1/1.4^{0.4} - 1)/(1 - \gamma) = 2094 \text{ J}$$

Thus the net work is $W_{net} = 2094 - 1663 = 431$ J.

(b) $Q_{ab} = 0$; $Q_{bc} = nC_p\Delta T = C_pP\Delta V/R = -5820$ J (in)

$\epsilon = W/Q_{in} = 431/5820 = 0.074$

Problem 7
A Carnot engine that uses one mole of an ideal gas ($\gamma = 5/3$) operates between 500 K and 300 K. The highest and lowest pressures at 500 kPa and 100 kPa. Find: (a) the net work done per cycle; (b) the efficiency.

Solution:
In Fig. 21.3, we have

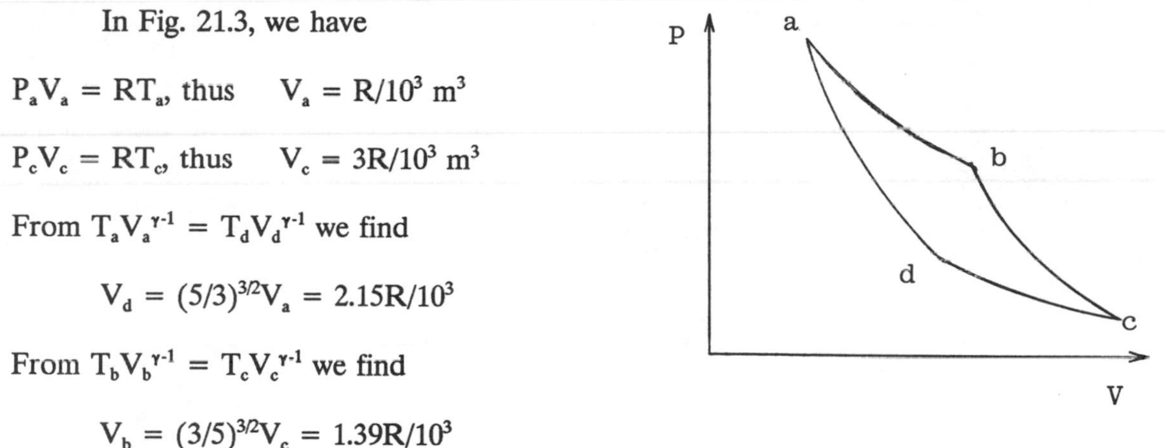

FIGURE 21.3

$P_aV_a = RT_a$, thus $V_a = R/10^3$ m^3

$P_cV_c = RT_c$, thus $V_c = 3R/10^3$ m^3

From $T_aV_a^{\gamma-1} = T_dV_d^{\gamma-1}$ we find

$V_d = (5/3)^{3/2}V_a = 2.15R/10^3$

From $T_bV_b^{\gamma-1} = T_cV_c^{\gamma-1}$ we find

$V_b = (3/5)^{3/2}V_c = 1.39R/10^3$

The work done in the segments are

$$W_{ab} = nRT_a\ln(V_b/V_a) = 500R \ln(1.39) = 1369 \text{ J}$$

$$W_{cd} = nRT_c\ln(V_d/V_c) = 300R \ln(2.15/3) = -831 \text{ J}$$

Thus, $W_{net} = 538$ J

(b) The thermal efficiency is

$$\epsilon_C = 1 - 300 \text{ K}/500 \text{ K} = 0.4$$

SELF-TEST

1. A refrigerator with a coefficient of performance of 4 freezes 2 kg of water at 0 °C to ice at 0 °C. What is the heat deposited into the room? The latent heat of fusion is 334 kJ.

2. An insulated copper jar of mass 120 g and initially at 20 °C, is filled with 100 g of water at 90 °C. Find: (a) the final temperature; (b) the change in entropy of the jar and the water. The specific heat of copper is 385 J/kg.K and of water it is 4190 J/kg.K.

3. One mole of an ideal monatomic gas initially at 0 °C and 1 atm absorbs 1.2 kJ of heat at constant pressure. What is the change in its entropy? Take C_p = 20.8 J/mol.K.

Chapter 22

ELECTROSTATICS

MAJOR POINTS

1. The properties of charge; its conservation, and quantization.
2. The distinction between conductors and insulators
3. Coulomb's law for the force between static, point charges
4. The principle of superposition used to determine the net force on one charge due to several other charges.

CHAPTER REVIEW

PROPERTIES OF CHARGE

There are two kinds of charge: positive and negative. Like charges repel each other and unlike charges attract each other. The SI unit of charge is the coulomb (C).

Conservation of charge: The net charge in an isolated system is constant.

Quantization of charge: Charge appears only in integer multiples of the elementary charge e:

$$q = \pm ne$$

where n is an integer and $e = 1.602 \times 10^{-19}$ C. The charge on an electron is -e and the charge on a proton is +e.

CONDUCTORS AND INSULATORS

A conductor has charges that are able to move throughout the medium. For example, a charge placed on a metal object will spread very quickly over the surface of the object. In an insulator all the charges are bound to particular sites; it does not allow the flow of charge.

COULOMB'S LAW

The magnitude of the electrostatic force between two point charges q and Q separated by a distance r is given by

$$F = kqQ/r^2$$

where $k \approx 9 \times 10^9$ N.m^2/C^2. The direction of each force is determined from the knowledge that unlike charges attract and like charges repel. Coulomb's law tells us the force between point charges at rest. (Moving charges also produce and experience magnetic effects.) In the special case of a continuous charge distribution that has spherical symmetry (a uniformly charged shell or sphere), one can use Coulomb's law with r taken to be the distance to the center.

The Superposition Principle

The net force on a charge q_1 exerted by charges q_2, q_3, etc., is given by the principle of linear superposition:

$$\mathbf{F}_1 = \mathbf{F}_{12} + \mathbf{F}_{13} + \mathbf{F}_{14} + ... + \mathbf{F}_{1N}$$

where, for example, $\mathbf{F}_{12}$ is the force exerted on q_1 by q_2. This principle says that one can calculate the force exerted by each charge in turn, ignoring any other charges present. The resultant is simply the vector sum of the individual forces.

HELPFUL HINTS

In order to find the net force on a charge q_1, keep the following points in mind.

1. Decide whether each force is an attraction or a repulsion, and then draw it in the appropriate direction with its tail at q_1.
2. Calculate the magnitudes of the forces. It helps to add the absolute value sign, for example

$$F_{12} = k|q_1q_2|/r^2$$

3. Choose a coordinate system and then find the components of the forces. Note: It is not correct to use the x component of the distance between to charges to find the x component of the force!

EXAMPLE 1

A charge $q_1 = -1\ \mu C$ is at the origin, $q_2 = -2\ \mu C$ is 10 cm from the origin at 37° to the y axis (see Fig. 22.1), and $q_3 = 3\ \mu C$ is at x = 15 cm. What is the net force on q_1?

Solution:

The directions of the forces on q_1 are already drawn in the figure. The magnitudes of these forces are

$F_{12} = k|q_1q_2|/r_1^2$

$= (9\text{x}10^9\ \text{N.m}^2/\text{C}^2)(2\text{x}10^{-12}\ \text{C}^2)/(0.10\ \text{m})^2$

$= 1.8$ N

$F_{13} = k|q_1q_3|/r_2^2$

$= (9\text{x}10^9\ \text{N.m}^2/\text{C}^2)(3\text{x}10^{-12}\ \text{C}^2)/(0.15\ \text{m})^2$

$= 1.2$ N

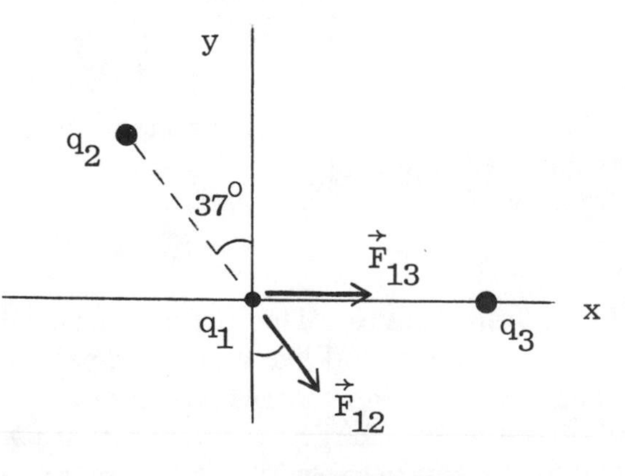

FIGURE 22.1

The components of the net force on q_1 are

$$F_{1x} = F_{12} + F_{13}\sin 37° = 2.28\text{ N}$$

$$F_{1y} = 0 - F_{13}\cos 37° = -1.44\text{ N}$$

Thus $\mathbf{F}_1 = 2.28\mathbf{i} - 1.44\mathbf{j}$ N.

SOLUTIONS TO SELECTED EXERCISES AND PROBLEMS IN THE TEXT

Exercise 11

Figure 22.2 shows five point charges on a straight line. The separation between the charges is 1 cm. For what values of q_1 and q_2 would the net force on each of the other three charges be zero?

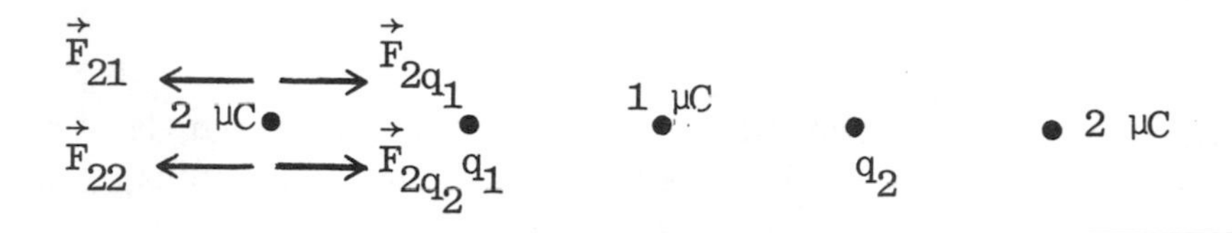

FIGURE 22.2

Solution:

From the symmetry of the arrangement we can immediately conclude that $q_1 = q_2$, and that they must be negative. The force on the 1 μC charge is clearly zero. If we let d = 0.01 m, the net force on the left 2 μC charge is (see Fig. 22.2):

$$F = k(2\ \mu C)[|q_1|/d^2 + |q_2|/9d^2 - (1\ \mu C)/4d^2 - (2\ \mu C)/16d^2] = 0$$

Since $|q_1| = |q_2|$, we find $q_1 = q_2 = -27/80\ \mu C$.

Exercise 17

In the quark model of elementary particlces a proton consists of two "up" (u) quarks, each with charge 2e/3, and a "down" (d) quark of charge -e/3. Suppose these particles lie equally spaced on a circle of radius 1.2×10^{-15} m, as in Fig. 22.3. Find the magnitude of the electrostatic force on each quark.

Solution:

From Fig. 22.3, we see that the distance between the quarks is

$$d = 2r\cos 30° = (3)^{1/2}r$$

Let $F_o = k(e^2/9)/3r^2 = 5.926$ N.

Force on the u quark:

$$F_x = 4F_o - 2F_o\cos60° = 3F_o$$

$$F_y = 2F_o\sin60° = (3)^{1/2}F_o$$

Thus $F_u = (12)^{1/2}F_o = 20.5$ N.

Force on the d quark:

$F_x = 0$; $F_y = 4F_o \sin60° = 20.5$ N.

Thus $F_d = 20.5$ N.

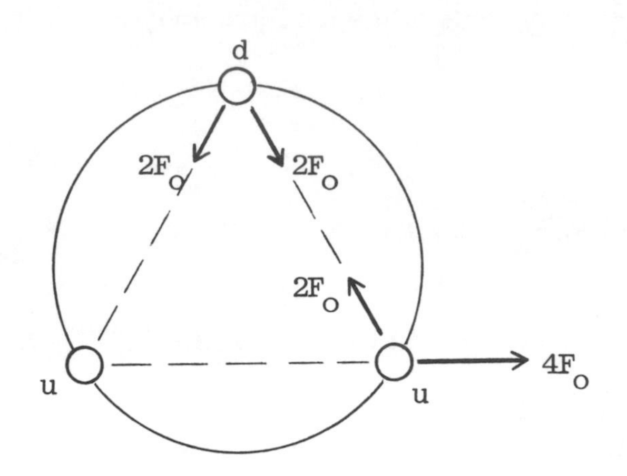

FIGURE 22.3

Problem 1
Three point charges, q_1, q_2, and q_3, lie at the corners of an equilateral triangle of side 10 cm. The forces between them are $F_{12} = 5.4$ N (attractive) $F_{13} = 15$ N (repulsive) and $F_{23} = 9$ N (attractive). Given that q_1 is negative, what are q_2 and q_3?

Solution:
The magnitudes of the forces are

$$k|q_1q_2|/r^2 = 5.4 \qquad (1)$$

$$k|q_1q_3|/r^2 = 15 \qquad (2)$$

$$k|q_2q_3|/r^2 = 9 \qquad (3)$$

Taking the ratio (3)/(2) we find $|q_2| = 9|q_1|/15$. When this is substituted into (1), with r = 0.1 m, we find $|q_2| = 1.90\ \mu C$. Since q_1 is negative and F_{12} is attractive we conclude that $q_2 = +1.90\ \mu C$. The other two charges are easily found: $q_1 = -3.16\ \mu C$ and $q_3 = -5.27\ \mu C$. (Check these values.)

SELF-TEST

1. A point charge $Q_1 = 2\ \mu C$ is at x = 0 while $Q_2 = 6\ \mu C$ is at x = d. (a) Where would you place a third charge q such that the net force on each of the three charges is zero? (b) What is the value of q?

2. Three point charges are located as follows: $Q_1 = 2\ \mu C$ is at (2cm, 1cm), $Q_2 = 5\ \mu C$ is at (-1cm, 3cm) and $Q_3 = 3\ \mu C$ is at the origin. What is the net force on Q_3?

Chapter 23

THE ELECTRIC FIELD

MAJOR POINTS

1. Definition of the electric field strength at a point
2. Calculate the electric field due to (a) a system of point charges, and (b) a continuous distribution of charge (which requires integration).
3. (a) Describe the properties of electric field lines
   (b) Draw electric field lines for simple charge distributions
4. Determine the motion of a point charge in a uniform electric field.
5. Dipoles: Dipole moment; the torque and potential energy in an electric field

CHAPTER REVIEW

THE ELECTRIC FIELD

The electric field strength at a given location is defined as the force per unit charge placed at that point:

$$\mathbf{E} = \mathbf{F}/q_t$$

The "test" charge q_t is assumed to be small enough not to disturb the "source" charges that produce the field. The direction of **E** is that of the force on a positive charge (although the test charge does not have to be positive). One can detect the presence of an electric field by noting whether a charge at rest experiences a force. The SI unit of **E** is N/C.

A charge q in an electric field **E** experiences a force

$$\mathbf{F} = q\,\mathbf{E}$$

Note that this is similar to the equation **W** = m **g**, where **g**, the gravtiational force per unit mass (N/kg), is the gravitational field strength.

The magnitude of the electric field created by a point charge Q can be determined from Coulomb's law, $F = kqQ/r^2$:

$$E = kQ/r^2$$

The direction of **E** is outward from a positive charge, and inward toward a negative charge.

If there are several charges present, we use the principle of linear superposition to determine the resultant field strength at a given position:

$$\mathbf{E}_T = \mathbf{E}_1 + \mathbf{E}_2 + \mathbf{E}_3 + \ldots + \mathbf{E}_N$$

Continuous Charge Distributions

If the distribution of charges is continuous, one must employ integration to sum the contributions of infinitesimal charge elements dq:

$$d\mathbf{E} = kdq/r^2\ \hat{\mathbf{r}}$$

The unit vector $\hat{\mathbf{r}}$ is directed from a given charge element to the point at which the field strength is being evaluated. However, in practice, we do not get involved with vector integration. We usually find an expression for **dE** and then integrate its components. Depending on the shape of the object, the charge distribution may be specified in several ways:

Volume charge density ρ (C/m^3)	$dq = \rho\ dV$
Surface charge density σ (C/m^2)	$dq = \sigma\ dA$
Linear charge density λ (C/m)	$dq = \lambda\ dx$

In order to perform an integration, the quantities dq and r must be expressed in terms of the same variable, such as x or θ.

EXAMPLE 1

Three point charges are located as follows: $q_1 = 2\ \mu C$ is 10 cm from the origin at 60° above the x axis, as shown in Fig. 23.1, $q_2 = -3\ \mu C$ is at (6 cm, -4 cm), and $q_3 = -1.5\ \mu C$ is at (0, 7 cm). (a) Determine the field strength at the origin. (b) What is the net force on $q_4 = -5\ \mu C$ at the origin?

Solution:

1. Draw arrows to represent the fields due to q_1 and q_3. Be careful with the directions.
2. Calculate the magnitudes of the fields.
3. Find the x and y components of the fields.
4. Add the components to find the resultant field.

The directions of the fields at the origin are indicated in Fig.23.1. The magnitudes are (check them)

$$E_1 = 1.8\text{x}10^6\ \text{N/C}$$

$$E_2 = 5.19\text{x}10^6\ \text{N/C}$$

$$E_3 = 2.76\text{x}10^6\ \text{N/C}$$

Next we need the components of each of these fields to calculate the components of the resultant field.

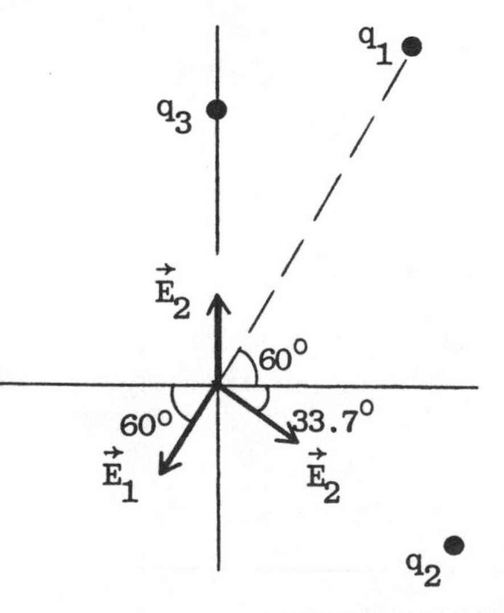

FIGURE 23.1

$$E_x = -E_1 \cos 60^\circ + E_2 \cos\alpha = 3.42\text{x}10^6 \text{ N/C}$$

$$E_y = -E_1 \sin 60^\circ - E_2 \sin\alpha + E_3 = -1.68\text{x}10^6 \text{ N/C}$$

The resultant field is $\mathbf{E} = (3.42\mathbf{i} - 1.68\mathbf{j})\text{x}10^6$ N/C.

(b) The resultant force on the charge at the origin is

$$\mathbf{F}_4 = q_4\mathbf{E} = (-5\ \mu\text{C})(3.42\mathbf{i} - 1.68\mathbf{j})\text{x}10^6 \text{ N/C}$$

$$= -17.1\mathbf{i} + 8.4\mathbf{j} \text{ N}$$

EXAMPLE 2
A thin semicircular rod of radius R has a uniform linear charge density λ C/m. Determine the electric field strength at the center.

Solution:
We start with the expression for the electric field due to a charge element,

$$dE = k\ dq/r^2.$$

In this case, the charge element is an infinitesimal length ds of arc with charge $dq = \lambda\ ds$ at the position shown in Fig. 23.2. It is convenient to express the length in terms of the angle θ: $ds = R\ d\theta$. The distance of all elements is fixed at $r = R$, thus,

$$dE_y = dE \sin\theta = k(\lambda R \sin\theta\ d\theta)/R^2$$

From the symmetry of the charge distribution we can say that there will be no net x component of the field.

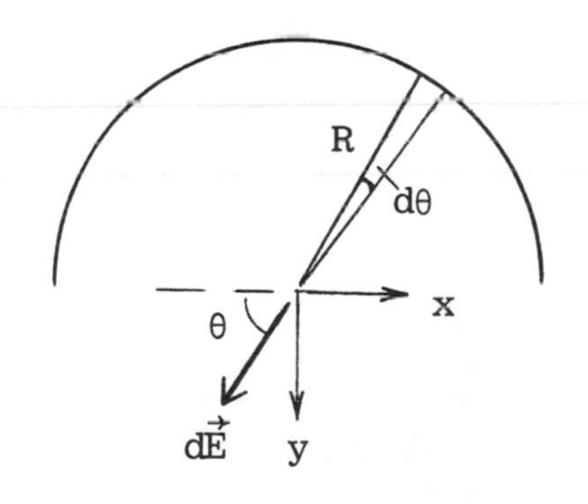

FIGURE 23.2

$$E_y = \int dE_y = (k\lambda/R)\int_0^{\pi} \sin\theta\ d\theta \quad = 2k\lambda/R$$

EXAMPLE 3

Use the expression $E = 2k\lambda/r$ given in Eq. 23.9 in the text for the field due to an infinite line of charge with linear charge density λ, to find the field due to an infinite plate with a surface charge σ C/m^2.

Solution:

We begin by chosing our infinitesimal element to be a line of width dx at a distance x from the origin, see Fig. 23.3. The charge in a length L is $\lambda L = \sigma dA = = \sigma(L\ dx)$, so $\lambda = \sigma dx$.

$$dE = 2k(\sigma\ dx)/r$$

Let us use θ as the variable. Note that $r = b/\cos\theta$, and since $x = b \tan\theta$, $dx = b \sec^2\theta\ d\theta$. Thus,

$$dE_y = 2k\sigma(b \sec^2\theta\ d\theta) \cos^2\theta/b$$

$$= 2k\sigma\ d\theta$$

For the whole plate, limits are $\theta = -\pi/2$ to $\pi/2$. Since $k = 1/4\pi\epsilon_o$,

$$E = \sigma/2\epsilon_o$$

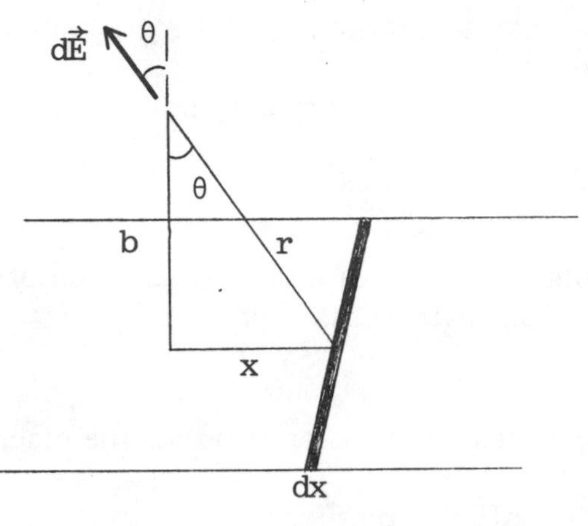

FIGURE 23.3

In Example 23.9 we use circular rings to obtain this result. Also, see Problem 9, where it is suggested that you use x as the variable.

FIELD LINES

Field lines are a useful way of visualizing electric fields. The properties of field lines and the information they provide are discussed in detail in the text.

1. The electric field at a point is directed along the tangent to the field line.
2. The strength of the field is proportional to the density of lines, that is, the number of lines crossing unit area normal to the lines.
3. Electrostatic field lines start on positive charges and end on negative charges.
4. The number of lines that enter or leave a charge is proportional to the magnitude of the charge.

In general, field lines do not represent the paths followed by charges released in the field. (There are two exceptions. Can you think of them?)

CONDUCTORS

If a conductor is in electrostatic equilibrium, then:

(a) The electric field is zero everywhere within the conductor.

(b) The electric field at the surface of the conductor is normal to the surface.

MOTION OF CHARGES IN UNIFORM FIELDS

When a charge q moves in a region where the field is **E** it experiences a force **F** = qE, so its acceleration is

$$\mathbf{a} = q\mathbf{E}/m$$

Note that in the case of an electron q = -e. As Example 23.1 in the text shows, the force of gravity may be ignored when we calculate the motion of charges in electric fields.

EXAMPLE 3

An electron enters the region between two horizontal plates moving at $v_o = 10^6$ m/s at 37° below the horizontal, see Fig. 23.4. The plates are 2 cm long and there is a uniform field E = 10^3 N/C directed vertically downward. Find: (a) the vertical position of the electron as it emerges on the other side of the plates; (b) the angle at which it emerges.

Solution:

The acceleration of the electron is

$$\mathbf{a} = -e\mathbf{E}/m = +\ 1.6 \times 10^{14}\mathbf{j}\ \ \text{m/s}^2$$

The time taken to reach the end of the plates depends on the component of the initial velocity along the x axis, thus

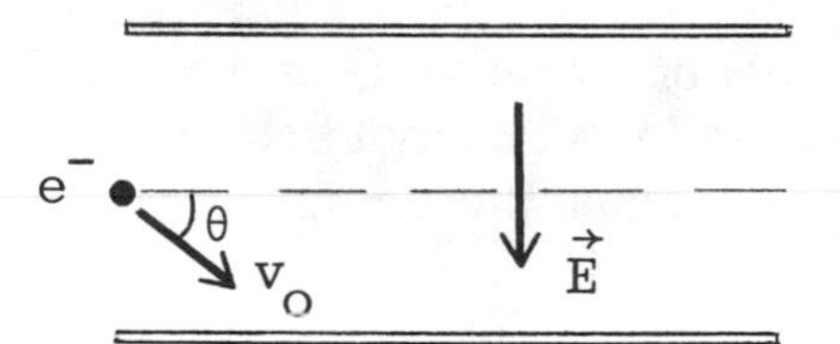

FIGURE 23.4

$$t = L/v_{ox} = L/(v_o \cos\theta)$$

$$= (0.02\text{ m})/(8\text{x}10^5) = 2.5\text{x}10^{-8}\text{ s}$$

(a) If we take the initial vertical coordinate $y_o = 0$, the final vertical coordinate as it emerges is given by

$$y = y_o + v_{oy}\,t + 1/2\ a\ t^2$$

$$= 0 - (v_o \sin\theta)t + (0.5)(1.6\text{x}10^{14}\text{ m/s}^2)(2.5\text{x}10^{-8}\text{ s})^2$$

$$= -0.015 + 0.05 = 0.035\text{ m}$$

Note that we used the vertical component of the initial velocity.

(b) The angle at which the electron emerges is given by the direction of its final velocity. Thus,

$$\tan\theta = v_y/v_x = (-v_o \sin\theta + at)/(v_o \cos\theta)$$

$$= (-6\text{x}10^5 + 4\text{x}10^6)/(8\text{x}10^5) = 4.25$$

Thus θ = 76.8° above the horizontal.

DIPOLES

Two equal and opposite point charges, $\pm Q$ separated by a distance d are said to form an electric dipole. The dipole moment is defined as

$$\mathbf{p} = Q\mathbf{d}$$

The vector **d** is directed from the negative charge to the positive charge.
In an external electric field, a dipole experiences a torque given by

$$\mathbf{\Gamma} = \mathbf{p} \times \mathbf{E}$$

The potential energy of a dipole in an external electric field is

$$U = -\mathbf{p}.\mathbf{E}$$

The minimum value of U occurs when the dipole is aligned along **E**.

EXAMPLE 4

A dipole consists of a charge -Q at the origin and +Q at 3 cm at 30° to the x axis. Take Q = 2 μC. There is a uniform external field E = 120 N/C along the y direction. Find: (a) the dipole moment; (b) the torque on the dipole; and (c) the potential energy of the dipole. See Fig. 23.5.

Solution:

(a) The magnitude of the dipole moment is

$$p = Qd = (2\ \mu C)(3x10^{-2}\ m)$$

$$= 6x10^{-8}\ C.m$$

Since it is directed at 30° to the x axis, the vector is

$$\mathbf{p} = (5.2\mathbf{i} + 3\mathbf{j})x10^{-8}\ C.m$$

(b) The vector torque is

$$\mathbf{\Gamma} = \mathbf{p}x\mathbf{E} = 6.24x10^{-6}\mathbf{k}\ N.m$$

The torque tends to align **p** along **E**.

(c) The potential energy is

$$U = -\mathbf{p}.\mathbf{E} = -pE\cos60° = -3.6x10^{-6}\ J$$

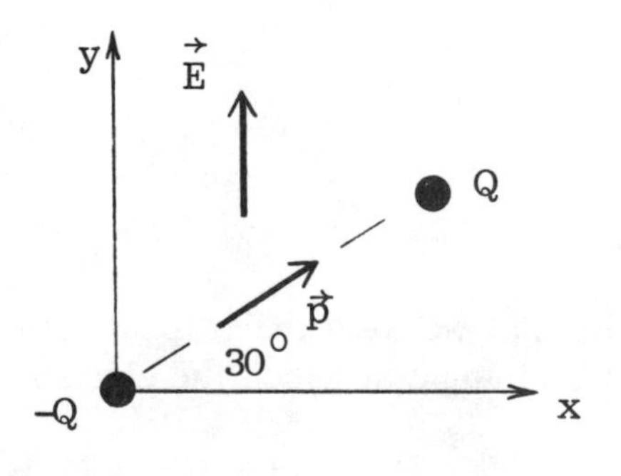

FIGURE 23.5

SOLUTIONS TO SELECTED TEXT EXERCISES AND PROBLEMS

Exercise 13
A point charge Q is at the origin. Show that the components of the electric field at a distance r are given by

$$E_\alpha = kQ\alpha/r3$$

where α = x, y, or z. (Hint: Note that $\mathbf{E} = E_x\,\mathbf{i} + E_y\,\mathbf{j} + E_z\,\mathbf{k} = (kQ/r^3)\,\hat{\mathbf{r}}$. Use the dot product.)

Solution:
The position vector is $\mathbf{r} = x\mathbf{i} + y\mathbf{j} + z\mathbf{k}$. We also know that $\mathbf{E} = kQ\mathbf{r}/r^3$.

$$E_x = \mathbf{E.i} = kQx/r^3; \quad E_y = \mathbf{E.j} = kQy/r^3; \qquad E_z = \mathbf{E.k} = kQz/r^3$$

Exercise 15
A point charge q is at x = 0, while -q is at x = 6 m. Calculate the field strength as a function of x for both positive and negative values of x at 0.5 m intervals. Make a rough sketch of E(x).

Solution:

$(x < 0)$

$$E = kq[-1/x^2 + 1/(6 - x)^2]$$

$(0 < x < 6)$

$$E = kq[1/x^2 + 1/(6 - x)^2]$$

$(x > 6)$

$$E = kq[1/x^2 - 1/(x - 6)^2]$$

A rough sketch is shown in Fig. 23.6.

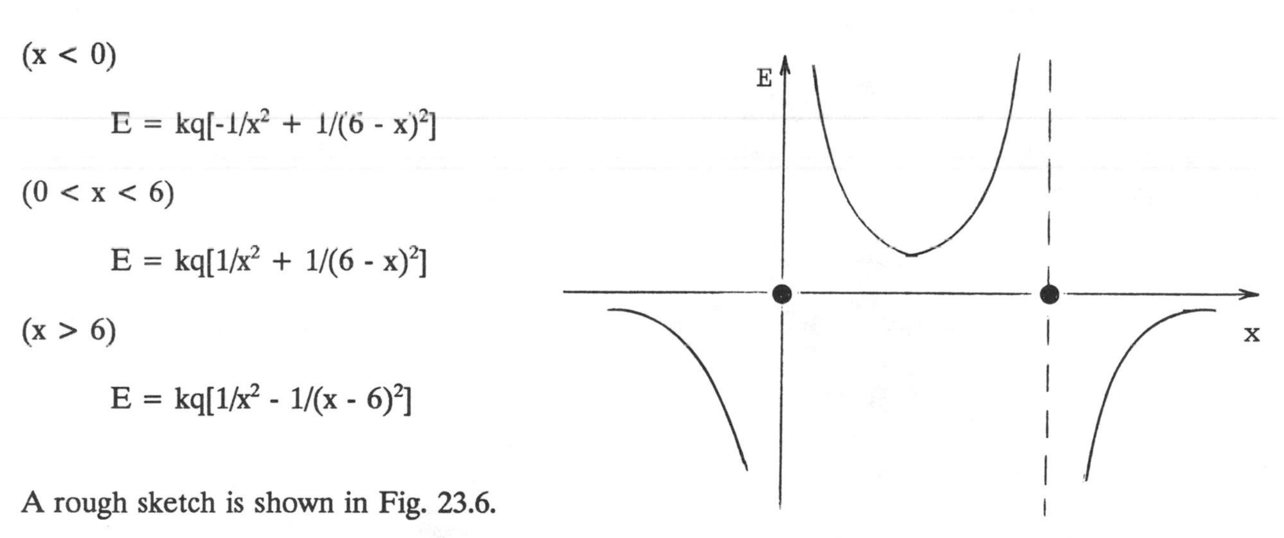

FIGURE 23.6

Exercise 19

Figure 23.7 shows a combination of charges called an electric quadrupole.
Find the electric field strength (a) at point A at (x, 0) and (b) at point B at (0, y). (c) Show that in either case E α $1/r^4$ for r >> a, where r is the distance from the origin. (Hint: Use the binomial expansion $(1 + z)^n \approx 1 + nz$, for small z.)

Solution:

(a) The directions of the fields are shown in Fig. 23.7. Their magnitudes are

$$E_1 = 2kq/x^2$$

$$E_2 = E_3 = kq/(a^2 + x^2)$$

From the symmetry of the system we see that there will be no net y component of the resultant field. The x component is

$$E_x = E_1 - (E_2 + E_3)\cos\theta$$

where $\cos\theta = x/(a^2 + x^2)^{1/2}$. Finally,

$$E_x = 2kq[1/x^2 - x/(a^2 + x^2)^{3/2}]$$

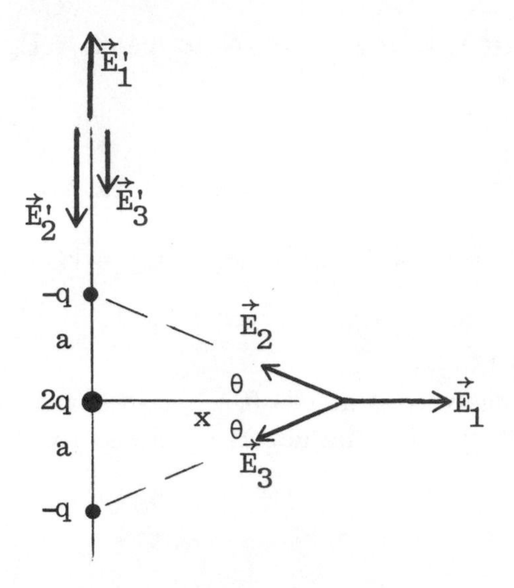

FIGURE 23.7

(b) The directions of the fields are shown in Fig. 23.7. There is clearly no x component. The y component is

$$E_y = 2kq/y^2 - kq/(y - a)^2 - kq/(y + a)^2$$

$$= 2kqa^2(a^2 - 3y^2)/y^2(y^2 - a^2)^2$$

(c) We can rewrite E_x in the form

$$E_x = (2kq/x^2)[1 - (1 + a^2/x^2)^{-3/2}]$$

Using the binomial expansion for a/x << 1, we have

$$(1 + a^2/x^2)^{-3/2} \approx 1 - 3a^2/2x^2 + ...$$

Thus E_x α $1/x^4$. For y >> a, we ignore a in comparison to y, so E_y α $1/y^4$.

Exercise 37
A point charge $q = 2\ \mu C$ is at a distance $d = 20$ cm from a uniformly charged infinite sheet with a surface charge density $\sigma = 20\ \mu C/m^2$. (a) What is the force on the point charge? (b) At what point(s) is the resultant field strength zero?

Solution:
(a) The field due to an infinite sheet is given in Eq. 23.10: $E = \sigma/2\epsilon_o$.

$$F = qE = q(\sigma/2\epsilon_o) = 2.26 \text{ N}$$

(b) The field at a distance r from a point charge is $E = kq/r^2$. The field strength is zero where the fields due to the sheet and the point charge are equal in magnitude and opposite in direction. Since both q and σ are positive, this cancellation can occur only to the left of q. Thus,

$$kq/r^2 = \sigma/2\epsilon_o$$

This leads to r = 12.6 cm to the left of q, as shown in Fig. 23.8.

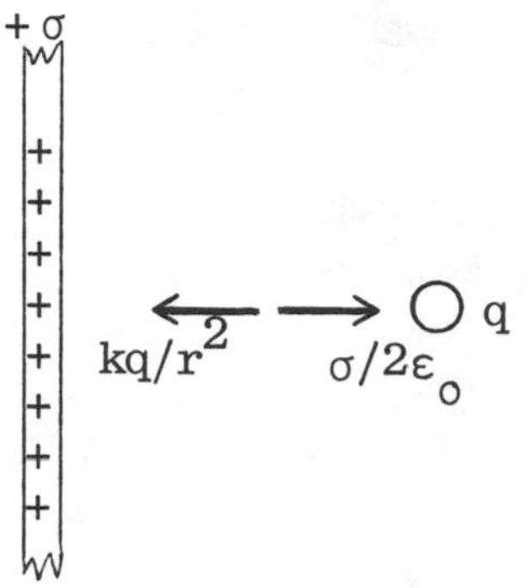

FIGURE 23.8

Problem 7
Show that the field strength at a distance y along the perpendicular bisector of a uniformly charged rod of length L, as in Fig. 23.9 is given by

$$E = 2kQ/y(L^2 + 4y^2)^{1/2}$$

(a) What is the form of this expression when y >> L? (b) What is the form when y << L? (Refer to the table of integrals in Appendix C.)

Solution:
The field due to an infinitesimal element of charge dq is

$$dE = k\ dq/r^2$$

We need to first obtain an expression for the field due to an arbitrary element of the rod. From the symmetry we see that there is no x component of the resultant field. The charge on an infinitiesimal length dx is $dq = \lambda dx$ and the distance $r = (x^2 + y^2)^{1/2}$. The y component of the field due to this element is

$$dE_y = dE\cos\theta = (k\lambda dx)y/r^3$$

where we have used $\cos\theta = y/r$. The total field is given by the integral of dE_y for all elements of the rod.

$$E = k\lambda y \int_{-L/2}^{L/2} dx/(x^2 + y^2)^{3/2}$$

$$= k\lambda y(1/y^2)\,|\,x/(x^2 + y^2)^{1/2}\,|$$

$$= 2kQ/y(L^2 + 4y^2)^{1/2}$$

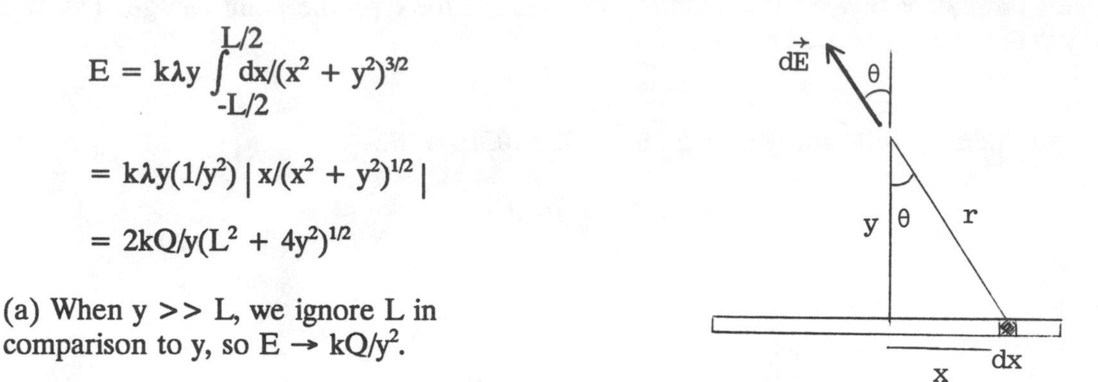

(a) When $y >> L$, we ignore L in comparison to y, so $E \rightarrow kQ/y^2$.

(b) When $y << L$, we ignore y in comparison to L, so $E \rightarrow 2kQ/yL$.

FIGURE 23.9

A more formal way of answering (a) and (b) is to make a binomial expansion in terms of L^2/y^2 for (a) and in terms of y^2/L^2 for (b). In either case retain only the first term. (Try this approach.)

Problem 15

In Millikan's oil drop experiment, the drops are first held motionless; by the application of a uniform field E. Next, the field is switched off and the drops are allowed to fall in air until they reach the terminal speed v_T. The fluid resistance is given by Stokes law, $F = 6\pi\eta r v_T$, where η is the coefficient of viscosity and r is the radius. The condition for fall at the terminal speed is

$$6\pi\eta r v_T = m_{eff} g$$

The effective mass of a drop is $m_{eff} = 4\pi r^3(\rho - \rho_A)/3$, where ρ is the density of the drop and ρ_A is the density of the air-which has a buoyant effect. Show that the charge on a drop is given by

$$q = 18\pi[\eta^3 v_T^3/2(\rho - \rho_A)g]^{1/2}$$

Solution:

Using the given expression for m_{eff}, the condition for the terminal speed is

$$4\pi r^3(\rho - \rho_A)g/3 = 6\pi\eta r v_T$$

so,

$$r^2 = 9\eta v_T/2g(\rho - \rho_A)$$

Next substitute this r into the condition for equilibrium when the field is turned on:

$$m_{eff} g = qE$$

Using the expression for m_{eff} again we obtain the given equation for q.

Problem 17

An electron is fired with an iniital speed $v_o = 3x10^6$ m/s midway between two 4-cm long horizontal plates, as shown in Fig. 23.10. For what initial angle will the electron emerge midway between the plates?

Solution:

The simplest appraoch is to use Eq. (iv) on p. 60 of the text for the range of a projectile when the initial and final positions are the same:

$$R = v_o^2 \sin 2\theta / a$$

where the acceleration is a = eE/m, thus,

$$\sin 2\theta = eER/mv_o^2$$

With R = 0.04 m, we find $\theta = 25.7°$.

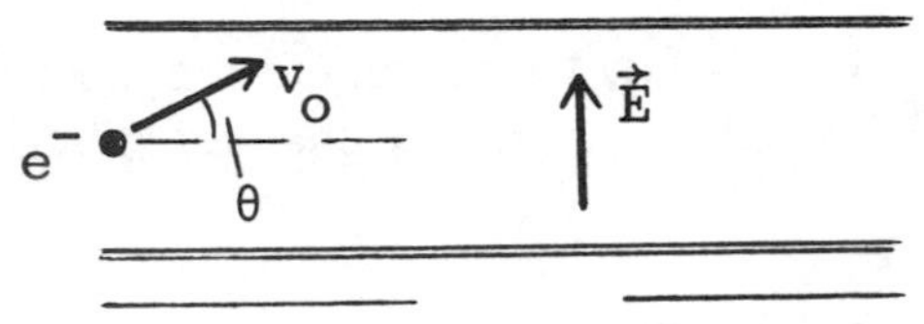

FIGURE 23.10

SELF-TEST

1. Three charges are in the positions shown in Fig. 23.11. Take q = -3 nC, Q_1 = -25 nC, and Q_2 = 40 nC. (a) What is the electric field due to Q_1 and Q_2 at the position of q? (b) If the sign of q is changed, what happens to the field calculated in part (a)?

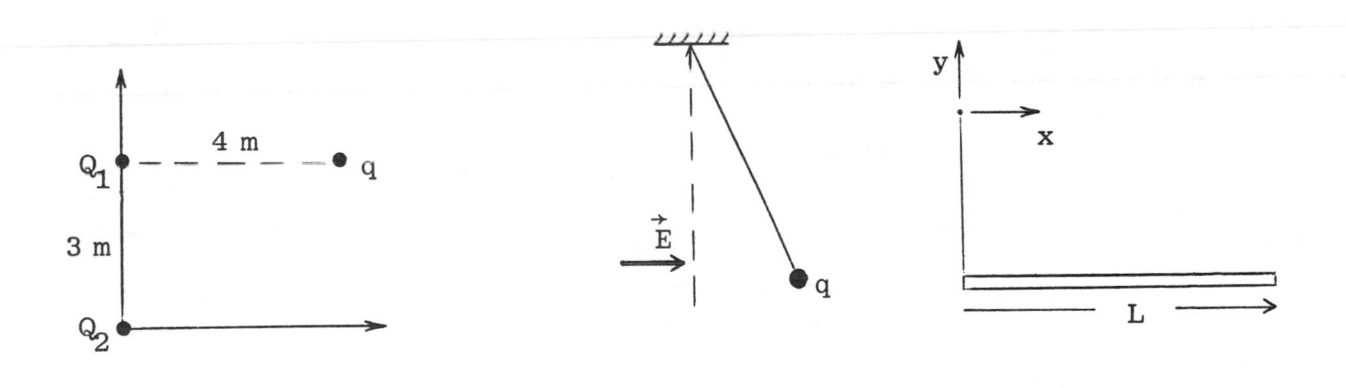

FIGURE 23.11 FIGURE 23.12 FIGURE 23.13

2. A particle of mass m = 1.5 g and charge q = 0.2 μC is suspended by a light string, as shown in Fig. 23.12. There is a uniform horizontal electric field E = $1.2x10^4$ V/m. At what angle to the vertical does the string rest?

3. A rod of length L has a uniform linear charge density λ (= Q/L) C/m. Consider a point that is at a perpendicular distance b from one end, as in Fig. 23.13. Find only the x component of the electric field at the point.

Chapter 24

GAUSS'S LAW

MAJOR POINTS

1. Define electric flux
2. State Gauss's law, which relates the flux through a closed surface to the charge enclosed by the surface.
3. Use Gauss's law to determine the electric field due to symmetrical charge distributions.
4. E = 0 inside a uniformly charged spherical shell

CHAPTER REVIEW

ELECTRIC FLUX

We have seen that the picture of field lines is useful in visualizing electric fields. Gauss imagined field lines as being analogous to streamlines in fluid flow and was able to apply the concept of "flux through a surface" to the electric field.

The electric flux through a plane area A in a uniform electric field **E** is defined as

$$\phi_E = \mathbf{E} \cdot \mathbf{A} = E\,A\cos\theta$$

where θ is the angle between the vectors **E** and **A**. The direction of the vector area **A** is perpendicular to the surface, as shown in Fig. 24.1. In the case of a closed surface, **A** is chosen in the direction of the outward normal. The SI unit of electric flux is $N.m^2/C$.

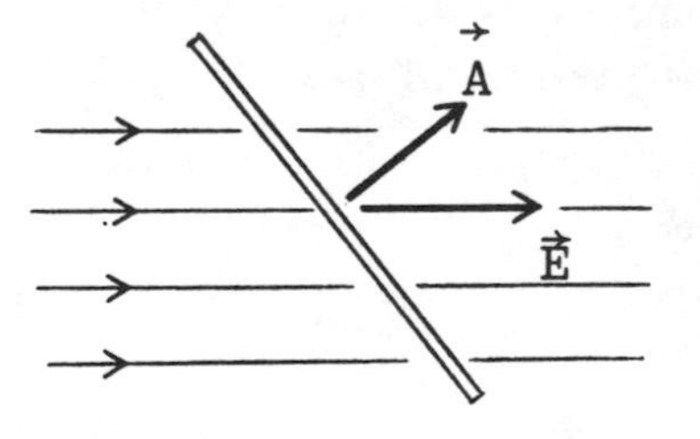

FIGURE 24.1

The concept of electric flux is not based on the use of field lines. However, one can think of the electric flux through a surface as being proportional to the number of lines crossing the surface.

If the field is not uniform or the area not plane, one must use integration to evaluate the flux:

$$\phi_E = \int \mathbf{E} \cdot d\mathbf{A}$$

In general this integral is difficult to perform, however, our problems are restricted to surfaces with planar, cylindrical or spherical symmetry for which **E** is either parallel to, or perpendicular to, d**A**.

GAUSS'S LAW

Gauss's law relates the net electric flux through a closed surface to the net enclosed charge:

$$\oint \mathbf{E} \cdot d\mathbf{A} = Q/\epsilon_0$$

The circle on the integral sign indicates that the integration must be carried out over a closed "Gaussian" surface. The vector **dA** is directed along the (local) outward normal from the closed surface.

HELPFUL HINTS

1. The symmetry of the charge distribution should allow you to specify the direction of the electric field.
2. The shape of the gaussian surface is determined by the symmetry of the field. Specifically, different regions of the surface should be chosen such that
   (a) The (vector) area elements are either parallel to, or perpendicular to, the local **E**. Thus **E.dA** equals E dA, or zero.
   (b) If **E** is parallel to **dA**, the electric field should be constant over the region so that the integral is simple:

$$\int \mathbf{E} \cdot d\mathbf{A} = \int E\, dA = E \int dA = E\, A$$

3. The charge Q is only the charge enclosed by the surface. It may be specified as a volume charge density ρ, a surface charge density σ or a linear charge density, λ. However, charges outside the Gaussian surface also make contributions to the field at a given point (even though their net flux is zero) and it is their presence that often allows us to make the useful simplifications based on symmetry.

EXAMPLE 1

An infinite cylinder of radius R has a nonuniform charge density $\rho = Cr$ where C is a constant. Find the field (a) at points outside the cylinder, and (b) within the cylinder. See Fig. 24.2.

Solution:

The symmetry of the charges distribution prompts use to choose a cylindrical Gaussian surface of radius r and length L. The electric field lines will be directed radially outward, perpendicular to the axis of the cylinder. The field is normal to the flat end faces of the Gaussian surface, so **E.dA** = 0 for S_1 and S_2. Along the curved surface S_3, **E** is parallel to **A** so **E.dA** = E dA. Furthermore, since all points of S_3 are at the same distance from the cylinder E is constant. Gauss's law takes the simple form

$$E\, A_3 = Q/\epsilon_0 \qquad \text{(i)}$$

(a) Outside the cylinder, Q is the total charge within a length L. In order to determine this we consider a thin cylindrical shell of radius x and thickness dx within the charged cylinder. (We do not use the symbol r since this is being used for the Gaussian surface.) The volume of this shell is

$$dV = (2\pi x\ dx)L$$

The charge within this shell is $dq = \rho\ dV = C\ x^2\ dV$ so the total charge within in the cylinder is

$$Q = \int (2\pi CL)\ x^2\ dx = 2\pi CLR^3/3 \quad \text{(ii)}$$

In (i) the area of the cylindrical surface S_3 is $A_3 = (2\pi rL)$. Using this and Q from (ii) in (i) we find

$$E\ (2\pi rL) = 2\pi CLR^3/3\epsilon_o$$

$$(r > R) \qquad E = CR^3/3\epsilon_o r \qquad \text{(iii)}$$

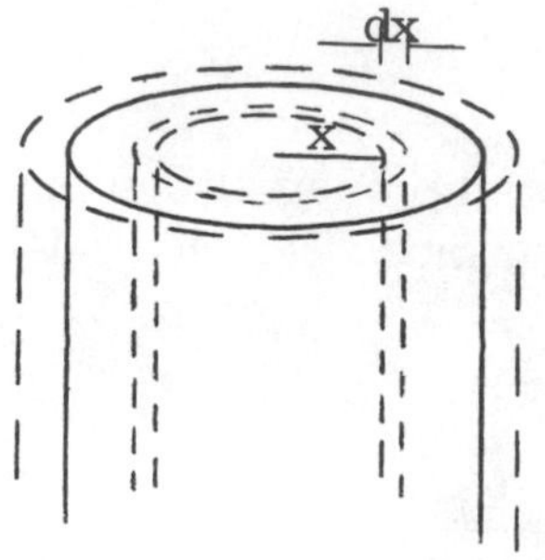

FIGURE 24.2

(b) For points within the cylinder, the Gaussian surface has the same shape. The charge enclosed within the cylindrical Gaussian surface of radius r is given by (ii) with R replaced by r, that is $Q = 2\pi CLr^3/3$. The expression for the area A_3 is the same, so from (i) we have

$$E\ (2\pi rL) = 2\pi CLr^3/3\epsilon_o$$

$$(r < R) \qquad E = Cr^2/3\epsilon_o$$

EXAMPLE 2

Show that the result for E obtained in the last example at points outside the cylinder is consistent with the result $E = 2k\lambda/r$ for an infinite line or cylinder with a linear charge density λ C/m.

Solution:

The linear charge density within the Gaussian surface of length L is

$$\lambda = Q/L = 2\pi CR^3/3.$$

Next we replace $CR^3/3$ in E in (iii) by $\lambda/2\pi$, so

$$E = \lambda/2\pi\epsilon_o r$$

Since $k = 1/4\pi\epsilon_o$, we find, as expected,

$$E = 2k\lambda/r$$

CONDUCTORS

(a) We know that $E = 0$ inside a conductor in electrostatic equilibrium. By drawing a Gaussian surface just inside a conductor, one can deduce that any net charge on a conductor must reside on its surface.

(b) If the surface charge density on the surface of a conductor is σ C/m^2, then the electric field just outside the surface is

$$E = \sigma/\epsilon_o$$

The field is directed along the local normal.

EXAMPLE 3
Two large conducting plates are parallel and have equal and opposite surface charge densities $\pm\sigma$ (Fig. 24.3). Find the field strength between the plates.

Solution:
We treat the plates as being essentially infinite so the field will be perpendicular to the plate. The appropriate Gaussian surface is a "Gaussian pill-box" through the surface of one plate. The base of the box can be square or circular. One flat face of the pill-box must be outside the plate for us to determine the field at that point. The field is normal to any area element on the curved surface, so there is no flux through this portion.

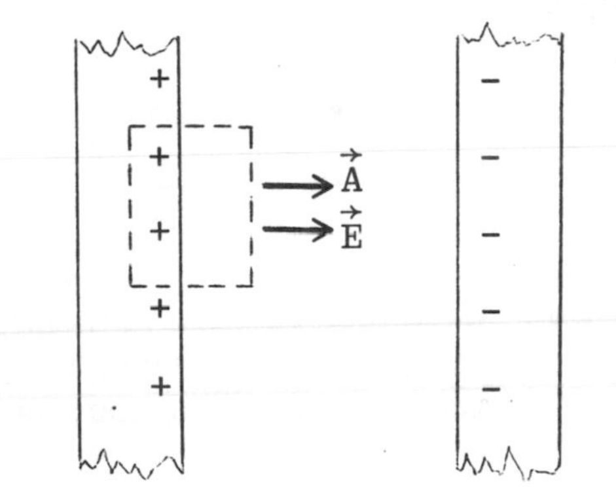

FIGURE 24.3

Because of the attraction between the charges on the two plates, the charges will lie on the inner surfaces, as shown in Fig. 24.3. We know that $E = 0$ within the plate, so this face will contribute nothing to the flux. If A is the area of the flat surface, the charge enclosed is σA. Thus Gauss's law tells us

$$E\,A = (\sigma\,A)/\epsilon_o$$

so $E = \sigma/\epsilon_o$. Note that we used the fact that $E = 0$ inside the plate and the knowledge that the charge appears on the inner faces to obtain this result.
However, these are not essential points. One could simply use the superposition principle to add the contributions of the two plates.

SOLUTIONS TO SELECTED EXERCISES AND PROBLEMS IN THE TEXT

Exercise 17
A long, straight coaxial cable, shown in Fig. 24.4, has an inner wire of radius a with a surface charge density σ_1 and an outer cylindrical shell of radius b with σ_2 C/m^2. Find the relationship between σ_1 and σ_2 for the field strength to be zero outside the cable, that is for $r > b$.

Solution:

The appropriate Gaussian surface is cylindrical in shape with radius r. For a wire of radius R, Gauss's law is

$$(2\pi rL)E = (2\pi RL)\sigma/\epsilon_o$$

thus $E = R\sigma/\epsilon_o r$.

For the resultant electric field to be zero, we need $a\sigma_1 + b\sigma_2 = 0$, that is, $\sigma_2 = -a\sigma_1/b$.

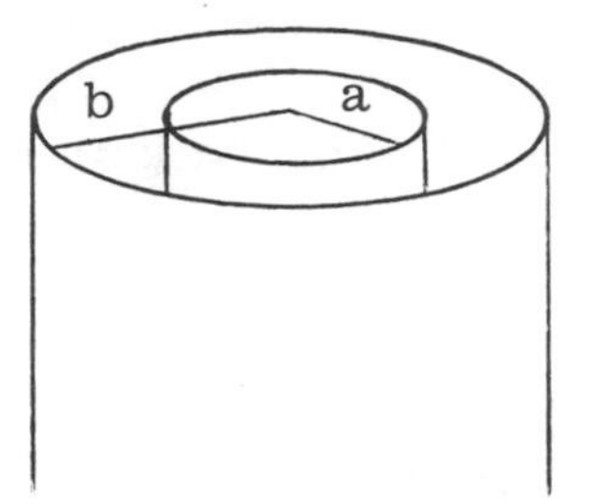

FIGURE 24.4

Exercise 21
A positively charged metal sphere of radius a is at the center of a metal shell of radius b (Fig. 24.5). The spheres carry equal and opposite charges $\pm Q$. Find the electric field as a function of the distance r from the common center for (a) $a < r < b$, and (b) $r > b$.

Solution:

(a) For $a < r < b$, the field due to a uniformly charged sphere is the same as that of a point charge at the center, thus $E = kQ/r^2$.

(b) The net field for $r > b$ is zero.

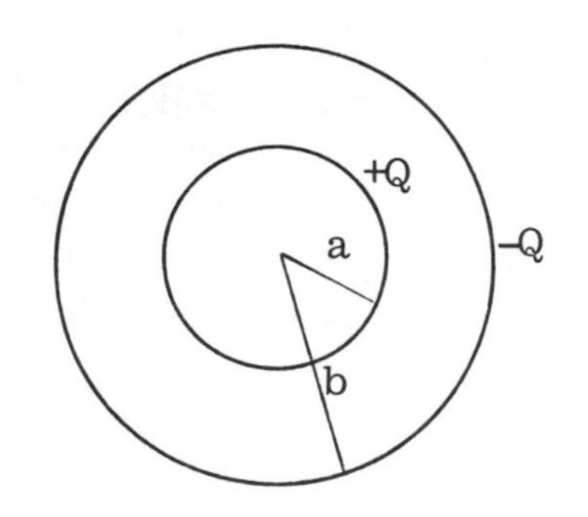

FIGURE 24.5

Problem 3
A conducting spherical shell has inner radius R_1 and outer radius R_2 with uniform charge densities σ C/m^2 on the inner and $-\sigma$ C/m^2 on the outer. (a) What can you say about the charge within the cavity? (b) What can you say about the net charge on the shell? (c) Find the field outside the shell.

Solution:
(a) The inner surface has a total charge $\sigma A_1 = 4\pi\sigma R_1^2$. All the lines from the charge within the cavity must end on the inner surface. Thus the charge within the cavity is $-4\pi\sigma R_1^2$

(b) The net charge on the shell is the difference between the outer and inner charges $-4\pi\sigma(R_2^2 - R_1^2)$

(c) The net charge enclosed by a Gaussian surface outside the sphere is simply the charge on the outer surface of the shell. From Gauss's law

$$4\pi r^2 E = (4\pi R_2^2 \sigma)/\epsilon_o$$

thus $E = \sigma R_2^2/\epsilon_o r^2$ (directed inward)

Problem 7
Consider a hydrogen atom to be a positive point charge e at the center of a uniformly charged sphere of radius R and with total charge -e. Determine the field as a function of the distance r from the nucleus.

Solution:
The field due to the positive point charge is $E_1 = ke/r^2$. To find the field due to the uniformly distributed charge, we consider a Gaussian sphere of radius r. From Gauss's law

$$4\pi r^2 E = \rho 4\pi r^3/3\epsilon_o$$

where $\rho = q/V = -3e/4\pi R^3$ is the charge density. We find

$$E_2 = \rho r/3\epsilon_o = -ker/R^3$$

The total field is therefore

$$E = ke(1/r^2 - r/R^3)$$

Problem 13
An infinite nonconducting slab of thickness t has a uniform charge density ρ C/m^3. Determine the electric field as a function of the distance from the central plane of symmetry.

Solution:

From the symmetry we can say that E will be the same at a given distance x from the central plane of symmetry. We could choose a Gaussian surface with one plane at the center or with both planes at some distance x from the center. With the latter choice, Gauss's law yields

$$2AE = 2Ax\rho/\epsilon_o$$

thus $E = \rho x/\epsilon_o$.

SELF-TEST

1. A conducting sphere of radius R has a charge density 3σ. There is a concentric conducting spherical shell of radius 2R with a charge density -2σ. Determine the field at a distance r from the common center for (a) $R < r < 2R$; (b) $r > 2R$.

2. A cable has a wire with a linear charge density $-\lambda$ along the central axis of a conducting sheath with a linear charge denisty 3λ. Determine the electric field (a) inside, and (b) outside thecable. (c) How is the charge distributed?

Chapter 25

ELECTRIC POTENTIAL

MAJOR POINTS

1. Definition of the scalar quantity electric potential, V
2. Determining the potential at a point due to (a) several point charges, and (b) a continuous distribution of charge.
3. Relating potential difference, ΔV, to the electric field
4. The electrostatic potential energy, U, of a system of charges
5. Obtaining the electric field from a given potential function

CHAPTER REVIEW

ELECTRIC POTENTIAL

The potential difference between two points is defined as the change in potential energy per unit charge:

$$\Delta V = \Delta U/q$$

The SI unit of V is the volt (V), where 1 V = 1 J/C. Electric potential is a scalar quantity, which, like the electric field, depends only on the "source" charges and not on any "test" charge. The external work required to move a charge q between two points without a change in kinetic energy is

(v = constant) $$W_{EXT} = +\Delta U = q(V_f - V_i)$$

If $V_i = 0$, we infer that $V_f = W_{EXT}/q$, which may be expressed as:

> The potential at a point is the external work need to bring a unit positive charge from infinity to that point with no change in kinetic energy.

Since the electrostatic field is conservative, we can relate the change in potential energy between two points to the work done by the (internal) conservative force. From Eq. 8.4 we have $\Delta U = - W_c$. The work done by the conservative electrostatic field on charge q is $W_c = \int q\mathbf{E}.d\mathbf{s}$, so the associated change in potential, $\Delta V = \Delta U/q$, from point A to point B is

$$V_B - V_A = - \int \mathbf{E}.d\mathbf{s}$$

The integration may be evaluated along <u>any</u> convenient path because the electrostaic field is conservative. The change in potential depends only on the initial and final points, not on the path taken. The sign of the integral is determined (a) by the signs of the components of E, and (b) by the direction of the path taken-which is indicated in the limits. The infinitesimal displacement ds is written without any negative sign, for example ds = dx**i** + dy**j**, or

$ds = dr\,\hat{r}$. In the special case of a uniform electric field, we have

$$\text{(Uniform E)} \qquad \Delta \mathbf{V} = -\mathbf{E} \cdot \Delta \mathbf{s}$$

This is often written as $\Delta V = \pm E\ell$, where $+\ell$ or $-\ell$ is the component of the displacement along the direction of **E**.

An equipotential is a surface on which the potential has a given value. It requires no net work to move a charge from one point to another on the same equipotential.

EXAMPLE 1

Given a uniform field $\mathbf{E} = 5\mathbf{i} + 4\mathbf{j}$ V/m, find: (a) The change in potential from point A at (-1 m, -1 m) to point B at (1 m, 2 m) along the path shown in Fig. 25.1; (b) the change in potential energy for a charge q = -0.4 C.

Solution:

(a) Since E is uniform we have

$$\Delta V = V_B - V_A = -\mathbf{E}.\Delta\mathbf{s}$$

where $\Delta s = \Delta x\mathbf{i} + \Delta y\mathbf{j} = 2\mathbf{i} + 3\mathbf{j}$ m, thus,

$$V_B - V_A = -(E_x\Delta x + E_y\Delta y)$$

$$= -22 \text{ V}$$

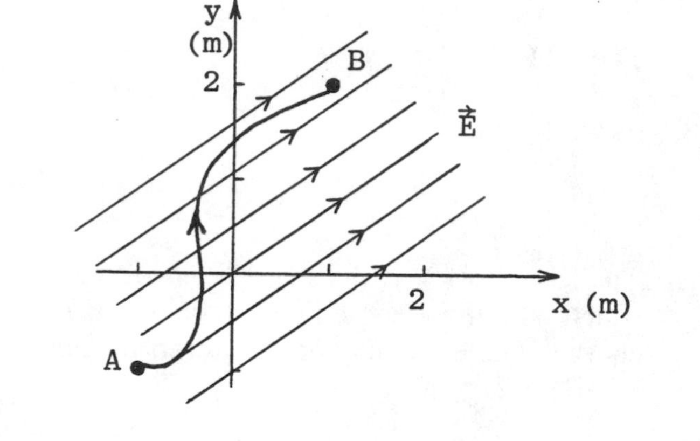

FIGURE 25.1

Note $V_B < V_A$: The potential decreases along the direction of the electric field.

(a) The change in potential energy is $\Delta U = q\Delta V$:

$$U_B - U_A = q(V_B - V_A)$$

$$= (-0.4 \text{ C})(-22 \text{ V}) = +8.8 \text{ J}$$

Since q is negative, its potential energy increases as V decreases.

POINT CHARGES

The potential at a distance r from a point charge Q is given by

$$V = kQ/r$$

Note that $V = 0$ at $r = \infty$. If several charges are present, the total potential is given by the superposition principle:

$$V = \Sigma\, k\, q_i/r_i$$

This is an <u>algebraic</u> sum (not a vector sum) and one must take into account the signs of the charges.

<u>Continuous distribution of charge</u>

The potential due to a continuous distribution of charge may be found using $\Delta V = -\int \mathbf{E} \cdot d\mathbf{s}$ if the field is known or easily determined. One might also make a direct calculation based on the superposition principle:

$$V = \int k\, dq/r$$

POTENTIAL ENERGY

The potential energy of a pair of point charges is

$$U = kqQ/r$$

In finding the potential energy of a system charges, it is convenient to use the form

$$U_{ij} = k\, q_i\, q_j/r_{ij}$$

It is understood that i cannot equal j and that the sum is taken over all distinct pairs. Thus we include either U_{12} or U_{21}, but not both. The potential energy of a system of charges is equal to the external work required to bring the charges to their given positions from an infinite separation.

EXAMPLE 2

A point charge $Q_1 = 4\ \mu C$ is at the origin and charge $Q_2 = -2\ \mu C$ is at $x = 1$ m. (a) Find the total potential at point A (0.5 m, 0) and at B (1 m, 0.75 m). (b) Where is the total potential zero? (c) What is the change in potential energy of a third charge $Q_3 = -3\ \mu C$ when it is moved from A to B? (d) What is the potential energy of the system of three charges when Q_3 is at B? See Fig. 25.2.

Solution:

(a) The total potential at any point is the scalar sum of the individual contributions. At point A, $r_1 = r_2 = 0.5$ m:

$$V_A = kQ_1/r_1 + kQ_2/r_2 = 3.6\text{x}10^4 \text{ V}$$

For point B, $r_1 = (1^2 + 0.75^2)^{1/2} = 1.25$ m, and $r_2 = 0.75$ m, so

$$V_B = kQ_1/r_1 + kQ_2/r_2 = 4.8\text{x}10^3 \text{ V}$$

(b) The total potential can be zero at two points: to the left and to the right of Q_2.

(Left) $Q_1/x = Q_2/(1 - x)$

This leads to x = 2/3 m.

(Right) $Q_1/x = Q_2/(x - 1)$

This leads to x = 2 m.

(c) The change in potential energy in moving from A to B is

$$U_B - U_A = Q_3(V_B - V_A)$$

$$= (-3\ \mu C)(-3.12x10^4\ V)$$

$$= 9.36x10^{-2}\ J$$

FIGURE 25.2

(d) The potential energy of the system of three charges when Q_3 is at B is

$$U = U_{12} + U_{13} + U_{23}$$

$$= (9x10^9)(-8/1 - 12/1.25 + 6/0.75)(10^{-12}) = -8.64x10^{-2}\ J$$

Notice that we count only distinct pairs. A simple interpretation of the negative sign is that the above value represents the external work that must be done to separate the charges and to place them at infinity.

EXAMPLE 3
At a certain distance r from a point charge q the potential is 54 V and the electric field strength is 21.6 V. Determine q and r

Solution:
We know that $V = kq/r$ and that $E = kq/r^2$. Taking the ratio,

$$V/E = r$$

$$= (54\ V)/(21.6\ V/m) = 2.5\ m$$

Since $V = kq/r$, we have

$$q = Vr/k = (54\ V)(2.5\ m)/(9x10^9\ N.m^2/C^2)$$

Thus $q = 15\ \mu C$.

CONSERVATION OF ENERGY

Since the electrostatic field is conservative, one can use the conservation of mechanical energy, $\Delta K + \Delta U = 0$, to deal with the motion of a point charge q in an external field. Since $\Delta U = q\,\Delta V$, we have

$$\Delta K = -q\,\Delta V$$

In this equation the signs of q and ΔV must be explicitly included.

EXAMPLE 4

An electron moves in a uniform field between two equipotential planes at 100 V and 36 V as in Fig. 25.3. Its initial velocity at the 100 V plane is $v_i = 5\text{x}10^6$ m/s in the direction of **E**. What is its final velocity at the other plane?

Solution:

The change in potential is
$\Delta V = 36 - 100 = -64$ V.

$$\Delta K = -q\,\Delta V$$

$$1/2\ mv_f^2 - 1/2\ mv_i^2 = -(-e)(-64\ \text{V})$$

Note carefully the appearance of three negative signs.

$$v_f^2 = v_i^2 - 128e/m$$

We find $v_f = 1.59\text{x}10^6$ m/s.

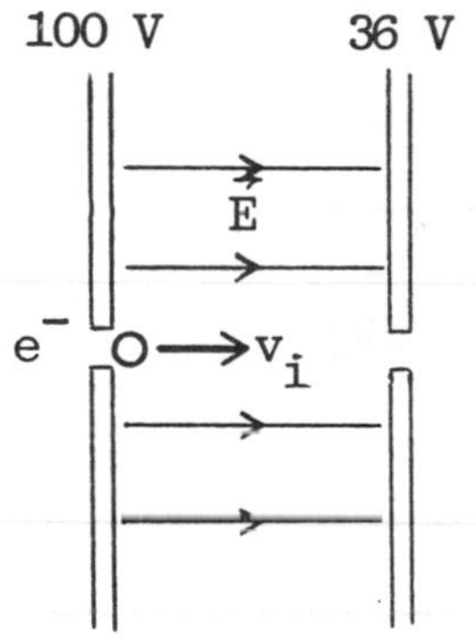

FIGURE 25.3

THE ELECTRIC FIELD DERIVED FROM THE POTENTIAL FUNCTION

If the scalar potential function is known, one can obtain each component of the electrostatic field from it:

$$\mathbf{E} = -\,\partial V/\partial x\ \mathbf{i} - \partial V/\partial y\ \mathbf{j} - \partial V/\partial z\ \mathbf{k}$$

In taking the partial derivative, say with respect to x, the other quantities, such as y and z, are constant. In general, the component of the field in the direction of the displacement ds is

$$E_s = -\,\partial V/\partial s$$

The lines of the electric field are perpendicular to the equipotential surfaces and point from higher potential to lower potential.

EXAMPLE 5
Given the potential function $V(x, y, z) = 3x^2y - 5xyz^2$ V, determine **E**.

Solution:

$$\partial V/\partial x = 6xy - 5yz^2; \quad \partial V/\partial y = 3x^2 - 5xz^2; \quad \partial V/\partial z = -10xyz$$

Thus,

$$\mathbf{E} = (5yz^2 - 6xy)\mathbf{i} + (5xz^2 - 3x^2)\mathbf{j} + 10xyz\mathbf{k}$$

SOLUTIONS TO SELECTED EXERCISES AND PROBLEMS IN THE TEXT

Exercise 5
An electric field is given by $\mathbf{E} = 2x\mathbf{i} - 3y^2\mathbf{j}$ N/C. Find the change in potential from the position $\mathbf{r}_A = \mathbf{i} - 2\mathbf{j}$ m to $\mathbf{r}_B = 2\mathbf{i} + \mathbf{j} + 3\mathbf{k}$ m.

Solution:
The change in potential is

$$V_B - V_A = -\int E_x\,dx - \int E_y\,dy$$

$$= -|x^2| - |y^3| = +6 \text{ V}$$

Exercise 17
What is the work required to move a particle of mass 2×10^{-2} g and a charge of -15 μC through a change in potential of -6000 V and also to increase its speed from zero to 400 m/s?

Solution:
The external work required is

$$W_{EXT} = \Delta K + \Delta U$$

$$= 1/2\ mv^2 + q(-6000)$$

$$= 1.6 \text{ J} + 0.09 \text{ J} = 1.69 \text{ J}$$

Exercise 31
Three point charges, $q_1 = 6\ \mu C$, $q_2 = -2\ \mu C$, and q_3 are located as shown in Fig. 25.4. For what value of q_3 will the potential at the origin be: (a) 0 V; (b) -400 kV?

Solution:
(a) The total potential is the sum

$$V = kq_1/r_1 + kq_2/r_2 + kq_3/r_3$$

Thus,

$$V = k(6\ \mu C/0.03\ m - 2\ \mu C/0.025\ m + q_3/0.025\ m)$$

Since $V = 0$ we find, $q_3 = -3\ \mu C$.

(b)
$$V = k(6\ \mu C/0.03\ m - 2\ \mu C/0.025\ m + q_3/0.025\ m)$$

Since $V = -4 \times 10^5$ V, we find, $q_3 = -4.11\ \mu C$

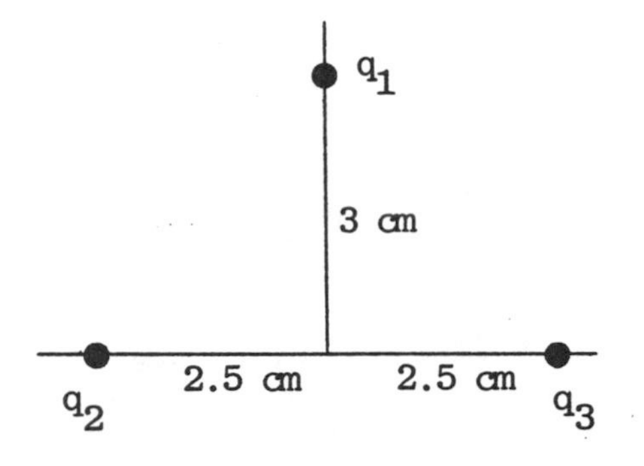

FIGURE 25.4

Exercise 35
A uniform electric field of 400 V/m is directed at 37° below the x axis, as shown in Fig. 25.5. Find the changes in potential: (a) $V_B - V_A$; (b) $V_B - V_C$.

Solution:
The field is

$$\mathbf{E} = E\cos 37°\mathbf{i} - E\sin 37°\mathbf{j}$$

$$= 319\mathbf{i} - 241\mathbf{j}\ \text{V/m}$$

Thus,

$$\Delta V = -\mathbf{E}.\Delta\mathbf{s} = -(319\mathbf{i} - 241\mathbf{j}).\Delta\mathbf{s}$$

(a) $\Delta\mathbf{s} = 0.03\mathbf{i}$ m, so $\Delta V = -9.57$ V;

(b) $\Delta\mathbf{s} = -0.03\mathbf{j}$ m, so $\Delta V = -7.23$ V

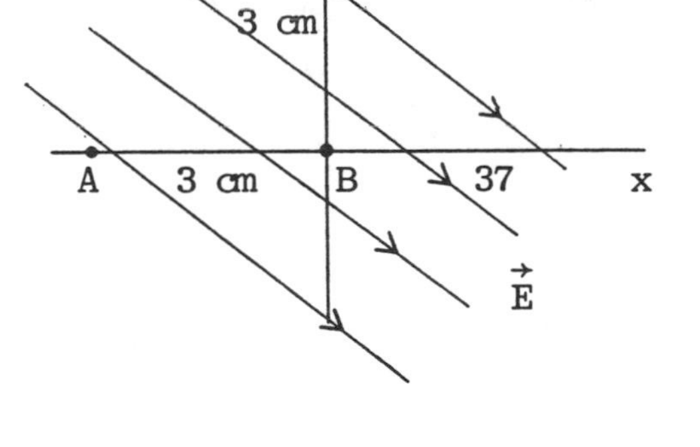

FIGURE 25.5

Exercise 43
Two equal positive charges Q are located at (0, a) and (0, -a), respectively. (a) Find the potential V(y) at a point (0, y) for y > a. (b) Use V(y) to find the electric field along the y axis.

Solution:
(a) The potential at P in Fig. 25.6 is

$$V = kQ[1/(y - a) + 1/(y + a)]$$

$$= 2kQy/(y^2 - a^2)\ .$$

(b) The field is given by

$$E_y = -\partial V/\partial y$$

$$= kQ[1/(y - a)^2 + 1/(y + a)^2]$$

$$= 2kQ(y^2 + a^2)/(y^2 - a^2)^2$$

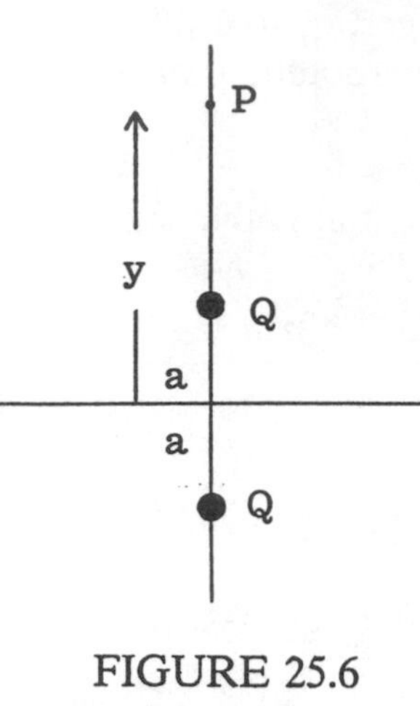

FIGURE 25.6

Problem 5
A metal sphere of radius R_1 has a charge Q_1. It is enclosed by a conducting spherical shell of radius R_2 that has a charge $-Q_2$; see Fig. 25.7. Determine: (a) the potential V_1 of the inner sphere; (b) the potential V_2 of the outer sphere; (c) the potential difference $V_1 - V_2$. (d) Under what condition is $V_1 = V_2$?

Solution:
The potential within a uniformly charged metal shell is equal to that at the surface, $V = kQ/R$. The net potential of each sphere must include the contribution from the other sphere. Outside a uniformly charged spherical shell, the potential is the same as that of a point charge at the center of the shell.

Keeping in mind these points, we find for Fig. 25.7:

(a) $V_1 = k(Q_1/R_1 - Q_2/R_2)$

(b) $V_2 = k(Q_1 - Q_2)/R_2$

(c) $V_1 - V_2 = kQ_1(1/R_1 - 1/R_2)$

(d) $Q_1 = 0$

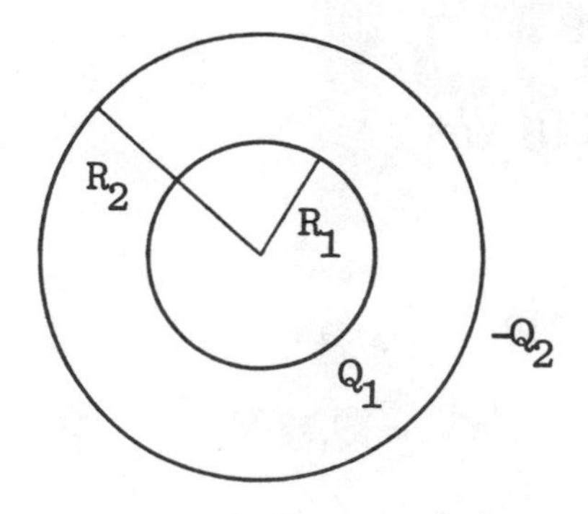

FIGURE 25.7

Problem 9
A rod of length L with a uniform linear charge density λ C/m lies along the x axis. Find he potential at a distance y from one end, perpendicular to the rod; see Fig. 25.8. The length of the rod is L. (See the table of integrals in Appendix C.)

Solution:
The rod must be divided up into infinitesimal elements of length dx. The contribution to the potential from such an element with charge $dq = \lambda\, dx$ is

$$dV = k\, dq/r$$

$$= k(\lambda dx)/(y^2 + x^2)^{1/2}$$

Using the table of integrals in Appendix C, we find (the integral is from $x = 0$ to $x = L$)

$$V = k\lambda \ln[1 + (1 + y^2/L^2)^{1/2}]$$

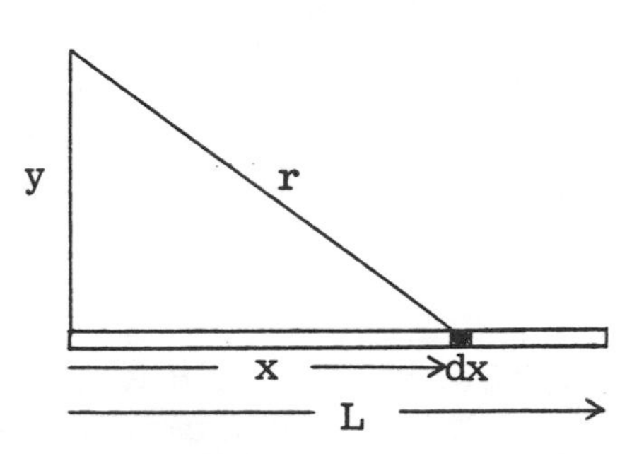

FIGURE 25.8

SELF TEST

1. Two equal positive charges Q are located at $y = \pm a$.
   (a) Find V(x), the potential at a point (x, 0).
   (b) From V(x) found in part (a), obtain E_x, the x component of the field at (x, 0).

2. An electron moves from a plate at which the potential is 20 V to another at which the potential is 6 V. If its initial speed is 4×10^6 m/s, what is its final speed?

3. Find the potential at a distance y from the center along the axis of a uniformly charged ring of radius a with total charge Q.

Chapter 26

CAPACITORS

MAJOR POINTS

1. The definition of the capacitance of a capacitor
2. Determining the equivalent capacitance of series and parallel combinations of capacitors.
3. (a) The energy stored in a capacitor
   (b) The energy density of the electric field
4. The effects of inserting a dielectric into a capacitor

CHAPTER REVIEW

A capacitor is a device that stores charge or electrical energy. It consists of two conductors, called the plates, usually with equal and opposite charges, separated by some distance. The capacitance C of a capacitor is defined as the ratio of the magnitude of the charge Q on one plate divided by the potential difference V between the plates:

$$C = Q/V$$

The SI unit of capacitance is the Farad (F), where 1 F = 1 C/V. The capacitance of a device depends on its size, shape and the medium between the plates; it is independent of Q or V.

<u>Parallel plate capacitor</u>

A frequently encountered geometry is that of two parallel plates of area A separated by some distance d. If the medium filling the space between the plates is air, the capacitance of this arrangement is

$$C = \epsilon_o A/d$$

<u>Series and Parallel Connections</u>

When two or more capacitors are connected in series (one-after the other), as in Fig. 26.1a, the equivalent capacitance of the combination is given by

$$1/C_{eq} = 1/C_1 + 1/C_2 + 1/C_3 + ...$$

The magnitude of the charge on each plate of the capacitors in series is the same. When capacitors are connected in parallel, as in Fig. 26.1b; the equivalent capacitance of the combination is

$$C_{eq} = C_1 + C_2 + C_3 + ...$$

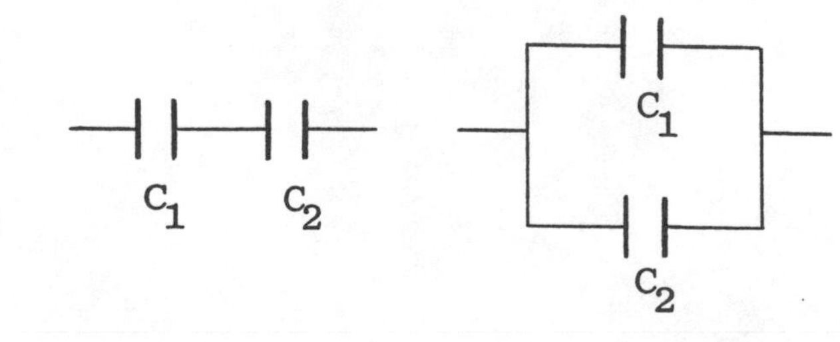

FIGURE 26.1

Capacitors in parallel have the same potential difference across their plates.

EXAMPLE 1
Four capacitors are connected to a 48 V battery as shown in Fig. 26.2. Find (a) the equivalent capacitance of the combination; (b) the potential difference across C_2; (c) the charge on C_3.

Solution:
We start with the simplest series or parallel combinations.

$$1/C_{12} = 1/C_1 + 1/C_2$$

leads to $C_{12} = 2\ \mu F$. This in parallel with C_3 gives $C_{123} = 12\ \mu F$. Finally, this in series with C_4 means that

$$1/C_{eq} = 1/C_{123} + 1/C_4 = 1/12 + 1/8$$

Thus, $C_{eq} = 4.8\ \mu F$.

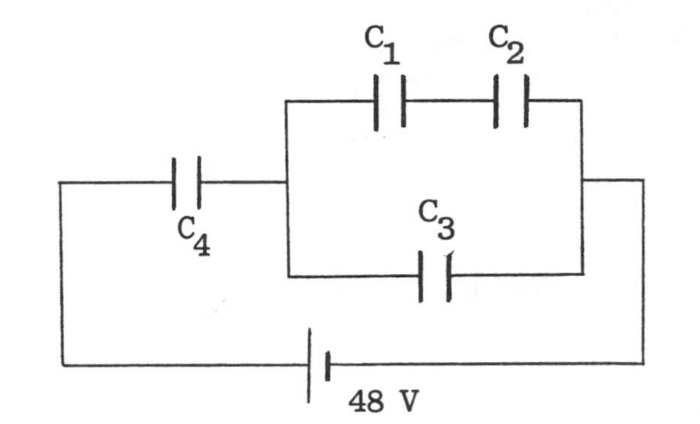

(b) The charge on C_{eq} is

$$Q = C_{eq}V = (4.8\ \mu F)(48\ V) = 230\ \mu C$$

FIGURE 26.2

This is equal to Q_4 and also to $(Q_1 + Q_3)$, or $(Q_2 + Q_3)$ since $Q_2 = Q_1$.

$$Q_1 + Q_3 = 230\ \mu C \qquad \text{(i)}$$

To determine the individual charges we note that the potential difference across C_{12} is equal to that across C_3: $Q_3/C_3 = Q_1/C_{12}$, or

$$Q_3 = 5Q_1 \qquad \text{(ii)}$$

From (i) and (ii) we find $Q_1 = 38.4\ \mu C = Q_2$ and $Q_3 = 191.6\ \mu C$. The potential difference across C_2 is $V_2 = Q_2/C_2 = 6.38$ V.

(c) In part (b) we found $Q_3 = 191.6\ \mu C$.

ENERGY STORED

The agent, such as a battery, responsible for charging a capacitor does working in separating the charges. This work is stored as potential energy in the capacitor. The energy stored in a capacitor is given by

$$U = 1/2\ CV^2 = Q^2/2C = 1/2\ QV$$

Note that U is not qV because both the charge and the potential difference built up gradually to their final values q and V. One can regard electrical energy as being stored in the electric field. The energy density (J/m^3) in a field E is given by

$$u_E = 1/2\ \epsilon_o E^2$$

DIELECTRICS

When an insulator is inserted to completly fill the space between the plates of a capacitor, the capacitance increases by a factor κ called the dielectric constant:

$$C = \kappa C_o$$

where C_o is the capacitance in vacuum (or to a good approximation air). If E_o is the field stength in vacuum, then the field strength in the dielectric is

$$E_D = E_o/\kappa$$

EXAMPLE 2

An air-filled capacitor has two parallel plates, each with an area 16 cm^2, separated by 3 mm. A 24-V potential difference is applied to the capacitor. Determine (a) the capacitance; (b) the electric field between the plates; (c) the surface charge density on each plate; (d) the energy stored, (e) the energy density. (f) If mica fills the space between the plates, what is the energy stored (the battery stays connected)?

Solution:

(a) The capacitance of a parallel plate

$$C = \epsilon_o A/d$$

$$= (8.85\text{x}10^{-12}\ F/m)(16\text{x}10^{-4}\ m^2)/(3\text{x}10^{-3}\ m) = 4.72\text{x}10^{-12}\ F$$

(b) Since the field is uniform, $E = V/d = 8000\ V/m$

(c) The charge density is

$$\sigma = Q/A = CV/A$$

$$= (4.72\text{x}10^{-12}\ F)(24\ V)/(16\text{x}10^{-4}\ m^2) = 7.08\text{x}10\text{-}8\ C/m^2$$

(d) The stored energy is

$$U = 1/2\ CV^2 = 1/2\ (4.72\text{x}10^{-12}\ F)(24\ V)^2 = 1.36\text{x}10^{-9}\ J$$

(e) The energy density of the electric field is

$$u_E = 1/2\ \epsilon_o\ E^2$$

$$= 1/2\ (8.85\text{x}10^{-12}\ \text{F/m})(8\text{x}10^3\ \text{V/m})^2 = 2.83\text{x}10^{-4}\ \text{J/m}^3$$

Equivalently we could have divided the total energy U by the volume fo the capacitor, Ad.

(f) In the presence of a dielectric the capacitance is $C = \kappa C_o$ where C_o is the capacitance with vacuum between the plates.

$$U = 1/2\ CV^2 = \kappa U_o$$

$$= 6.8\text{x}10^{-9}\ \text{J}$$

EXAMPLE 3

Two capacitors, $C_1 = 2.5\ \mu F$ and $C_2 = 4.5\ \mu F$, are connected in series as shown in Fig. 26.3a. The 20 V battery is removed and the plates of like sign are connected together. For each capacitor, determine (a) the initial charge and (b) the final charge.

Solution:

(a) The equivalent capacitance is given by

$$1/C_{eq} = 1/C_1 + 1/C_2$$

which leads to, $C_{eq} = 1.6\ \mu F$. The initial charge on each capacitor is $Q_1 = Q_2 = 32\ \mu C$.

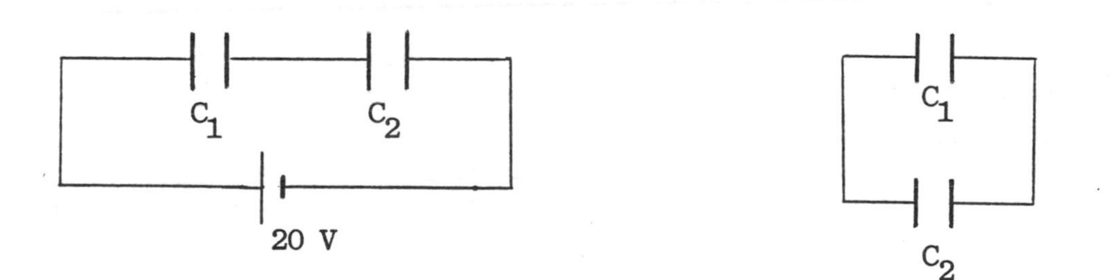

FIGURE 26.3

(b) When they are reconnected, the total charge is shared but their potential differences are equal. Thus,

$$Q_1' + Q_2' = 64\ \mu C \qquad \text{(i)}$$

$$Q_1'/C_1 = Q_2'/C_2 \qquad \text{(ii)}$$

We substitute $Q_2' = 1.8Q_1'$ from (ii) into (i) to obtain $Q_1' = 22.9\ \mu C$, then $Q_2' = 41.1\ \mu C$.

SOLUTIONS TO SELECTED TEXT EXERCISES AND PROBLEMS

Exercise 5

Assume that the earth (radius 6400 km) is surrounded by a conducting sphere 50 km above the surface and that there is a constant electric field of 100 N/C directed vertically downward. (a) What is the surface charge density on the earth's surface? (b) What is the capacitance of the system? (c) Compare the answer to part (b) with the capacitiance of the earth treated as a n isolated sphere.

Solution:

(a) The electric field due to a uniformly charged sphere of radius R is kQ/R^2. With $Q = \sigma A = \sigma(4\pi R^2)$ and $k = 1/4\pi\epsilon_o$, we have $E = \sigma/\epsilon_o$ so

$$\sigma = \epsilon_o E = 8.85 \text{ x } 10^{-10} \text{ C/m}^2$$

(b) From Eq. 26.5 the capacitance of a spherical capacitor is

$$C = R_1R_2/k(R_2 - R_1) = 91.7 \text{ mF}$$

(c) From Eq. 26.4, the capacitance of an isolated sphere is $C = 4\pi\epsilon_o R$.
For the earth we find $C = 712\ \mu F$.

Exercise 15

The three capacitors in Fig. 26.4 have an equivalent capacitance of 12.4 μF. Find C_1.

Solution:

The equivalent capacitance of C_1 and 4 μF is $(1/4 + 1/C_1)^{-1}$. This is in parallel with 10 μF. Therefore,

$$12.4 = 10 + (1/4 + 1/C_1)^{-1}$$

We find

$$1/C_1 = 1/4 - 1/2.4$$

so, $C_1 = 6\ \mu F$.

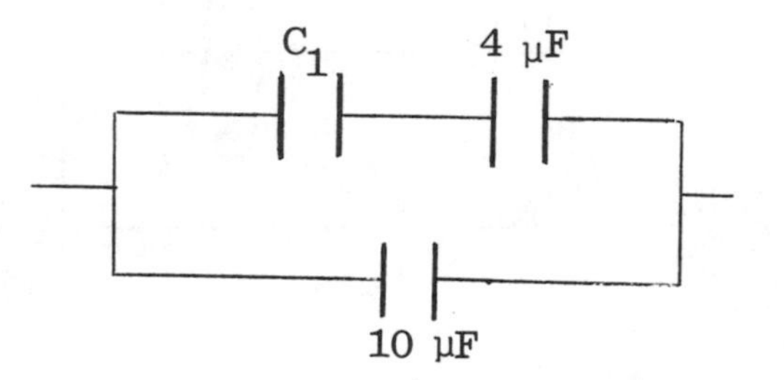

FIGURE 26.4

Exercise 35

A parallel-plate capacitor with a plate separation d is initially connected to a battery with a potential difference V. The battery is then removed. The plates are pulled apart till the separation is 2d. What is the change in each of the following quantities: (a) the potential difference; (b) the charge on each plate; (c) the energy stored in the capacitor.

Solution:

(a) For a uniform field, the potential difference is $V_i = Ed$. Thus, $V_f = E(2d) = 2V_i$.

(b) Since the battery has been removed there is no change in the charge on each plate.

(c) Since $d' = 2d$, the new capacitance is $C' = 0.5C$. The new energy is $U' = 1/2\ C'\ V'^2 = 1/2\ (0.5C)(2V)^2 = CV^2 = 2U$. The stored energy doubles. The energy comes from the work done to separate the plates.

Exercise 41

The space between the plates of a parallel-plate capacitor is filled with two dielectrics of equal size, as shown in Fig. 26.5. What is the resulting capacitance in terms of κ_1, κ_2, and C_o, the capacitance with a vacuum between the plates?

Solution:

We can treat the system as two capacitors in parallel. In the presence of a dielectric $C = \kappa C_o$. Since each dielectric fills only half of the area, we have

$$C = C_o(\kappa_1 + \kappa_2)/2$$

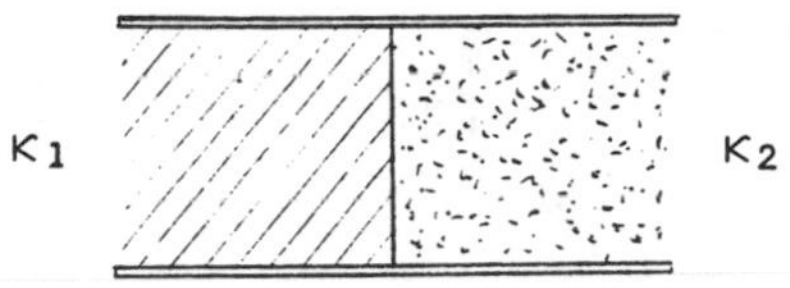

κ_1 κ_2

FIGURE 26.5

Problem 3

A parallel-plate capacitor has plates of area A separated by a distance d and is connected to a battery with a potential difference V. A metal block of thickness ℓ is midway between the plates, as shown in Fig. 26.6. What is the work required to remove the block given that the battery is disconnected before the block is removed?

Solution:

The electric field in the spaces between the plates and the block is $E = \sigma/\epsilon_o$. The potential difference is $V = E(d - \ell)$ since the potential difference across the metal block is zero. The capacitance of the system is then

$$C = \sigma A/V = \epsilon_o A/(d - \ell)$$

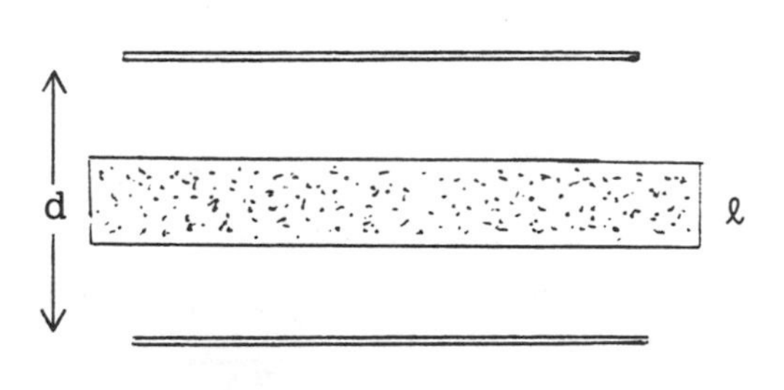

The initial energy stored is

$$U_1 = 1/2\ CV^2 = \epsilon_o AV^2/2(d - \ell)$$

FIGURE 26.6

When the battery is removed, the charges on the plates remain, and so the electric field is unchanged. When the block is removed the potential difference between the plates increases to

$V' = Ed = dV/(d - \ell)$. The capacitance is simply $C' = \epsilon_o A/d$. Thus the stored energy is

$$U_2 = 1/2\, C'V'^2 = 1/2\, (\epsilon_o A/d)[dV/(d - \ell)]^2$$

$$= \epsilon_o A d V^2/2(d - \ell)^2$$

The change in energy associated with the removal of the block is equal to the work required:

$$W = \Delta U = \epsilon_o A \ell V^2/2(d - \ell)^2$$

Problem 11

A spherical capacitor consists of concentric spheres of radii R_1 and R_2. (a) Show that when $R_2 - R_1 << R_2$, the capacitance becomes that of a parallel-plate capacitor. (b) Show that in the appropriate limit the capacitance of the spherical capacitor reduces to that of an isolated sphere.

Solution:

(a) From Eq. 26.5 the capacitance of a spherical capacitor is

$$C = R_1R_2/k(R_2 - R_1)$$

If $R_2 >> (R_2 - R_1) = d$, then $R_1R_2 \simeq R^2$, and $C = 4\pi\epsilon_o R^2/d = \epsilon_o A/d$. This is the capacitance of a parallel plate capacitor.

(b) When $R_2 >> R_1$, $R_2/(R_2 - R_1) \simeq 1$, thus $C = R_1/k = 4\pi\epsilon_o R$. This the capacitance of an isolated sphere.

SELF-TEST

1. Consider Fig. 26.7 where $C_1 = 4\ \mu F$, $C_2 = 2.4\ \mu F$, $C_3 = 5\ \mu F$ and $\mathcal{E} = 15$ V. What is the charge and potential difference for each capacitor.

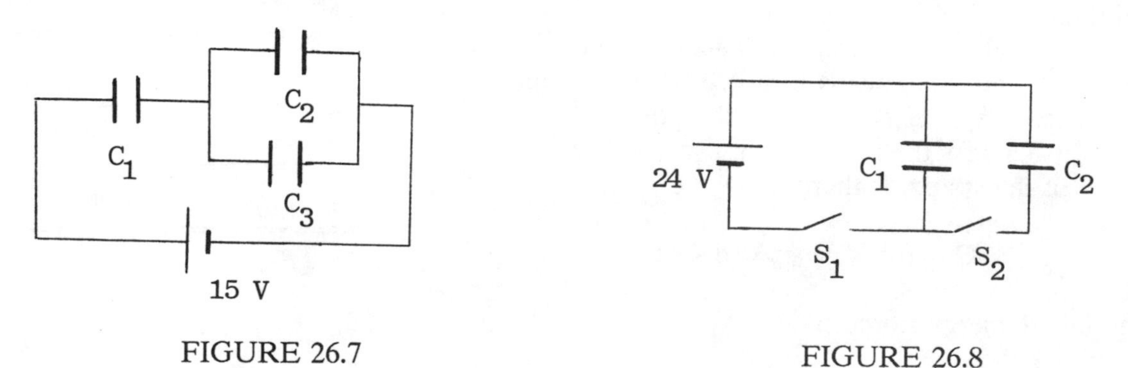

FIGURE 26.7

FIGURE 26.8

2. Capacitors $C_1 = 5\ \mu F$ and $C_2 = 3\ \mu F$ are connected, as in Fig. 26.8. Initially switch S_1 is closed and S_2 is open. Then S_1 is opened and S_2 is closed. For each find the final (a) charge and (b) stored energy.

Chapter 27

CURRENT AND RESISTANCE

MAJOR POINTS

1. Definitions of current and current density
2. Definition of resistivity and resistance
3. Temperature dependence of resistivity
4. Ohm's law
5. Electrical power

CHAPTER REVIEW

CURRENT

An electrical current is defined as the rate of flow of charge through a surface

$$I = dQ/dt$$

The SI unit of current is the ampere, where 1 A = 1 C/s. The direction of the current is the direction in which positive charges flow. In a metal, the electrons flow opposite to the direction of the current.

CURRENT DENSITY AND RESISTIVITY

The magnitude of the (average) current density J is the current flowing per unit area normal to the flow of charge:

$$J = I/A$$

If the number of charge carriers per unit volume is n and each carries a charge q, we find

$$\mathbf{J} = n\, q\, \mathbf{v}_d$$

where $\mathbf{v}_d$ is the drift velocity of the charges. The drift velocity (typically 10^{-4} m/s) is superimposed on the random thermal velocities (typically 10^6 m/s) of the conduction electrons in a metal. Whereas I is a scalar measured on a macroscopic scale, J is a vector that can vary from point to point.

The current density in a material is proportional to the electric field

$$\mathbf{J} = \sigma\, \mathbf{E} = (1/\rho)\, \mathbf{E}$$

where σ, measured in Ω.m, is the resistivity, and ρ, is the conductivity. Resistivity and conductivity are properties of a medium.

EXAMPLE 1

A current of 1.5 A flows in an copper wire of radius 0.5 mm. Find: (a) the current density; (b) the drift speed; (c) the electric field within the wire. Take the free electron density to be $8.4\text{x}10^{28}$ m^{-3}.

Solution:

(a) The current density is

$$J = I/A = I/\pi r^2 = 1.91\text{x}10^6 \text{ A/m}^2$$

(b) Since $J = nev_d$, the drift speed is

$$v_d = J/ne = 1.42\text{x}10^{-4} \text{ m/s}$$

(c) The electric field is

$$E = \rho J = (1.7\text{x}10^{-8} \ \Omega.\text{m})(1.91\text{x}10^6 \text{ A/m}^2) = 3.25\text{x}10^{-2} \text{ V/m}$$

RESISTANCE

If a current I flows when a potential difference V is applied across a sample, its resistance is defined to be

$$R = V/I$$

This holds for <u>any</u> material since it is merely a definition. For a sample in the form of a wire of length L and cross-sectional area A, the resistance is

$$R = \rho L/A$$

The resistivity ρ, and hence also R, depend on temperature:

$$\rho = \rho_o[1 + \alpha(T - T_o)]$$

$$R = R_o[1 + \alpha(T - T_o)]$$

where α is the temperature coefficient of resistivity. The values ρ_o and R_o are the values at the "reference" temperature T_o. Note that resistivity is a characteristic of a material, whereas resistance is a property of a particular sample of the material.

OHM'S LAW

According to Ohm's law, the current flowing through a conductor is directly proportional to the potential difference across it. This relation is usually written in the form

$$V = IR$$

where R is a constant, independent of V or I. (The relation $\mathbf{J} = \mathbf{E}/\rho$ is the microscopic form of Ohm's law, if ρ is constant.)

POWER

When charges move through a potential difference V the power required is given by

$$P = IV$$

For a conductor with a resistance R, the power dissipated as heat may also be expressed in the forms

$$P = I^2 R = V^2/R$$

EXAMPLE 2

A copper wire has a radius of 0.8 mm and a length of 14 m. It carries a 6 A current. Find: (a) the potential difference between the ends of the wire; (b) the resistance of the wire; (c) the power dissipated in the wire.

Solution:

(a) The current density is

$$J = I/\pi r^2$$

$$= (6\ \mathrm{A})/\pi(8\mathrm{x}10^{-4}\ \mathrm{m})^2 = 2.98\mathrm{x}10^6\ \mathrm{A/m^2}$$

From the microscopic form of Ohm's law

$$E = \rho J$$

$$= (1.7\mathrm{x}10^{-8}\ \Omega.\mathrm{m})(2.98\mathrm{x}10^6\ \mathrm{A/m^2}) = 5.07\mathrm{x}10^{-2}\ \mathrm{V/m}$$

For a uniform electric field the potential difference is

$$V = E\ell = (5.07\mathrm{x}10^{-2}\ \mathrm{V/m})(14\ \mathrm{m}) = 0.71\ \mathrm{V}$$

(b) The resistance of the length of wire of length L and area $A = \pi r^2$ is

$$R = \rho L/\pi r^2$$

$$= (1.7\mathrm{x}10^{-8}\ \Omega.\mathrm{m})(14\ \mathrm{m})/\pi(8\mathrm{x}10^{-4}\ \mathrm{m})^2 = 0.12\ \Omega$$

A simpler approach would be to use Ohm's law $R = V/I = 0.71\ \mathrm{V}/6\ \mathrm{A} = 0.12\ \Omega$.

(c) The power dissipated is $P = I^2R = (6\ \mathrm{A})^2(0.12\ \Omega) = 4.3\ \mathrm{W}$

EXAMPLE 3
The current in a wire varies according to I = $2t^2 - 3t + 5$ A. What is the charge that flows in the interval t = 3 s to t = 5 s?

Solution:
Since I = dQ/dt, the infinitesimal charge that flows in the interval dt is dQ = I dt. Over a finite time interval we have

$$Q = \int_3^5 I dt = \int_3^5 (2t^2 - 3t + 5)\, dt$$

$$= \left| 2t^3/3 - 3t^2/2 + 5t \right| = 46.8 \text{ C}$$

EXAMPLE 4
What is the fractional change in the resistance of a copper wire from 30 °C to 65 °C?

Solution:
We can rewrite the relation $R = R_o(1 + \alpha\, \Delta T)$ in terms of $\Delta R = R - R_o$:

$$\Delta R/R_o = \alpha\, \Delta T$$

$$= (3.9\text{x}10^{-3}\ ^\circ\text{C}^{-1})(35\ ^\circ\text{C}) = 13.7\%$$

SOLUTIONS TO SELECTED TEXT EXERCISES AND PROBLEMS

Exercise 5
The starter motor on a car draws 80 A through a copper cable of radius 0.3 cm. (a) What is the current density? (b) Find the electric field within the wire?

Solution:
(a) The current density is

$$J = I/A = I/\pi r^2$$

$$= 2.83 \text{ x } 10^6 \text{ A/m}^2$$

(b) The electric field is

$$E = \rho J$$

$$= (1.7\text{x}10^{-8}\ \Omega\text{.m})(2.83\text{x}10^6 \text{ A/m}^2) = 4.81 \text{ x } 10^{-2} \text{ V/m}$$

Exercise 23
A wire of length 10 m and 1.2 mm diameter has a resistance of 1.4 Ω. What is the resistance if the length is 16 m and the diameter is 0.8 mm?

Solution:
The resistance of a length L of a wire with a cross-sectional area A, is

$$R = \rho L/A$$

If the resistivity is unchanged, we haev

$$R_2/R_1 = (L_2/A_2)(A_1/L_1) = 3.6$$

Thus $R_2 = 5.04\ \Omega$

Exercise 37
A power station supplies 100 kW to a load via cables whose total resistance is 5 Ω. Find the power loss in the cables if the potential difference across the load is (a) 10^4 V; (b) $2\text{x}10^5$ V.

Solution:
(a) First we need to find the current in the load:

$$I = P/V = 10\ \text{A}$$

The power dissipated in the cables would be

$$P = I^2R = 500\ \text{W}$$

(b) In this case the current in the load is I = 0.5 A, and the power dissipated in the cables is P = 1.25 W.

These values show that the power loss in transmission cables can be reduced by supplying power at a high potential difference. (In practice the current in the load is not the same as that in the cables because of a device called a transformer (see Ch. 33) that connects them. However, the logic in favour of high potential difference is still correct.)

Exercise 39
Water is contained in a glass tube of radius 1 cm and length 20 cm. What potential difference would be required to heat the water by 30 C° in 4 min? The resistivity of water is 10^{-2} Ω.m.

Solution:

The resistance fo the water column is

$$R = \rho L/A = 6.37\ \Omega$$

The mass of water is $m = (1\ \text{g/cm}^3)(2\pi rL) = 62.8$ g. The heat energy required to change the temperature by 30 °C is

$$\Delta E = mc\Delta T$$

$$= (62.8 \text{ g})(4.186 \text{ J/g})(30 \text{ °C}) = 7890 \text{ J}$$

Since this energy must be supplied in 4 min the power required is

$$P = \Delta E/\Delta t = 7890 \text{ J}/240 \text{ s} = 32.9 \text{ W}$$

The potential difference across the ends of the column may be found from $P = V^2/R$. We find $V = (PR)^{1/2} = 14.5$ V.

Problem 7
An electroplating cell uses $AgNO_3$ to deposit silver (108 u) on an electrode. If a current of 0.2 A is equally shared by the Ag^+ and NO_3^- ions, what mass of Ag is deposited in 10 min?

Solution:
The charge that flows in 10 min is $Q = It = 120$ C. Since each silver atom carries a single elementary charge, we have

$$\text{Number of atoms } N = Q/e = (120 \text{ C})/e = 75\text{x}10^{19}$$

The mass of one Ag atom is $m = 108 \text{ u} = 1.79\text{x}10^{-25}$ kg. Thus the total mass deposited is

$$\text{Mass deposited} = Nm = (75\text{x}10^{19})(1.79\text{x}10^{-25} \text{ kg})$$

$$= 1.34\text{x}10^{-4} \text{ kg} = 134 \text{ mg}$$

SELF-TEST

1. A wire of length 5 m and radius 0.6 mm has a resistance of 0.3 Ω and carries a current of 2.5 A. Find: (a) the resistivity of the material; (b) the current density; (c) the electric field within the wire; (d) the power dissipation.

2. The resistance of a wire changes from 2.4 Ω at 20 °C to 4.1 Ω at a higher temperature. What is the higher temperature if the temperature coefficient of resistivity is $3.9\text{x}10^{-3}$ °C^{-1}.

Chapter 28

DIRECT CURRENT CIRCUITS

MAJOR POINTS

1. (a) The definition of emf
   (b) Terminal potential difference, internal resistance
2. Kirchhoff's junction rule and loop rule
3. The equivalent resistance of series and parallel combinations of resistors.
4. (a) The variation of charge and potential differences in RC circuits.
   (b) The meanings of the time constant, and the half-life

CHAPTER REVIEW

EMF

The emf of a device is the work it can perform in moving a unit charge around a closed loop. A real source of emf is treated as an ideal source of emf with an internal resistance in series with it. The potential difference "available" for an external circuit is reduced by the potential difference across the internal resistance.
From point a to point b in Fig. 28.1

$$V_{ba} = V_b - V_a = \mathscr{E} - Ir$$

This terminal potential difference is numerically equal to the emf only if I = 0 (open circuit) or r = 0 (ideal source).

FIGURE 28.1

EXAMPLE 1
A battery with an emf of 12 V and an internal resistance of 0.15 Ω is being charged by a current of 2 A. Determine (a) the terminal potential difference; (b) the power supplied by the emf.

Solution:
(a) In Fig. 28.2, we start at a and add the changes in potential:

$$V_b - V_a = \mathscr{E} + Ir$$

$$= 12 + (2\ \text{A})(0.15\ \Omega)$$

$$= 12.3\ \text{V}$$

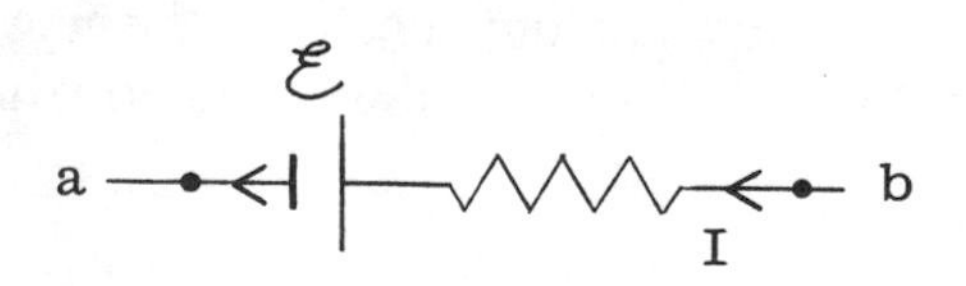

FIGURE 28.2

When a battery is being charged, the terminal potential difference is greater than the emf.

(b) The power supplied by the emf is $P = -\mathcal{E}I = -24$ W. The minus sign indicates that power is being supplied to the battery.

KIRCHHOFF'S RULES

Junction rule:

The algebraic sum of the currents entering a junction must equal the sum of the currents leaving.

$$\Sigma I = 0$$

For example, in Fig. 28.3, we could write

$$I_1 + I_2 - I_3 - I_4 = 0$$

In a multiloop ciruit, the directions of the currents in each branch may be assigned arbitrarily. This rule is an example of the conservation of charge.

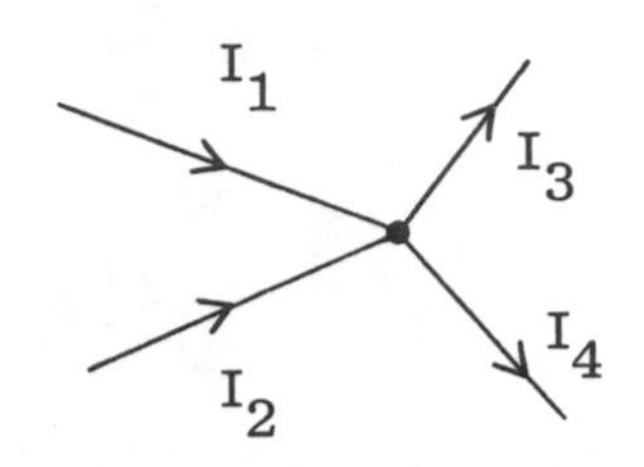

FIGURE 28.3

Loop rule:

The algebraic sum of the changes in potential around a closed loop is zero.

This rule is an example of the conservation of energy.

HELPFUL HINTS:

1. Current flows downhill in potential, thus: The sign of ΔV across a resistor is negative if your are tracing a branch of the circuit along the current; ΔV is positive if you are going against the current.
2. The sign of ΔV for a battery depends on the order in which the terminals of the source of emf are encountered; it does not depend on the direction of the current.
3. If you obtain negative currents do not redraw your arrows in the circuit. Keep your original directions but substitute the negative values in your equations.
4. The number of independent loop equations is equal to the number of loops.

SERIES AND PARALLEL CONNECTIONS

The equivalent resistance of a series or a parallel combination of resistors is given by

Series: $R_{eq} = R_1 + R_2 + R_3 + ...$

Parallel: $1/R_{eq} = 1/R_1 + 1/R_2 + 1/R_3 + ...$

In a series connection, I is the same for all elements; in a parallel connection, V is the same for all elements.

EXAMPLE 2

Consider the combination of resistors shown in Fig. 28.4. Determine (a) the current through each resistor, and (b) the power dissipated by each resistor.

Solution:

(a) First we must find the equivalent resistance. For the three resistors in parallel

$$1/R' = 1/R_2 + 1/R_3 + 1/R_4$$

$$= 1/2\ \Omega + 1/3\ \Omega + 1/4\ \Omega$$

Thus $R' = 0.923\ \Omega$. This in series with R_1 gives us $R_{eq} = R' + 1.5\ \Omega = 2.42\ \Omega$. The current supplied by the 10 V battery is

$$I_1 = \mathscr{E}/R_{eq} = 4.13\ \text{A}.$$

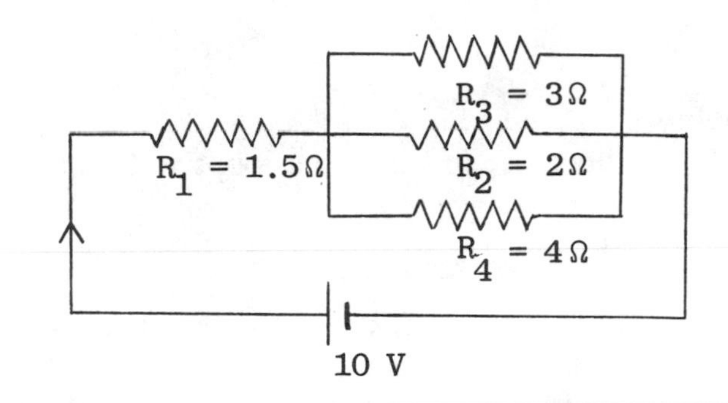

FIGURE 28.4

This is the current through R_1 and the potential difference across it is $V_1 = I_1R_1 = 6.20$ V.

Since the potential differences across the parallel resistors are the same, and equal to $\mathscr{E} - V_1 = 3.8$ V. Thus,

$$I_2R_2 = I_3R_3 = I_4R_4 = 3.8\ \text{V}$$

This leads immediately to $I_2 = 1.9$ A, $I_3 = 1.27$ A, and $I_4 = 0.95$ A. Note that $I_1 = I_2 + I_3 + I_4$, as it should.

(b) The power dissipated by each resistor is given by $P = I^2R$. We find $P_1 = 25.58$ W, $P_2 = 7.22$ W, $P_3 = 4.84$ W, and $P_4 = 3.61$ W. Note that the power supplied by the battery $P = \mathscr{E}I = 41.3$ W, is equal to $(P_1 + P_2 + P_3 + P_4)$.

EXAMPLE 3
Consider the circuit shown in Fig. 28.5. Find (a) the currents, (b) the change in potential $V_A - V_B$. Take $R_1 = 3\ \Omega$, $R_2 = 2\ \Omega$, $R_3 = 4\ \Omega$, $r_1 = r_2 = r_3 = 1\ \Omega$, $\mathscr{E}_1 = 9$ V, $\mathscr{E}_2 = 5$ V, $\mathscr{E}_3 = 7$ V.

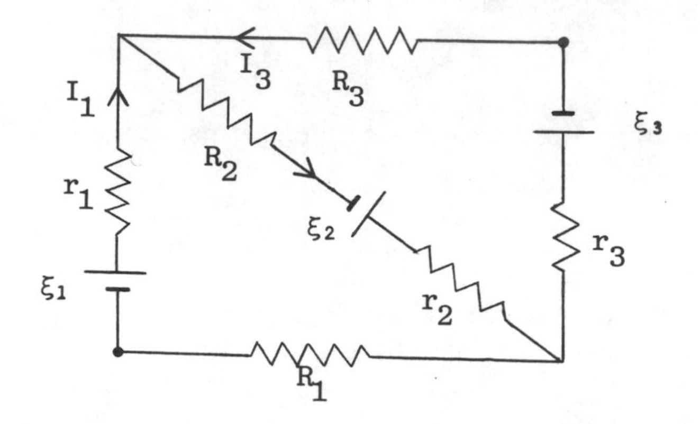

FIGURE 28.5

Solution:

The sense in which the loops will be traced is indicated.

(Junction) $I_2 = I_1 + I_3$ (i)

(Left loop) $-I_1R_1 + \mathscr{E}_1 - I_1r_1 - I_2R_2 + \mathscr{E}_2 - I_2r_2 = 0$ (ii)

(Right loop) $+I_2r_2 - \mathscr{E}_2 + I_2R_2 + I_3R_3 + \mathscr{E}_3 + I_3r_3 = 0$ (iii)

When values are substituted into (ii) and (iii) and (i) is used, we find

$$7I_1 + 3I_3 = 14$$

$$3I_1 + 8I_3 = -2$$

Solving these leads to $I_1 = 2.51$ A, $I_3 = -1.19$ A and then $I_2 = 1.32$ A.
Having found that I_3 is negative do not go back and change the direction on the original figure, because the equations will then appear to be wrong; simply substitute correctly in any subsequent calaculation.

(b) To find $V_A - V_B$ note that B is the initial point and A is the final point. Starting at B we add the changes in potential along <u>any</u> path to A. For example, going clockwise around the perimeter:

$$V_A - V_B = \mathscr{E}_3 + I_3r_3 - I_1R_1$$

$$= 7 + (-1.19\ \text{A})(1\ \Omega) - (2.51\ \text{A})(3\ \Omega) = -1.72\ \text{V}$$

Since this is negative, we see that $V_A < V_B$.

RC CIRCUITS

Discharging

In a discharge circuit, the potential difference across the resistance and the capacitance are equal are vary in time according to (see Fig. 28.6)

$$V_C = V_R = V_o \exp(-t/RC)$$

where V_o is the initial potential difference at $t = 0$. Note that $Q = CV_C$. The quantity

$$\tau = RC$$

is called the time constant of the circuit.

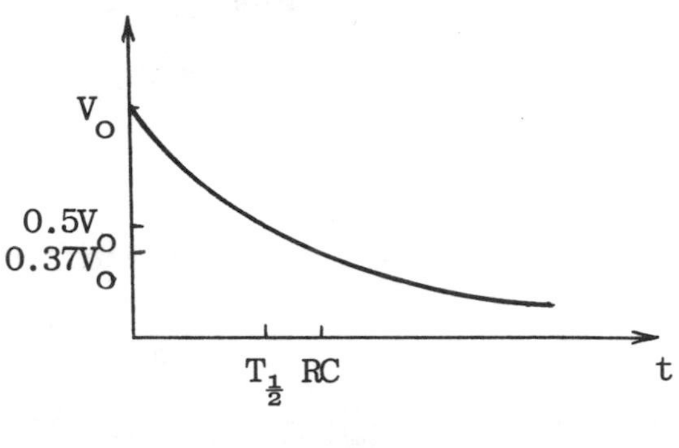

FIGURE 28.6

$$\text{At } t = RC, \quad V = V_o e^{-1} = 0.37\ V_o$$

It takes one time constant to fall to 37% of any starting value. Another time of interest is the half-life, $T_{1/2}$, which is the time required to fall to 50% of any starting value.

Charging

In a charging circuit, the potential differences across R and C are given by (see Fig. 28.7)

$$V_R = V_o \exp(-t/RC)$$

$$V_C = V_o[1 - \exp(-t/RC)]$$

Note that in this case,

$$V_R + V_C = V_o$$

In both the charge and the discharge cases, the current decays:

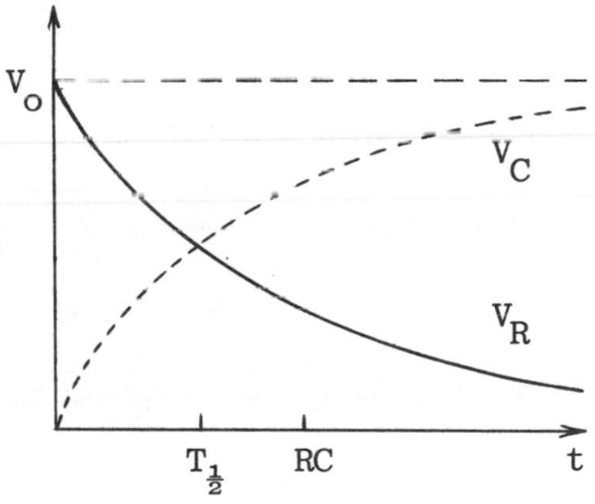

FIGURE 28.7

$$I = I_o \exp(-t/RC)$$

EXAMPLE 4

Consider an RC charging circuit. Take $C = 50\ \mu F$, $R = 2\ k\Omega$ and $V_o = 100$ V. The capacitor is initially uncharged and the switch is closed at $t = 0$. Find the time required for each of the following quantities to reach 75% of its maximum value; (a) V_C; (b) V_R; (c) U_C; (d) P_R

Solution:

(a) The potential difference across the capacitor is

$$V_C = V_o[1 - \exp(-t/RC)]$$

where RC = 10 s. We are given $V_C = 0.75V_o$, so

$$\exp(-t/10) = 0.25$$

thus,

$$t = -10 \ln(0.25) = 13.9 \text{ s}$$

(b) The potential difference across the resistor is

$$V_R = V_o \exp(-t/RC)$$

With $V_R = 0.75V_o$, we find exp(-t/10) = 0.75, thus t = 2.88 s.

(c) The energy stored in the capacitor, $U_C = CV_C^2/2$, is

$$U_C = U_o(1 - e^{-t/10})^2$$

where $U_o = 0.5CV_o^2$. If $U_C = 0.75U_o$, we have

$$(1 - e^{-t/10}) = (0.75)^{1/2}$$

This leads to exp(-t/10) = 0.134, and then t = 20.1 s.

(d) The power dissipated, $P_R = I^2R$, is

$$P_R = P_o \exp(-2t/10)$$

where $P_o = I_o^2R$. The condition $P_R = 0.75P_o$, occurs when exp(-t/5) = 0.75, thus t = 1.44 s.

SOLUTIONS TO SELECTED EXERCISES AND PROBLEMS IN THE TEXT

Exercise 5

A 16-V battery delivers 50 W to a 4-Ω external resistor. (a) Find the internal resistance of the battery. (b) For what value of external resistor would the delivered power be 100 W?

Solution:

The current in the circuit is $\mathscr{E}/(R + r)$, and the power dissipated in R is $P = I^2R$.

(a) $P = [\mathscr{E}/(R + r)]^2R$, so $50 = 256 \times 4/(4 + r)^2$, thus r = 0.525 Ω

(b) $100 = 256R/(R + 0.525)^2$, so $R^2 - 1.51R + 0.276 = 0$. The solutions of this quadratic are R = 0.215 Ω, 1.295 Ω

Exercise 17
Given three 4-Ω resistors rated at 20 W, find the maximum potential difference that can be applied if they are connected (a) all in parallel; (b) as in Fig. 28.8.

Solution:
(a) The maximum potential difference for each resistor is

$$V = (PR)^{1/2} = 8.94 \text{ V}$$

In parallel, each has the same potential difference, so we can apply only 8.94 V.

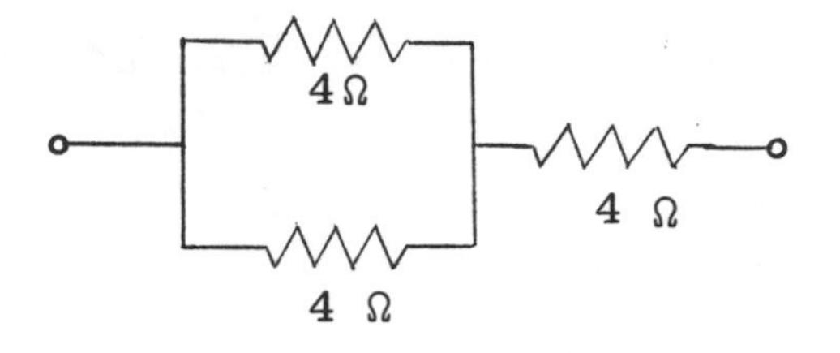

FIGURE 28.8

(b) The maximum current through the single resistor is I = 8.94 V/4 Ω = = 2.236 A. The current through each R in parallel would then be 1.118 A. The potential difference across the combination is

$$V = (1.118 \text{ A})(4\ \Omega) + 8.94 \text{ V} = 13.4 \text{ V}$$

Exercise 21
A single battery is connected to three resistors as shown in Fig. 28.9. Find the current in each resistor.

Solution:

The ends of the 7 Ω resistor are connected by a wire and so there can be no potential difference across it, thus $I_7 = 0$. For the other two resistors,

$$I_3 = 10 \text{ V}/3\ \Omega = 3.33 \text{ A}$$

$$I_4 = 10 \text{ V}/4\ \Omega = 2.5 \text{ A}$$

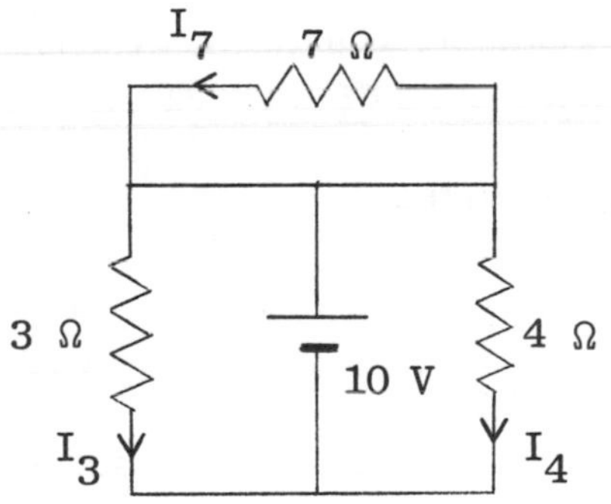

I_7 7 Ω
3 Ω 10 V 4 Ω
I_3 I_4

FIGURE 28.9

Exercise 41
In an RC discharge circuit the current drops to 10% of the initial value in 5 s. (a) How long does it take to drop to 50%? (b) Express the current flowing at 10 s as a percentage of the initial current.

Solution:
The current varies according to $I = I_o \exp(-t/\tau)$. Since $I = 0.1I_o$,

$$0.1 = \exp(-5/\tau)$$

Thus $\tau = -5/\ln(0.1) = 2.17$ s.

(a) Since $I/I_o = 0.5 = \exp(-t/2.17)$, we find $t = \tau \ln 2 = 1.50$ s.

(b) $I/I_o = \exp(-10/2.17) = 0.01$ or 1 %.

Exercise 49
In the circuit of Fig. 28.10, the internal resistances of the voltmeter and ammeter are $R_V = 1$ kΩ, $R_A = 0.1$ Ω. The resistor has a 10-Ω resistance. (a) What are the true values for the current through and potential difference across the resistor? (b) What are the current and potential differnece as measured by the meters?

Solution:
(a) The resistance of the ammeter and the resistor is 10.1 Ω. Thus the current through these two elements is

$$I = 100 \text{ V}/10.1 \text{ Ω} = 9.9 \text{ A}$$

The potential difference across the resistance is $V = (10 \text{ Ω})(9.9 \text{ A}) = 99$ V.

(b) 9.9 A, 100 V

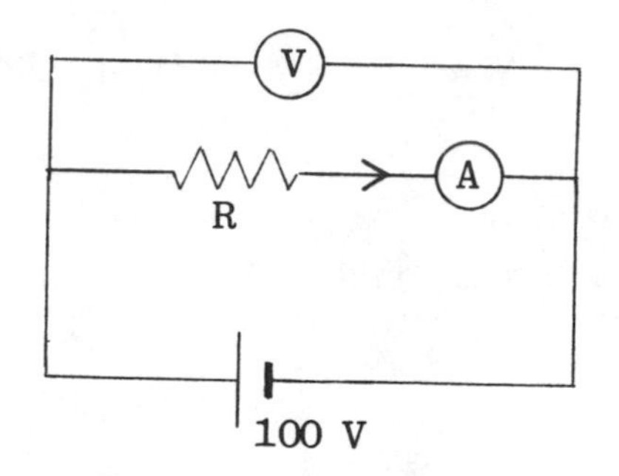

FIGURE 28.10

Problem 13
The switch in Fig. 28.11 is closed at $t = 0$. What are the currents throught the resistors at (a) $t = 0$, and (b) $t = \infty$. (c) Show that the current charging the capacitor is given by

$$i_C = \mathscr{E}/R_1 \; \exp(-t/\tau)$$

where $\tau = R_1R_2C/(R_1 + R_2)$. (Hint: Write the loop equation for each loop and relate the currents via the junction rule. Obtain a differential equation for i_C and integrate it.)

Solution:
(a) At $t = 0$, there is no charge on C and hence no potential difference across C or R_2. Thus $I_1 = \mathscr{E}/R_1$ and $I_2 = 0$.

(b) At $t = \infty$, C is fully charged and so the same current flows in both resistors, thus $I = \mathscr{E}/(R_1 + R_2)$.

(c) Applying the loop rule to the left loop:

$$\mathscr{E} - IR_1 - Q/C = 0$$

The potential difference across C and R_2 are equal, that is, $Q/C = I_2R_2$. Also, $I = I_C + I_2$, where $I_C = dQ/dt$ is the current charging the capacitor (not flowing through it!). The loop equation becomes

$$\mathscr{E} - (I_C + Q/R_2C)R_1 - Q/C = 0$$

Taking the time derivative of this equation we find

$$-R_1\, dI_C/dt - I_C(R_1/R_2 + 1)/C = 0$$

Let $\tau = R_1R_2C/(R_1 + R_2)$ and rearrange to integrate

$$dI_C/I_C = -dt/\tau$$

which yields $\ln I_C = -t/\tau + k$. At $t = 0$, $I_C = I = \mathscr{E}/R_1$, thus $k = \ln(\mathscr{E}/R_1)$ and finally

$$I_C = (\mathscr{E}/R_1)\exp(-t/\tau)$$

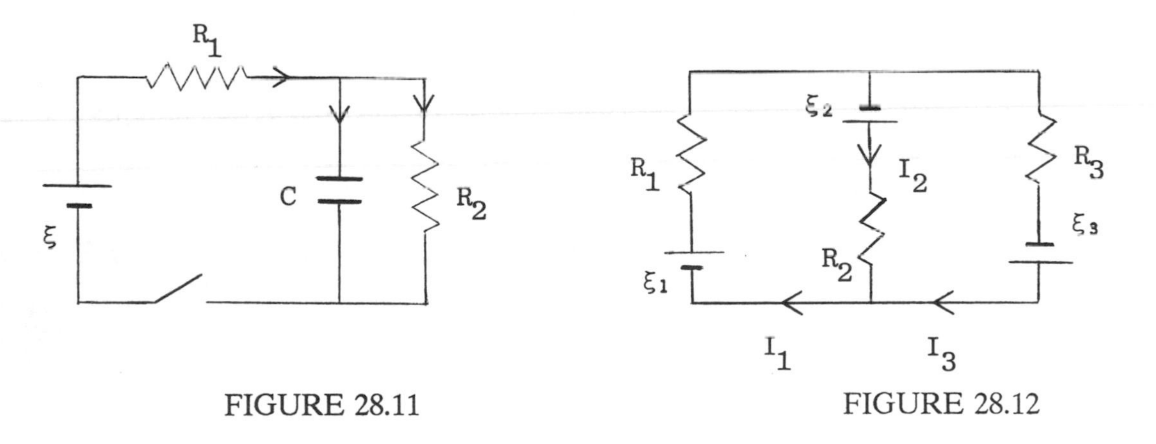

FIGURE 28.11 FIGURE 28.12

SELF-TEST

1. A 5 Ω resistor, connected to a 12 V source of emf, dissipates 25 W. Find the internal resistance of the source.

2. (a) Write two loop equations for the left and right loops in the circuit in Fig. 28.12. (b) Find the three currents given $R_1 = 2\ \Omega$, $R_2 = 3\ \Omega$, $R_3 = 1\ \Omega$, and $\mathscr{E}_1 = 5$ V, $\mathscr{E}_2 = 2$ V, $\mathscr{E}_3 = 11$ V.

3. In an RC charging circuit $C = 10^{-4}$ F and $R = 10^5\ \Omega$ are in series with a source of emf $\mathscr{E} = 100$ V. $Q = 0$ when the switch is closed at $t = 0$. (a) Write expressions for the charge on C and the current through R as functions of time. (b) What is the power loss in R at $t = 5$ s ? (c) How much energy is stored in the capacitor after one time constant?

Chapter 29

THE MAGNETIC FIELD

MAJOR POINTS

1. The magnetic force on a moving charge or current element
2. (a) The torque on a current loop in a magnetic field
   (b) The definition of magnetic dipole moment

CHAPTER REVIEW

THE MAGNETIC FORCE

The force exerted by a magnetic field **B** on a charge q moving with velocity **v** is

$$\mathbf{F} = q\,\mathbf{v} \times \mathbf{B}$$

(You should review the cross-product, Sec. 2.6.) This magnetic force is perpendicular to both **v** and **B**. The sign of the charge must be explicitly included. The magnitude of the force is

$$F = q\,v\,B\sin\theta$$

Thus $F = 0$ if either $v = 0$ or $\theta = 0$. Since **F** is normal to **v**, the magnetic force on a particle does no work on it. The magnetic force changes the direction of the velocity but not the speed.

EXAMPLE 1

An electron moves in a uniform magnetic field $\mathbf{B} = -B\mathbf{i}$. What is the force on it for the following velocities: (a) $\mathbf{v}_1$ at 30° to the +x axis; (b) $\mathbf{v}_2 = -v_2\mathbf{j}$; (c) $\mathbf{v}_3 = v_3\mathbf{k}$ (out of the page).

Solution:

We use $\mathbf{F} = -e\mathbf{v} \times \mathbf{B}$ (see Fig. 29.1)

(a) $\mathbf{v}_1 = v_1\cos30°\,\mathbf{i} + v_1\sin30°\,\mathbf{j}$

thus,

$\mathbf{F}_1 = -ev_1(0.866\mathbf{i} + 0.5\mathbf{j}) \times (-B\mathbf{i})$

$= -0.5ev_1B\mathbf{k}$

(b) $\mathbf{F}_2 = -e(-v_2\mathbf{j}) \times (-B\mathbf{i}) = ev_2B\mathbf{k}$.

(c) $\mathbf{F}_3 = -e(v_3\mathbf{k}) \times (-B\mathbf{i}) = ev_3B\mathbf{j}$.

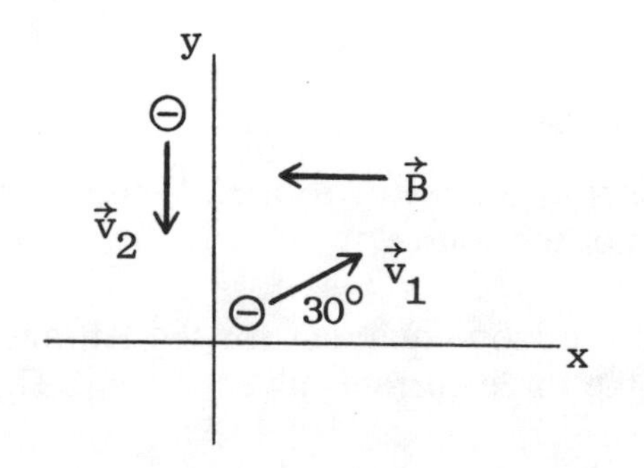

FIGURE 29.1

Magnetic force on a current

The force exerted by a uniform magnetic field on a straight current-carrying wire of length ℓ is

$$\mathbf{F} = \mathrm{I}\, \boldsymbol{\ell} \times \mathbf{B}$$

The direction of $\boldsymbol{\ell}$ is along the current. If the wire is not straight or the field is not uniform, one must calculate the force on a current element I $d\boldsymbol{\ell}$:

$$d\mathbf{F} = \mathrm{I}\, d\boldsymbol{\ell} \times \mathbf{B}$$

The net force on a closed loop in a uniform field is zero.

EXAMPLE 2

Figure 29.2 shows a straight wire of length ℓ carrying a current I at 30° to the +x axis. Find the force on the wire for each of the magnetic fields indicated.

Solution:

In each case we must find $\mathbf{F} = \mathrm{I}\, \boldsymbol{\ell} \times \mathbf{B}$, where

$\boldsymbol{\ell} = \ell \cos30°\mathbf{i} + \ell \sin30°\mathbf{j}$

$\mathbf{F}_1 = \mathrm{I}\boldsymbol{\ell} \times (-B_1\mathbf{i}) = \mathrm{I}\ell B_1 \sin30°\, \mathbf{k}$

$\mathbf{F}_2 = \mathrm{I}\boldsymbol{\ell} \times (B_2\mathbf{j}) = \mathrm{I}\ell B_2 \cos30°\, \mathbf{k}$

$\mathbf{F}_3 = \mathrm{I}\boldsymbol{\ell} \times (B_3\mathbf{k})$

$= \mathrm{I}\ell B_3(-\cos30°\, \mathbf{j} + \sin30°\, \mathbf{i})$

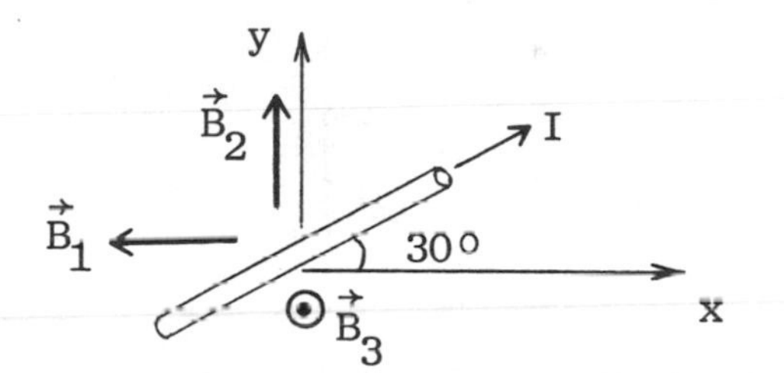

FIGURE 29.2

You should verify the direction of each of the these forces with the right-hand rule.

TORQUE ON A COIL

When a plane loop of area A with N turns carries a current I, it has a magnetic dipole moment

$$\boldsymbol{\mu} = \mathrm{N\, I\, A}\, \hat{\mathbf{n}}$$

The direction of the unit vector $\hat{\mathbf{n}}$ is normal to the plane of the loop and given by a right-hand rule: Curve the fingers in the sense of the current; the thumb indicates the direction of $\hat{\mathbf{n}}$.

When a current-carrying (plane) loop is placed in a magnetic field, it experiences a torque given by

$$\boldsymbol{\tau} = \boldsymbol{\mu} \times \mathbf{B}$$

The torque tends to align μ along **B**. The potential energy of a current-carrying coil in a magnetic field is

$$U = -\boldsymbol{\mu} \cdot \mathbf{B}$$

Compare these with the expressions for an electric dipole: $\boldsymbol{\tau} = \mathbf{p} \times \mathbf{E}$ and $U = -\mathbf{p} \cdot \mathbf{E}$.)

EXAMPLE 3
A rectangular coil lies diagonally across two sides of a square of side d = 6 cm; see Fig. 29.3. It has 8 turns and carries a current of 3 A. There is a uniform external field $\mathbf{B} = 2\mathbf{i}$ T. Determine: (a) the force on each side; (b) the torque on the coil; (c) the potential energy of the coil

Solution:
(a) To find the force on each side we may use $F = I\ell B\sin\theta$ and apply the right-hand rule for the direction, or $\mathbf{F} = I\boldsymbol{\ell} \times \mathbf{B}$, which requires us to express ℓ in unit vector notation.

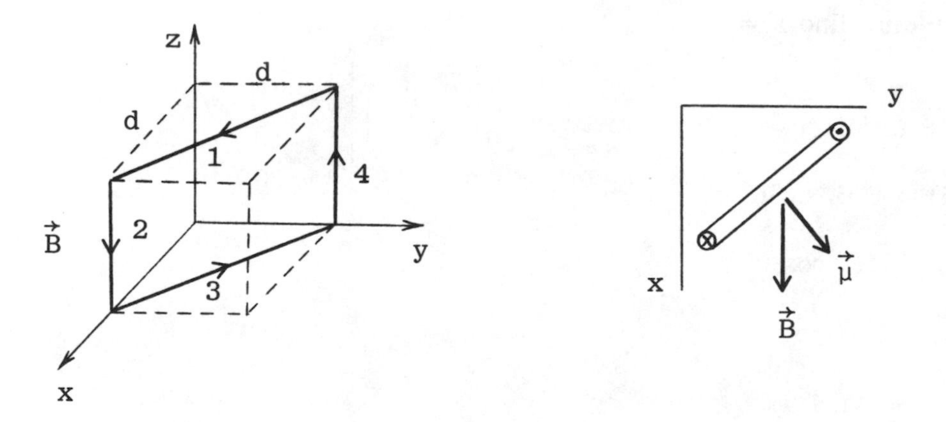

FIGURE 29.3

The angle between ℓ_1 and **B** is 45° and $\ell = (2)^{1/2}d = 8.49\times10^{-2}$ m.

$$F_1 = NI\ell B\sin\theta$$

$$= 8(3\text{ A})(8.49\times10^{-2}\text{ m})(2\text{ T})(0.707) = 2.88\text{ N}$$

Using the right-hand rule the force is along the +z axis. In unit vector notation, $\ell_1 = d\mathbf{i} - d\mathbf{j}$ (go from the tail to the tip along two sides of the square), thus

$$\mathbf{F}_1 = NI\,\boldsymbol{\ell} \times \mathbf{B} = NI(d\mathbf{i} - d\mathbf{j}) \times B\mathbf{i}$$

$$= NIdB\,\mathbf{k} = 2.88\mathbf{k}\text{ N}$$

Clearly $\mathbf{F}_3 = -\mathbf{F}_1 = -2.88\mathbf{k}$ N. These two froces tend to stretch the coil. The force on side 2 is

$$\mathbf{F}_2 = NI\,(-d\mathbf{k}) \times B\mathbf{i} = -NIdB\mathbf{j} = -2.88\mathbf{j}\ \text{N}$$

Clearly, $\mathbf{F}_4 = -\mathbf{F}_2 = 2.88\mathbf{j}$ N. Confirm these directions with the right-hand rule. These two forces tends to rotate the coil.
(b) To determine the torque on the coil it is advisable to redraw the coil from a different perspective--looking down the z axis. When this is done, as in Fig. 29.3b, we see that the angle between $\boldsymbol{\mu}$ and $\mathbf{B}$ is 45°.

$$\tau = \mu B \sin\theta = NIAB \sin 45°$$

$$= 8(3\ \text{A})(36\text{x}10^{-4}\ \text{m}^2)(2\ \text{T})(0.707) = 0.122\ \text{N.m}$$

The sense of τ is clockwise; its vector direction is along -z, so $\boldsymbol{\tau} = -0.122\mathbf{k}$ N.m.
(c) The potential energy of the coil is

$$U = -\boldsymbol{\mu}.\mathbf{B} = -\mu B\cos\theta = -0.122\ \text{J}$$

MOTION OF CHARGED PARTICLES

When a charged particle enters a uniform magnetic field perpendicular to the lines, it moves in a circular path. Since $\theta = 90°$, from Newton's second law we have:

$$qvB = mv^2/r$$

The cyclotron angular frequency, $\Omega = v/r$,

$$\Omega_c = qB/m$$

does not depend on either the speed of the particle or the radius of the path. The period and frequency are given by

$$T = 1/f = 2\pi/\Omega$$

EXAMPLE 4

A proton with a kinetic energy of 120 keV moves in a circle of radius 24 cm perpendicular to a uniform magnetic field. (a) What is the magnitude of the field? (b) What is the frequency of the orbit?

Solution:

(a) From Newton's second law, $qvB = mv^2/r$ we find $v = qrB/m$. Thus, the kinetic energy is

$$K = 1/2\ mv^2 = (qrB)^2/2m$$

where $K = 120$ keV $= 1.92\text{x}10^{-14}$ J. The magnetic field is given by

$B = (2mK)^{1/2}/rq$

$= [2(1.67\text{x}10^{-27}\text{ kg})(1.92\text{x}10^{-14}\text{ J})]^{1/2}/(0.24\text{ m})(1.6\text{x}10^{-19}\text{ C}) = 0.209\text{ T}$

(b) The cyclotron frequency is

$$f = \Omega/2\pi = qB/2\pi m$$

$$= (1.6\text{x}10^{-19}\text{ C})(0.209\text{ T})/2\pi(1.67\text{x}10^{-27}\text{ kg}) = 3.2\text{ MHz.}$$

LORENTZ FORCE

When a charged particles is subject to both electric and magnetic fields, the force on the particle is called the Lorentz force:

$$\mathbf{F} = q(\mathbf{E} + \mathbf{v} \times \mathbf{B})$$

The only instance we will encounter with both fields occurs in the velocity selector.

Velocity Selector

Suppose a uniform electric and magnetic field are perpendicular and a charged particle enters the region of the fields perpendicular to both fields. Then, the net force is zero if

$$\mathbf{E} = -\mathbf{v} \times \mathbf{B}$$

In terms of the magnitudes, only those particles whose speed is $v = E/B$ will pass through undeflected. This arrangement of fields is useful in selecting velocities and measuring them.

EXAMPLE 5

An electron moving moving with velocity $\mathbf{v} = -10^6\mathbf{j}$ m/s enters a region between two plates where the electric field is $\mathbf{E} = 10^3\mathbf{i}$ V/m. What magnetic field would allow the electron to pass through undeflected?

Solution:

The given information is shown in Fig. 29.4. The condition, $\mathbf{E} = -\mathbf{v}\text{x}\mathbf{B}$, for no deflection takes the form

$$E\mathbf{i} = -(-v\mathbf{j}) \times \mathbf{B} = v\mathbf{j} \times \mathbf{B}$$

Since $\mathbf{i} = \mathbf{j} \times \mathbf{k}$, we see that $\mathbf{B} = B\mathbf{k}$ (out of the page). Note that the sign of the charge is not relevant, although you would have to take it into account if you were to consider the magnetic and electric forces.

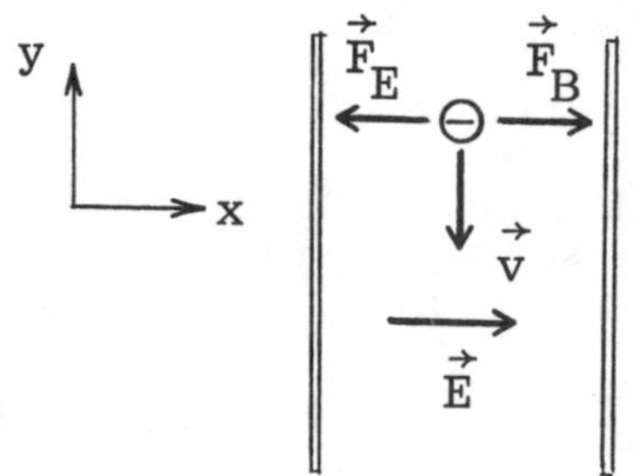

FIGURE 29.4

SOLUTIONS TO SELECTED TEXT EXERCISES AND PROBLEMS

Exercise 11

When a proton has a velocity $\mathbf{v} = (2\mathbf{i} + 3\mathbf{j})\times 10^6$ m/s, it experiences a force $\mathbf{F} = -1.28\times 10^{-13}\ \mathbf{k}$ N. When its velocity is along the +z axis, it experiences a force along the +x axis. What is the magnetic field?

Solution:

When **v** is in the xy plane, **F** is along z. Since the magnetic force is always perpendicular to to both **v** and **B**, it follows that $B_z = 0$. When **v** is along the +z axis, we are told **F** is along the +x axis. Thus, $B_x = 0$. Next we use $\mathbf{F} = q\mathbf{v} \times \mathbf{B}$ with the given data:

$$-1.28\times 10^{-13}\ \mathbf{k} = (1.6\times 10^{-13})(2\mathbf{i} + 3\mathbf{j}) \times (B_y\mathbf{j})$$

$$= B_y\ (3.2\times 10^{-13}\ \mathbf{k})$$

Therefore, $\mathbf{B} = -0.4\mathbf{j}$ T

Exercise 15

The current-carrying triangular wire loop shown in Fig. 29.5 is in a uniform magnetic field $\mathbf{B}_3 = +B_3\mathbf{j}$. What is the force on each side of the loop?

Solution:

The magnetic force on a straight wire is $\mathbf{F} = I\ \boldsymbol{\ell} \times \mathbf{B}$. The three sides of the triangle are $\ell_1 = d\mathbf{i}$; $\ell_2 = d\mathbf{j}$ and $\ell_3 = -d\mathbf{i} - d\mathbf{j}$.

$\mathbf{F}_1 = IdB_3\mathbf{k}$; $\mathbf{F}_2 = 0$; $\mathbf{F}_3 = -IdB_3\mathbf{k}$

Confirm the directions with the right-hand rule. Note that $\mathbf{F}_1 + \mathbf{F}_2 + \mathbf{F}_3 = 0$; the net magnetic force on a closed current loop is zero.

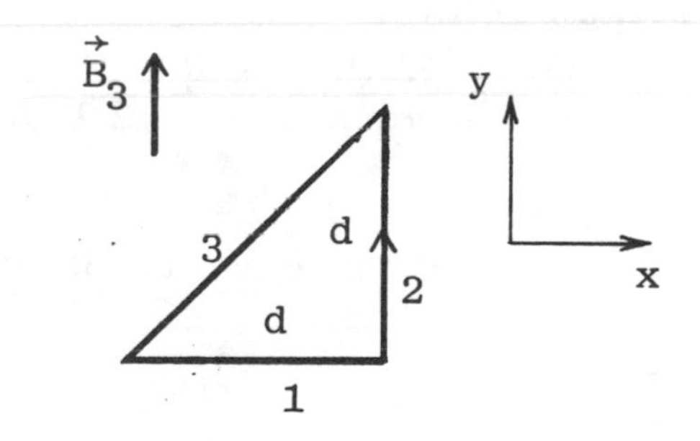

FIGURE 29.5

Exercise 25

A circular loop of radius 4 cm carries a 2.8-A current. The magnetic moment of the loop is directed along $\hat{\mathbf{n}} = 0.6\mathbf{i} - 0.8\mathbf{j}$. The magnetic field is $\mathbf{B} = 0.2\mathbf{i} - 0.4\mathbf{k}$ T. Find: (a) the torque on the loop; (b) the potential energy of the loop.

Solution:

(a) The torque is given by $\boldsymbol{\tau} = \boldsymbol{\mu} \times \mathbf{B} = NIA\ \hat{\mathbf{n}} \times \mathbf{B}$, where $A = \pi r^2$. Thus

$$\boldsymbol{\tau} = IA(0.6\mathbf{i} - 0.8\mathbf{j}) \times (0.2\mathbf{i} - 0.4\mathbf{k})$$

$$= (4.5\mathbf{i} + 3.4\mathbf{j} - 2.25\mathbf{k})\times 10^{-4}\ \text{N.m}$$

(b) The potential energy is $U = -\boldsymbol{\mu} \cdot \mathbf{B}$:

$$U = -IA\mathbf{n} \cdot \mathbf{B}$$

$$= -IA(0.6\mathbf{i} - 0.8\mathbf{j}) \cdot (0.2\mathbf{i} - 0.4\mathbf{k}\ T) = -1.69\times10^{-3}\ J$$

Exercise 31

Suppose that an electron and a proton both move perpendicular to the same uniform magnetic field. Find the ratio of the radii of their orbits given that they have the same value for the following quantities: (a) speed; (b) kinetic energy.

Solution:

(a) From Newton's second law $qvB = mv^2/r$, we find $v = qrB/m$. If v, B and q are the same for both particles, we have that r/m = constant, or

$$r_e/r_p = m_e/m_p = 5.45 \times 10^{-4}$$

(b) The kinetic energy is $K = 1/2\ mv^2 = (qrB)^2/2m$. If K, q and B are the same for both particles, then r^2/m = constant, or

$$r_e/r_p = (m_e/m_p)^{1/2} = 2.33 \times 10^{-2}$$

Exercise 45

An ion ($m = 1.2\times10^{-25}$ kg, $q = 2e$) is accelerated from rest by a voltage of 200 V and then eneters a uniform field $B = 0.2$ T normal to the lines. Find the radius of its path.

Solution:

When the particle is accelerated, from the conservation of energy we obtain

$$1/2\ mv^2 = qV \qquad \text{(i)}$$

When the particle enters the magnetic field its path is governed by Newton's second law:

$$mv^2/r = qvB \qquad \text{(ii)}$$

From (i) and (ii) we obtain expressions for v^2 and equate them:

$$2qV/m = (qrB/m)^2$$

therefore,

$$r = (2mV/q)^{1/2}/B$$

Inserting the given values we find r = 6.12 cm. (Check it.)

SELF-TEST

1. A rectangular loop with sides a = 20 cm and c = 50 cm carries a current I = 3 A and pivots about the z axis, as shown in Fig. 29.6. There is a uniform field **B** = 0.5**j** T. Find: (a) The force on each side; (b) The torque on the loop.

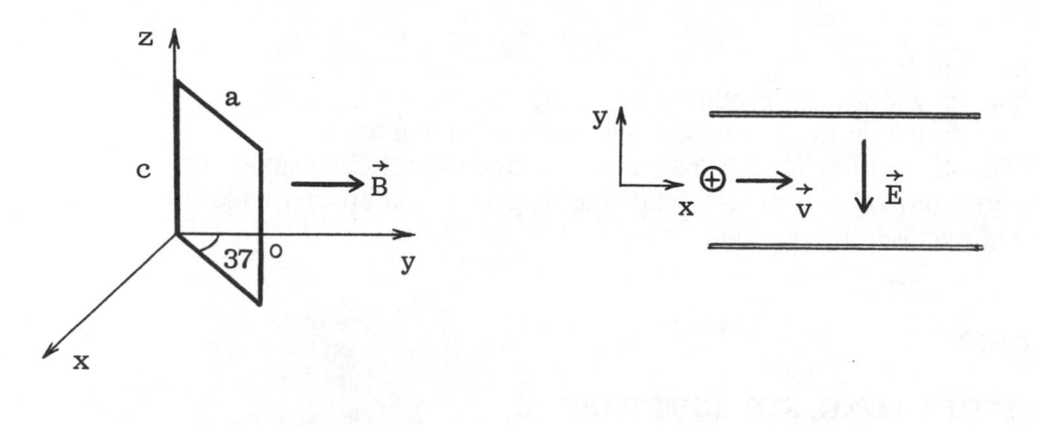

FIGURE 29.6

FIGURE 29.7

2. A proton moving at **v** = 10^6**i** m/s enters a region where the electric field is **E** = - 10^5**j** V/m, see Fig. 29.7. What magnetic field would allow the charge to pass through the region undeflected? Explain your logic clearly.

3. An electron moves perpendicular to a 3 mT magnetic field in a circular path of radius 8 cm. Determine (a) the period; (b) the linear momentum; (c) the kinetic energy

Chapter 30

SOURCES OF THE MAGNETIC FIELD

MAJOR POINTS

1. (a) The magnetic field produced by a long straight wire
   (b) The magnetic force between two current-carrying wires
2. The Biot-Savart law for the magnetic field due to an infinitesimal current element
3. Ampere's law relates an integral of the magnetic field around a closed loop to the net current enclosed by the loop

CHAPTER REVIEW

FIELD DUE TO A LONG, STRAIGHT WIRE

A long straight wire carrying a current I produces a magnetic field whose lines are circles centered on the wire. At a perpendicualr distance R from the wire, the magnetic field is

$$B = \mu_o I/2\pi R$$

The direction of B is found with a right-hand rule: If you "hold" the wire with your right hand, with the thumb pointing along the current, the curled fingers indicate the direction or sense of B.

Magnetic force between wires

Consider two long parallel wires carrying currents I_1 and I_2 and separated by a perpendicular distance d, as shown in Fig. 30.1. The force on a length ℓ_1 of the wire carrying I_1 due to the wire carrying I_2 is

$$\mathbf{F}_{12} = I_1\, \boldsymbol{\ell}_1 \mathbf{x} \mathbf{B}_2$$

The direction of B_2 at the position of I_1 is indicated. Using the right-hand rule for cross products, this force is directed toward the left. From Newton's third law, we know that the force per unit length

$$F/\ell = \mu_o I_1 I_2/2\pi d$$

is the same for both wires. Currents in the same direction attract, currents in opposite directions repel.

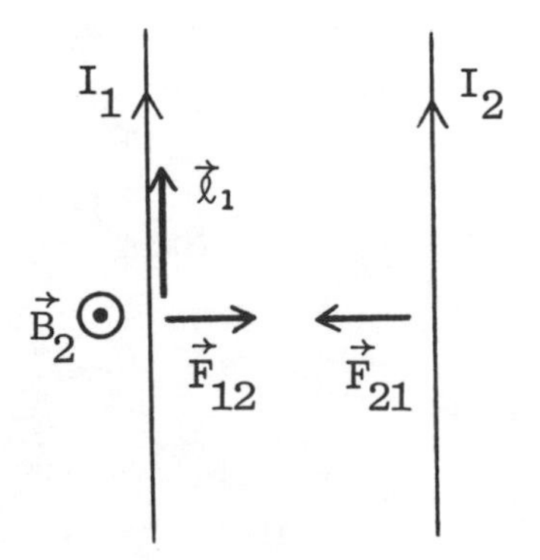

FIGURE 30.1

EXAMPLE 1
Two straight parallel wires carry currents $I_1 = 3$ A and $I_2 = 10$ A, as shown in Fig. 30.2. (a) Find the resultant magnetic field at point P; (b) Where is the resultant field zero? (c) What is the force per unit length on the wires?

Solution:
(a) The directions of the fields are indicated in the figure. Note that each field is perpendicular to the line joining the current to P. The magnitudes of the fields at P are

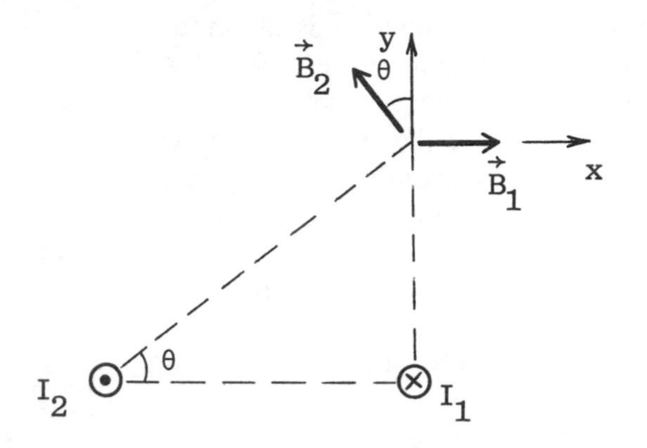

FIGURE 30.2

$$B_1 = \mu_o I_1/2\pi r_1$$

$$= (2\text{x}10^{-7}\ \text{T.m/A})(3\ \text{A})/(0.03\ \text{m})$$

$$= 2\text{x}10^{-5}\ \text{T}$$

Similarly, $B_2 = 4\text{x}10^{-5}$ T. The components of the resultant field are

$$B_x = B_1 - B_2 \sin\theta = -0.4\text{x}10^{-5}\ \text{T}$$

$$B_y = B_2 \cos\theta = 3.2\text{x}10^{-5}\ \text{T}$$

Thus, $\mathbf{B} = (-0.4\mathbf{i} + 3.2\mathbf{j})\text{x}10^{-5}$ T.

(b) The resultant field will vanish at some point at a distance d to the right of I_1 (why?). The condition $I_1/r_1 = I_2/r_2$ yields

$$3/d = 10/(4 + d)$$

Thus d = 12/7 cm = 1.71 cm.

(c) The force per unit length on each wire is

$$F/\ell = \mu_o I_1 I_2/2\pi d$$

$$= (2\text{x}10^{-7}\ \text{T.m/A})(30\ \text{A}^2)/(0.04\ \text{m})$$

$$= 1.5\text{x}10^{-4}\ \text{N/m}.$$

This force is attractive. (Confirm this by determining the direction of B_2 at the position of I_1 and then using the right-hand rule.)

BIOT-SAVART LAW

The magnetic field produced by a current element I $d\ell$ at a distance r is given by the Biot-Savart law

$$d\mathbf{B} = (\mu_o/4\pi)\, I\, d\boldsymbol{\ell} \times \hat{\mathbf{r}} / r^2$$

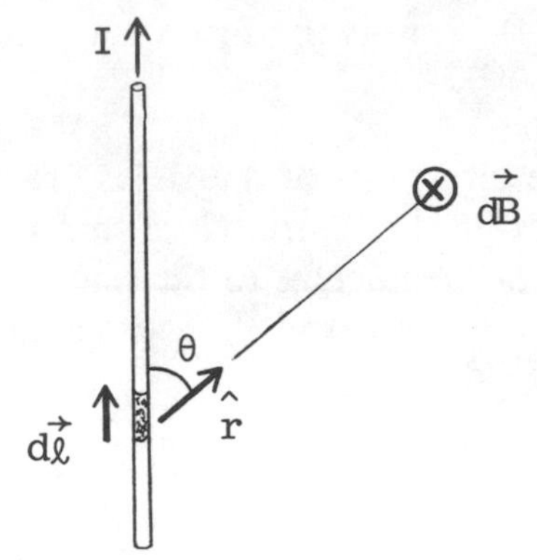

FIGURE 30.3

where the unit vector $\hat{\mathbf{r}}$ is directed from the element to the point at which the field is being calculated, see Fig. 30.3. Note that the direction of the field is perpendicular to both $\boldsymbol{\ell}$ and $\hat{\mathbf{r}}$. The magnitude of the field is

$$dB = (\mu_o/4\pi)\, I\, d\ell \sin\theta / r^2$$

where θ is the angle between $d\boldsymbol{\ell}$ and $\hat{\mathbf{r}}$.

HELPFUL HINTS

1. Choose an arbitrary current element
2. Determine the direction of **dB** from $d\boldsymbol{\ell} \times \hat{\mathbf{r}}$
3. The three variables ℓ, r and θ must related to each other so that the final expression for dB involves only one variable.

EXAMPLE 2

Determine the magnetic field at the center of a circular loop of radius R carrying a current I.

Solution:

We choose an arbitrary current element as shown in Fig. 30.4. The unit vector $\hat{\mathbf{r}}$ is directed toward the center. From the right-hand rule we see that $d\boldsymbol{\ell} \times \hat{\mathbf{r}}$ is upward, along the axis of the loop. After determining the direction of **dB** we are no longer conerned with the vector nature of the integral.

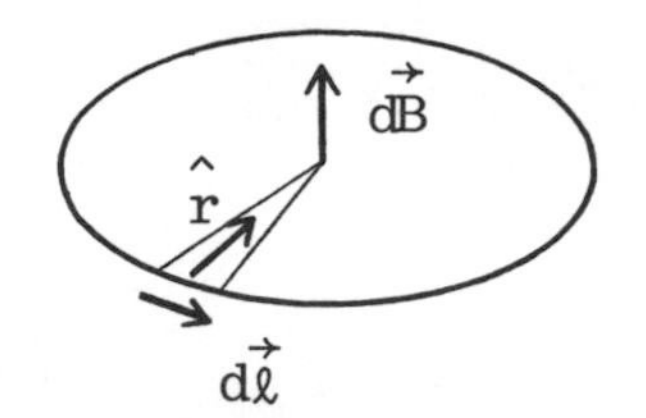

$$|d\boldsymbol{\ell} \times \hat{\mathbf{r}}| = d\ell \sin 90^\circ = d\ell$$

Since r = R is fixed, the Biot-Savart law takes the form

FIGURE 30.4

$$B = (\mu_o I/4\pi R^2) \oint d\ell$$

The integral around the closed loop is simply the circumference, $2\pi R$.
Finally,

$$B = \mu_o I/2R$$

AMPERE'S LAW

Ampere's law relates a path integral of the magnetic field around a closed loop to the net steady current passing through the area of the loop:

$$\oint \mathbf{B} \cdot d\boldsymbol{\ell} = \mu_o I$$

In cases with a high degree of symmetry, Ampere's law provides a very simple way to calculate the magnetic field. In such cases **B** is either parallel or perpendicular to $d\boldsymbol{\ell}$ so $\mathbf{B}.d\boldsymbol{\ell}$ is either zero or B $d\ell$. Ampere's law is analogous to Gauss's law for the electric field.

EXAMPLE 3

The current density in a cylinder of radius R is given by J = Kr, where K is a constant. Determine the magnetic field within the cylinder (Fig. 30.5).

Solution:

The cylindrical symmetry indicates that the appropriate path for the integration is a circle. The current enclosed in a circle of radius r is

$$I = \int \mathbf{J} \cdot dA$$

where $dA = 2\pi r\, dr$ is the area of an infinitesimal ring of radius r and thickness dr. The magnetic field is tangent to the circle, so at all points we have $\mathbf{B} \cdot d\boldsymbol{\ell} = B\, d\ell$. Ampere's law takes the form

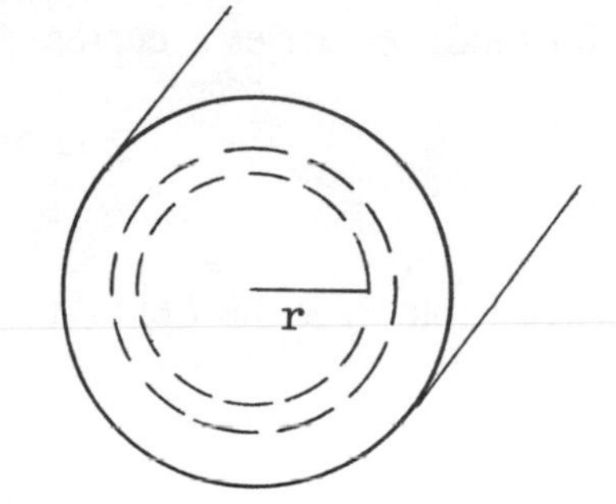

FIGURE 30.5

$$\oint B\, d\ell = \mu_o \int (Kr)\, 2\pi r\, dr$$

$$B\,(2\pi r) = 2\pi\mu_o K\,(r^3/3)$$

Thus, $B = \mu_o K r^2/3$.

SOLUTIONS TO SELECTED TEXT EXERCISES AND PROBLEMS

Exercise 7

A dc power line 20 m above the ground carries a 600-A current due north. If the horizontal component of the earth's field is 0.5 G due north, in what direction does a compass needle, on the ground directly below the line, point?

Solution:

We are given the earth's field is $B_e = 0.5$ G due north. The field due to the wire is $B_w = \mu_o I/2\pi d = 0.06$ G due west (confirm this with the right-hand rule). A compass needle with point along the resultant field, as shown in Fig. 30.6. Thus,

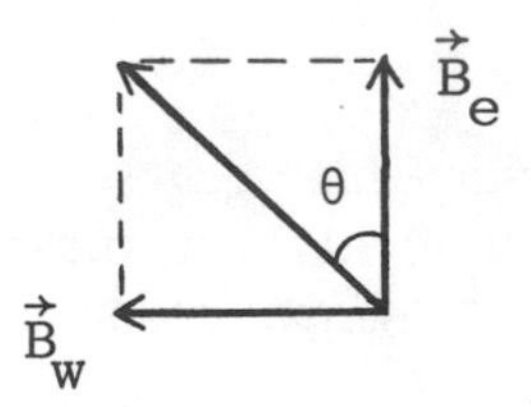

$$\tan\theta = B_w/B_e$$

Since $\tan\theta = 0.12$, we find $\theta = 6.8°$ W of N.

FIGURE 30.6

Exercise 13

A square coil of side ℓ carries a current I. Show that the field strength at its center is

$$B = 2(2)^{1/2}\mu_o I/\pi\ell$$

Solution:

From Example 30.2 the field due to a finite straight wire is

$$B = (\mu_o I/2\pi R)(\sin\alpha_1 + \sin\alpha_2)$$

where the angles are measured relative to the line perpendicular to the wire. For the field at the center of the square, Fig. 30.7, $R = \ell/2$, and $\alpha_1 = \alpha_2 = 45°$, therefore $\sin\alpha_1 = \sin\alpha_2 = 1/(2)^{1/2}$. The fields due to the four wires are in the same direction, so the resultant field is four times as large:

$$B = 2(2)^{1/2}\mu_o I/\pi\ell$$

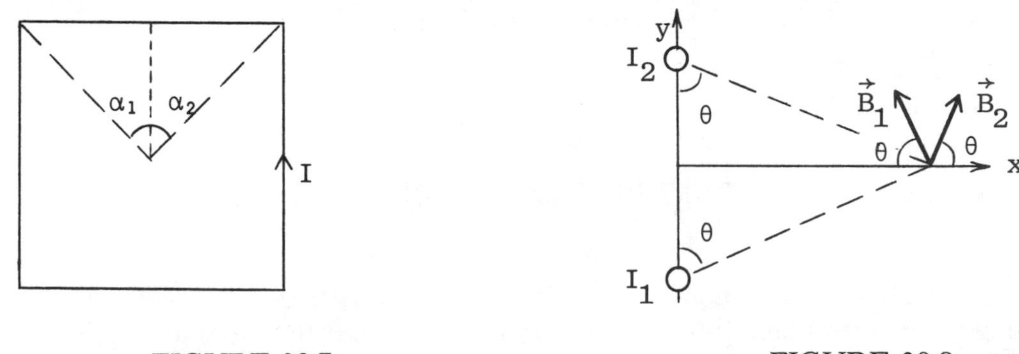

FIGURE 30.7

FIGURE 30.8

Exercise 19

Two long wires are parallel to the z axis and are located at $x = 0$, $y = \pm a$. (a) Find B(x) at a point (x, 0) in the xy plane, given that the wires carry equal currents in the same direction. (b) At what point is B a maximum?

Solution:
(a) Let us assume that both currents are coming out of the page. The direction of each field is perpendicular to the radial line from the wire to given point, as shown in Fig. 30.8.
From symmetry we can say immediately that $B_x = 0$. The magnitudes of the fields are equal

$$B_1 = B_2 = \mu_o I/2\pi r$$

where $r = (a^2 + x^2)^{1/2}$. The y component of the total field $\mathbf{B}_T = \mathbf{B}_1 + \mathbf{B}_2$ is

$$B_{Ty} = (B_1 + B_2)\sin\theta$$

where $\sin\theta = x/r$. Thus

$$B_{Ty} = 2(\mu_o I/2\pi r)(x/r) = \mu_o Ix/\pi(a^2 + x^2)$$

(b) To find where the field is a maximum set $dB/dx = 0$ to find $x = \pm a$.

Exercise 29
An infinite metal plate of thickness t, as in Fig. 30.9, carries a uniform current density J. (a) Use the right-hand rule and symmetry to determine the direction of the field above and below the plate. (b) Determine the magnetic field at a distance a from the plate.

Solution:
(a) By considering two segments at equal distances on either side of a given point it is easy to see that the resultant field must be parallel to the plate. Using the right-hand rule the direction is indicated in the Fig. 30.9.

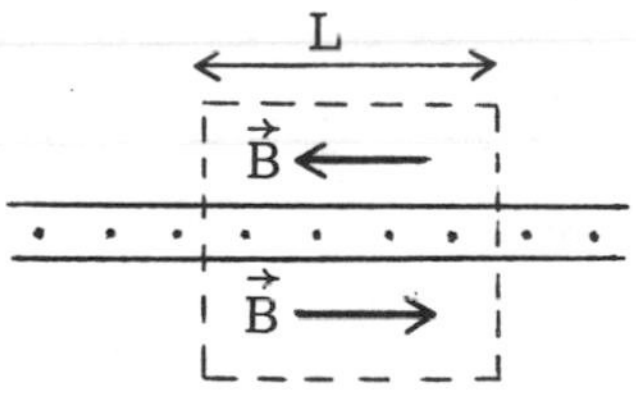

FIGURE 30.9

(b) We apply Ampere's law to the dashed rectangular path of length L. Only the lengths parallel to the plate contribute to the integral. The current flowing through the path is $I = JA = JLt$. Thus Ampere's law

$$\oint \mathbf{B} \cdot d\boldsymbol{\ell} = \mu_o I$$

takes the form

$$B(2L) = \mu_o(JLt)$$

which yields $B = \mu_o Jt/2$.

Problem 9

A long straight solid wire of radius R contains a cavity of radius r along its length. As shown in Fig. 30.10 the centers of the wire and of the cavity are a distance a apart. The current is uniformly distributed throughout the rest of the wire. (a) Show that the field in the cavity is uniform. (b) What is its value? (Hint: Use the superposition principle. Add the field due to a completely solid wire of radius R to that of a wire of radius a carrying a current in the opposite direction.)

Solution:

From Eq. (ii) of Example 30.5, the field inside a wire is

$$B = \mu_o Ir/2\pi R^2$$

where **B** is perpendicular to **r**. The current that would have flowed through the cavity is $I' = (a^2/R^2)I$. Thus the field associated with the cavity would be

$$B' = \mu_o I'r'/2\pi a^2 = \mu_o Ir'/2\pi R^2$$

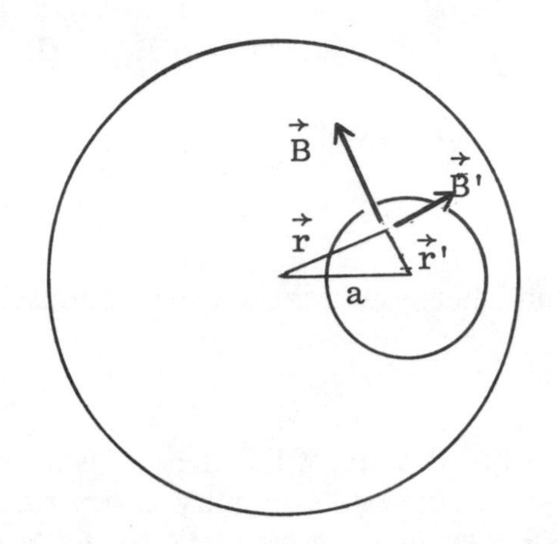

FIGURE 30.10

The total field is the sum of the contributions of a wire of radius R carrying a current I and a wire of radius a with a current I' flowing in the opposite direction. Let **s**, **s'** and **D** be equal in magnitude but normal to **r**, **r'** and **d** respectively. The resultant field then is

$$\mathbf{B}_T = \mathbf{B} + \mathbf{B}' = (\mu_o I/2\pi R^2)(\mathbf{s} - \mathbf{s}')$$

$$= (\mu_o I/2\pi R^2)\mathbf{D}$$

This is constant within the cavity.

SELF-TEST

1. Two straight infinite wires are parallel and carry currents $I_1 = 8$ A and $I_2 = 12$ A in the directions shown in Fig. 30.11. (a) What is the resultant magnetic field at point P? (b) What is the force per unit length exerted by I_2 on I_1?

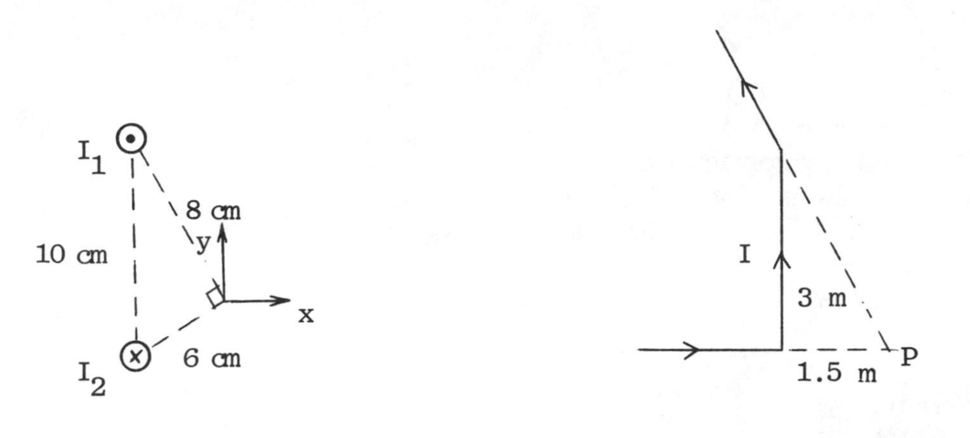

FIGURE 30.11

FIGURE 30.12

2. A long straight wire carries a current in the -z direction. What is the direction of the force on a proton that is on the +x axis and moving (a) in the -y direction; (b) in the -x direction?

3. Use the Biot-Savart law for the field due to a current element to find the field strength at point P in Fig. 30.12.

4. A metal cylinder of radius b has a concentric hole of radius a. It carries a uniformly distributed current I. Determine the field for $a < r < b$.

Chapter 31

ELECTROMAGNETIC INDUCTION

MAJOR POINTS

1. Definition of magnetic flux
2. The production of induced emfs
3. Faraday's law of electromagnetic induction
4. Lenz's law for the direction of the induced emf
5. The operation of dc and ac generators
6. Motional emfs
7. Induced electric fields

CHAPTER REVIEW

MAGNETIC FLUX

The magnetic flux through a plane (flat) area A in a uniform magnetic field B is defined to be

$$\phi_B = \mathbf{B} \cdot \mathbf{A} = B A \cos\theta$$

where the vector **A** is normal (perpendicular) to the plane, as in Fig. 31.1. The SI unit of ϕ_B is the weber (Wb). As is the case with electric flux (Ch. 24), one can think of magnetic flux as being proportional to the number of field lines that intercept a surface. The quantity B is also called the magnetic flux density; thus 1 T = 1 Wb/m^2. If the area is not flat or the field is nonuniform, the magnetic flux is found by integration:

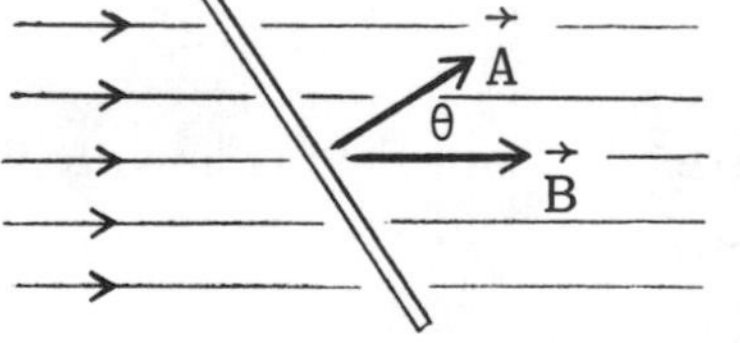

FIGURE 31.1

$$\phi_B = \int \mathbf{B} \cdot d\mathbf{A}$$

FARADAY'S LAW

A source of emf is an agent that is able to move charges around a closed loop. Familiar sources of emf are batteries and solar cells. Faraday discovered that an emf is created, "induced", when the magnetic flux through a loop changes. The induced emf in any closed loop is proportional to the rate of change of flux through it:

$$\mathcal{E} \propto d\phi_B/dt$$

The flux can change because of a change in the magnetic field B, a change in the area of the loop A, a change in the orientation of the loop θ, or a combination of these. Note that the value of the flux, large or small, is not relevant; it is the rate of change of flux that determines the induced emf.

LENZ'S LAW

The direction of the induced emf is such that the induced current creates a magnetic flux that opposes the change in flux that produced it. If $\Delta\phi_B$ is negative, the induced magnetic field is in the same direction as the external field; it tries to prevent the decrease in ϕ_B, as in Fig. 31.2b.

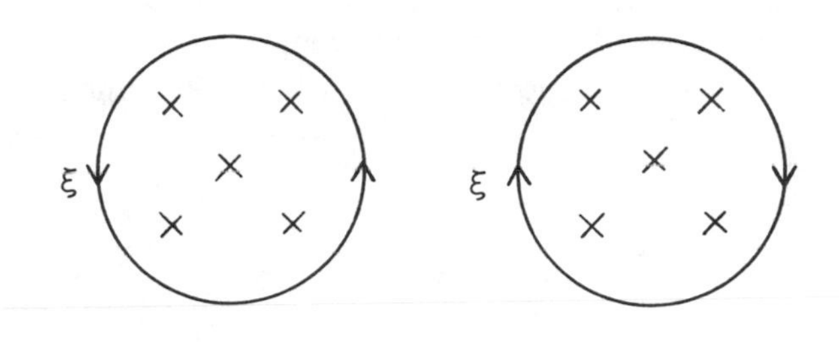

FIGURE 31.2

Lenz's law can be incorporated into Faraday's statement by the simple inclusion of a negative sign. If the loop is actually a flat coil with N turns, then the emf's induced in the turns will be in series. In this case, Faraday's law of induction is

$$\mathcal{E} = - N\, d\phi_B/dt$$

where ϕ_B is the flux through each turn. Positive sense of $\mathcal{E}$ is determined by a right-hand rule with the thumb in the direction of A - which points in the original direction of **B** (see p.621).

When an induced emf arises from the relative motion of a conductor and a magnetic field, it is called a motional emf. One can infer the existence of a motional emf by considering the magnetic force, $\mathbf{F} = q\mathbf{v}\mathbf{x}\mathbf{B}$, on moving charges.

EXAMPLE 1

A plane rectangular coil with sides 10 cm x 15 cm has 25 turns is oriented with its plane at θ = 40° to **B**, see Fig. 31.3. Find average induced emf if B changes from 0.12**i** T to -0.3**i** T in 0.02 s. Indicate the sense of the induced emf.

Solution:

The given angle θ is <u>not</u> the required angle. We need the angle between the normal to the plane and the magnetic field. This angle is $\alpha = 50°$. Thus, the magnetic flux is $\phi_B = BA \cos\alpha$ and the average induced emf is

$$\mathscr{E}_{av} = \Delta\phi/\Delta t = (\Delta B/\Delta t)\, A \cos\alpha$$

$$= -0.202 \text{ V}$$

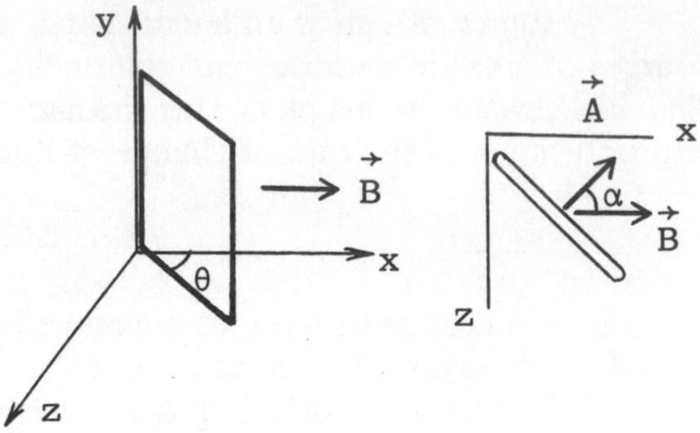

We take the initial flux to be positive, thus the final flux is negative. Since the flux change is negative the induced magnetic field is in the same direction as the external field. From the right hand rule, the induced current flows up along the +y axis, as shown in Fig. 31.3.

FIGURE 31.3

EXAMPLE 2

A conducting rod of length R, pivoted at one end, rotates at ω rad/s with the other end in contact with a circular ring, as shown in Fig. 31.4. Obtain an expression for the induced emf.

Solution:

The area enclosed within the sector of the circle is

$$A = \pi R^2(\theta/2\pi) = R^2\theta/2$$

The rate of change of flux is

$$d\phi_B/dt = B\, dA/dt = BR^2\omega/2$$

where $\omega = d\theta/dt$ is the angular velocity. Since the flux is increasing, the induced field opposes this increase and is directed opposite to the external field. Thus the induced current flows as shown.

FIGURE 31.4

INDUCED ELECTRIC FIELDS

An induced emf has an electric field associated with it. The emf is defined as $\mathcal{E} = W/q$, the work per unit charge in moving a charge around a closed loop. Since $\mathbf{F} = q\mathbf{E}$, we have $W = \int q\mathbf{E}.\, d\boldsymbol{\ell}$. Faraday's law takes the form

$$\oint \mathbf{E}.\, d\boldsymbol{\ell} = -d\phi_B/dt$$

Unlike the electrostatic field, which is conservative, the induced electric field is nonconservative and is not associated with a scalar potential function. The induced electric can exist in free space.

EXAMPLE 3

A ring of radius R lies perpendicular to a uniform magnetic field that varies according to $B = B_o \exp(-t/\tau)$. What is the induced electric field in the ring?

Solution:

By symmetry the induced electric field will have the same value at all points on the ring, Fig. 31.5. Furthermore, the field is tangential to the circular loop and so $\mathbf{E}.d\ell = E\, d\ell$. The integral around the ring is simply

$$\oint \mathbf{E}.d\ell = E \oint d\ell = E(2\pi R)$$

The flux through the loop is

$$\phi_B = \mathbf{B.A} = BA = B(\pi R^2)$$

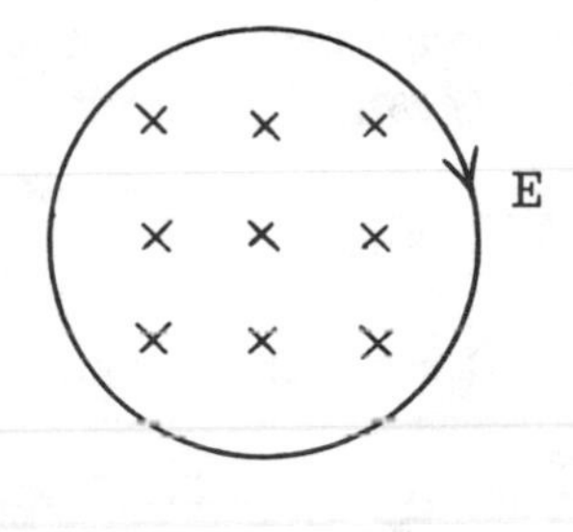

FIGURE 31.5

Faraday's law becomes

$$E(2\pi R) = -(\pi R^2)dB/dt$$

Since $dB/dt = -(B_o/\tau)\exp(-t/\tau)$, we find

$$E = -(R/2)dB/dt = (RB_o/2\tau)\exp(-t/\tau)$$

The positive result means that the induced current produces an induced field that reinforces the external field. It is important to realize, however, that the induced electric field would exist even if the ring were not present.

SOLUTIONS TO SELECTED TEXT EXERCISES AND PROBLEMS

Exercise 15
A rectangular loop of mass m, width ℓ, and resistance R falls vertically through a uniform horizontal field B, as shown in Fig. 31.6. (a) Show that the loop reaches a terminal velocity $v_T = mgR/(B\ell)^2$. (b) Show that at v_T the rate at which gravitational energy is being lost is equal to the electrical power dissipation.

Solution:
(a) The flux through the portion of the loop within the field is $\phi_B = BLy$. The magnitude of the induced emf is

$$\mathscr{E} = d\phi_B/dt = BL\, v$$

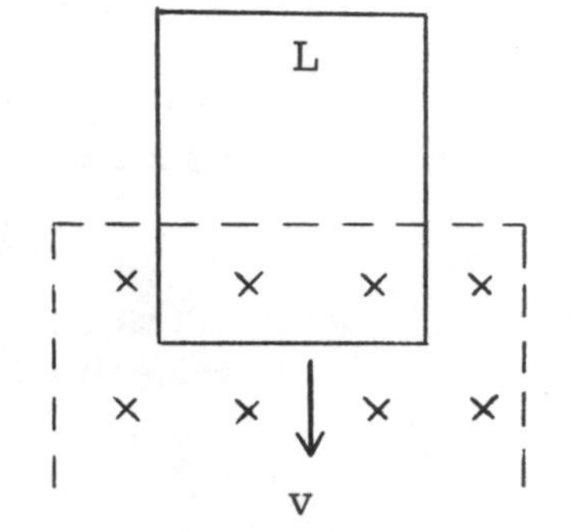

FIGURE 31.6

therefore the induced current is $I = \mathscr{E}/R = BLv/R$. The flux through the loop is increasing and so the induced current flows in a direction such that the induced magnetic field is opposite to the external field. The net vertical force on the horizontal wire within the field is zero.

$$\Sigma F_y = mg - ILB = ma$$

When a = 0, the loop has reached the terminal speed $v_T = mgR/(BL)^2$.
(b) The gravitational potential energy is $U_g = mgy$ and $v_T = -dy/dt$, thus the rate of change is

$$dU_g/dt = -mgv_T = -(mg/BL)^2R$$

The electrical power dissipation is

$$P_{elec} = I^2R = (mg/BL)^2R$$

We see that $-dU_g/dt = P_{elec}$.

Exercise 17
A magnetic field given by $B(t) = 0.2t - 0.5t^2$ T is directed perpendicular to the plane of a circular coil containing 25 turns of radius 1.8 cm and whose total resistance is 1.5 Ω. Find the power dissipation at 3 s.

Solution:
The flux is simply $\phi_B = BA$, so the magnitude of the induced emf is

$$\mathscr{E} = NA\, dB/dt = N(\pi r^2)(0.2 - t)$$

$$= 25\pi(1.8\text{x}10^{-2}\text{ m})^2(2.8\text{ s}) = 71.3\text{ mV}$$

The electrical power dissipated is $P = \mathcal{E}^2/R = 3.38$ mW.

Exercise 23

An electron is at a distance d from the axis of a solenoid, as shown in Fig. 31.7. The uniform magnetic field in the solenoid is changing according to B = Ct. Obtain an expression for the electric force on the electron.

Solution:

In terms of the induced electric field, Faraday's law is

$$\oint \mathbf{E}.d\boldsymbol{\ell} = -d\phi_B/dt$$

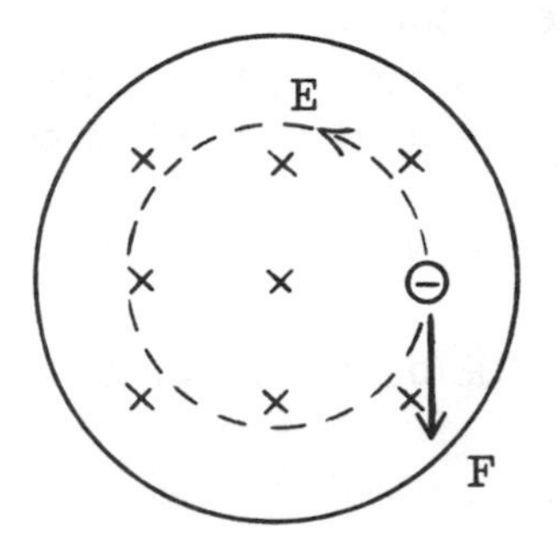

FIGURE 31.7

For a circular loop, **E** is constant in magnitude and tangential to the loop, that is, $\mathbf{E}.d\boldsymbol{\ell} = E\, d\ell$. The integral in Faraday's law becomes $E \int d\ell = (2\pi d)E$, thus

$$(2\pi d)E = A\ dB/dt = (\pi d^2)(C)$$

thus E = dC/2. The magntiude of the force on an electron is F = eE = eCd/2.

Problem 5

An elasticized conducting band is around a spherical balloon. Its plane passes through the center of the balloon. A uniform magnetic field of magnitude 0.4 T is directed perpendicular to the plane of the band. Air is let out of the balloon at 100 cm^3/s at an instant when the radius of the balloon is 6 cm. What is the induced emf in the band?

Solution:

We need to find the rate at which the cross-sectional area of the band is changing. Since the volume of the sphere is $V = 4\pi r^3/3$, we are given

$$dV/dt = (4\pi r^2)(dr/dt) = 10^{-4}\text{ m}^3/\text{s}$$

When r = 0.06 m, we find $dr/dt = 2.21\text{x}10^{-3}$ m/s. The flux through the circular loop is $\phi = \pi r^2 B$. Therefore, the magnitude of the induced emf is

$$\mathcal{E} = d\phi/dt = 2\pi r B\ dr/dt$$

$$= 2\pi(0.06\text{ m})(0.4\text{ T})(2.21\text{x}10^{-3}\text{ m/s}) = 333\ \mu\text{V}$$

Problem 9

Figure 31.8 shows a square loop of side L perpendicular to the uniform field of a solenoid. (a) Show that at any point on a side the component of the induced electric field along the side is

$0.25L\, dB/dt$. (b) Evaluate $\int \mathbf{E}.d\ell$ around the loop.

Solution:

(a) The induced electric field is given by $\int \mathbf{E}.d\boldsymbol{\ell} = -d\phi_B/dt$, where $d\phi_B/dt = A\, dB/dt$. For a circular loop of radius r we obtain (omitting the sign)

$$E(2\pi r) = (\pi r^2)\, dB/dt$$

thus, $E = (r/2)dB/dt$. The electric field lines are circular.

The component of **E** along a side of the rectangular path in the figure is

$$E \cos\theta = E(L/2r) = (L/4)\, dB/dt$$

(b) For one side of the rectangle,

$$\oint \mathbf{E}.d\boldsymbol{\ell} = (E \cos\theta)(L) = (L^2/4)dB/dt$$

For the four sides we have $L^2\, dB/dt$.

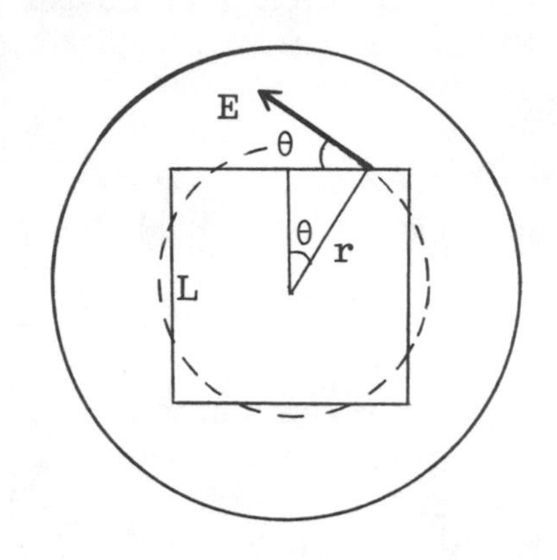

FIGURE 31.8

SELF-TEST

1. A circular loop of radius 15 cm is made of thin wire. It lies with its plane normal to a magnetic field 0.3 T directed into the page. The wire is pulled into a square shape in 0.05 s. What is the average induced emf?

2. A rod slides on two metal rails separated by 0.5 m and that terminate in a resistor R = 1.25 Ω, as shown in Fig. 31.9. The speed of the rod is 25 m/s. The external field of 0.6 T is directed into the page. (a) What is the magnetic force on the rod? (b) What is the mechanical power required to keep the rod moving at constant speed? (c) What is the electrical power dissipated in the rod?

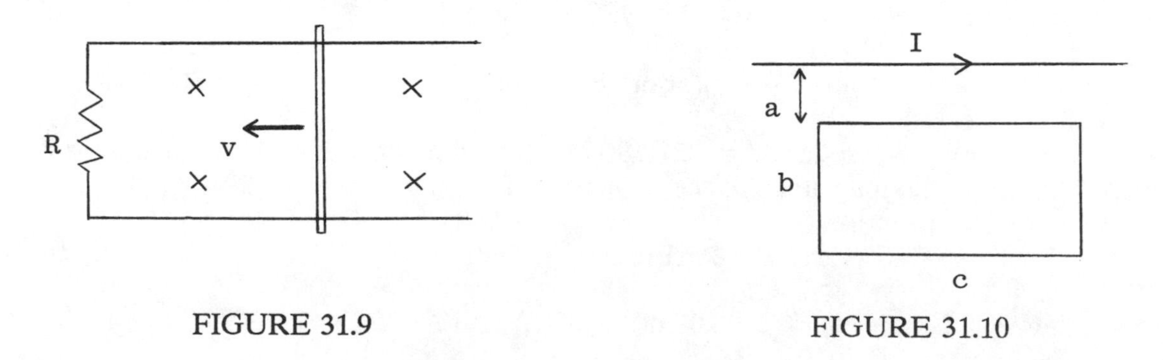

FIGURE 31.9

FIGURE 31.10

3. A current I flows in a long straight wire. A rectangular loop is placed as shown in Fig. 31.10. (a) Find the magnetic flux through the rectangular loop; (b) If the current in the wire varies according to $I = 3t^2 + 7t$ A, what is the induced emf in the loop?

Chapter 32

INDUCTANCE AND MAGNETIC MATERIALS

MAJOR POINTS

1. The definitions of self-inductance and mutual inductance
2. The rise and decay of current in an LR circuit
3. (a) The energy stored in an inductor
   (b) The energy density of the mangetic field
4. Oscillations in an LC circuit.

CHAPTER REVIEW

SELF-INDUCTANCE

When the magnetic field produced by a coil changes, the change in flux causes a "self-induced" emf in the coil. This self-induced emf tends to oppose the change in flux that produces it.

The magnetic field, and hence the magnetic flux, ϕ, produced by a device, such as a coil with N turns, is proportional to the current flowing. The self-inductance, L, of the device may be defined by the relation

$$N \phi = L I$$

where ϕ is the magnetic flux through each turn. The quantity $N\phi$ is called the "flux linkage". The SI unit of L is the henry (H). Alternatively, one can express the self-induced emf,

$$\mathcal{E} = - N\, d\phi/dt = - d(N\phi)/dt$$

in the form

$$\mathcal{E} = - L\, dI/dt$$

Thus L can be determined from a measurement of $\mathcal{E}$ and dI/dt. The polarity of the self-induced emf is determined by the rate of change of current (see Fig. 32.4 in the text), not by the direction of the current. Like capacitance, self-inductance is a quantity that depends on the size and shape of a device and the presence of (magnetic) materials. A device designed to have a specific amount of inductance is called an inductor.

There are two ways in which you can calculate the self-inductance of a device. The first is based on Eq. 32.3, $L = N\phi/I$, which requires a calculation of flux, the second is based on Eq. 32.12, $U = 1/2\, LI^2$, which requires a calculation of the total energy stored in the device.

EXAMPLE 1
When the current through a coil with 40 turns changes at 150 A/s, the self-induced emf is 9 V. (a) What is its self-inductance? (b) If the current is 2.4 A at this instant, what is the magnetic flux through each turn?

Solution:
(a) Ignoring the negative sign, the self-inductance is given by

$$L = \mathcal{E}/(dI/dt)$$

$$= (9\ V)/(150\ A/s) = 0.06\ H$$

(b) The flux through each turn is given by

$$\phi = LI/N$$

$$= (0.06\ H)(2.4\ A)/40 = 3.6\ mWb.$$

Note that we did not assume that the field is uniform, that is, we have not assumed that $\phi = BA$.

MUTUAL INDUCTANCE

When changes in the current flowing in one coil produce an induced emf in a nearby coil, the mutual inductance M of the coils is defined by

$$\mathcal{E}_1 = -M\ dI_2/dt; \qquad \mathcal{E}_2 = -M\ dI_1/dt$$

The value of M depends on the geometries of the coils and their relative positions. Mutual inductance can also be determined from the relations

$$N_1\ \phi_{21} = M\ I_1; \qquad N_2\ \phi_{12} = M\ I_2$$

where ϕ_{21} is the flux through coil 2 produced by the field of coil 1.
The net emf in either coil arises from both mutual inductance and self-inductance.

EXAMPLE 2
When the current in one coil changes at 45 A/s the mutually-induced emf in a second coil is 0.32 V. What is the mutual inductance of the coils?

Solution:
Ignoring the negative sign the mutual inductance is given by

$$M = \mathcal{E}_2/(dI_1/dt)$$

$$= (0.32\ V)/(45\ A/s) = 7.1\ mH.$$

ENERGY

The energy stored in the magnetic field of an inductor is

$$U = 1/2\ LI^2$$

This has the same form as $U = 1/2\ CV^2$ for a capacitor. The energy density of the magnetic field (J/m^3) is

$$u_B = B^2/2\mu_o$$

EXAMPLE 3

A solenoid with 120 turns has a length of 40 cm and a radius of 1.5 cm. It carries a 3 A current. (a) What is the magnetic energy density inside the solenoid? (b) What is the total energy stored? (c) Find the self-inductance. Ignore end corrections.

Solution:

(a) The magnetic field in an infinite solenoid with n turns per unit unit is $B = \mu_o nI$. Thus the energy density is

$$u_B = B^2/2\mu_o = \mu_o\ n^2\ I^2/2$$

$$= (4\pi x10^{-7}\ H/m)(120/0.4\ m)^2\ (3\ A)^2/2 = 0.509\ J/m^3$$

(b) The total energy stored is

$$U = u_B V = u_B(\pi r^2 \ell)$$

$$= (0.509\ J/m^3)(\pi)(1.5x10^{-2}\ m)^2(0.4\ m) = 1.44x10^{-4}\ J$$

(c) Since the total energy is $U = 1/2\ LI^2$, the self-inductance is

$$L = 2U/I^2$$

$$= 2(1.44x10^{-4}\ J)/(9\ A^2) = 32\ \mu H.$$

LR CIRCUITS

Consider the circuit in Fig. 32.1. Switch S_2 is open. When S_1 is closed, the "back emf" of the inductor prevents the current from rising instantaneously. Applying Kirchhoff's loop rule we obtain

$$\mathscr{E} - IR - L\ dI/dt = 0$$

From this equation we find that the current in the circuit will rise according to

$$I = I_o\ [1 - \exp(-t/\tau)]$$

where $I_o = \mathcal{E}/R$ and the time constant $\tau = L/R$. Now if S_2 is closed and S_1 is opened, the current will decay according to

$$I = I_o \exp(-t/\tau)$$

In this case, the self-induced emf opposes the decrease in current and prevents it from decreasing to zero instantaneously. These equations resemble the corresponding equations for the charge on a capacitor in an RC circuit.

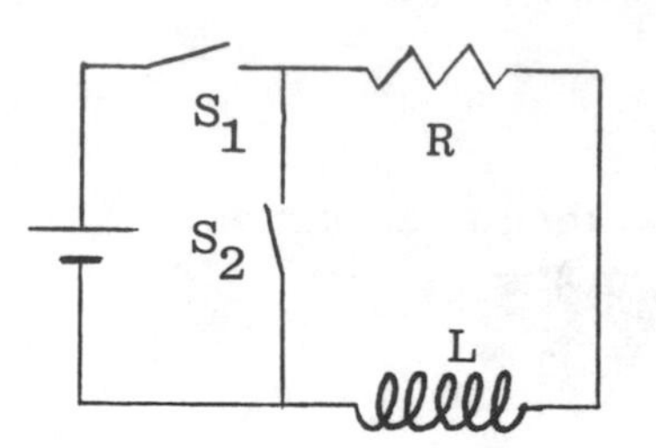

FIGURE 32.1

EXAMPLE 4

An inductor, L = 120 mH, a resistor, R = 4 Ω, and a source of emf $\mathcal{E}$ = 24 V are in series. The switch is closed at t = 0. When the current is 2 A, what is the power supplied (a) to the resistor; (b) to the inductor; and (c) by the source of emf?

Solution:

The current in the circuit rises according to

$$I = I_o [1 - \exp(-t/\tau)]$$

With $I_o = \mathcal{E}/R = 6$ A, and $\tau = L/R = 30$ ms. When I = 2 A, we find $\exp(-t/\tau) = 0.667$, thus $t = -\tau \ln(0.667) = 12.1$ ms.

(a) The power dissipated in the resistor is $P_R = I^2R = 16$ W.

(b) The power supplied to the inductor is (ignoring the negative sign)

$$P_L = I\,\mathcal{E}_L = L\,I\,dI/dt$$

$$= LI\,(I_o/\tau)\exp(-t/\tau)$$

$$= (0.12\text{ H})(2\text{ A})(6\text{ A}/0.03\text{ s})\exp(-12.1/30) = 32\text{ W}$$

(c) The power supplied by the source of emf is

$$P_{\mathcal{E}} = I\mathcal{E} = 48\text{ W}$$

Clearly, $P_{\mathcal{E}} = P_R + P_L$, as we would expect.

LC OSCILLATIONS

Consider an inductor and capacitor connected as shown in Fig. 32.2. By applying Kirchhoff's loop rule we obtain

$$d^2Q/dt^2 + (1/LC)Q = 0$$

This is the differential equation for simple harmonic oscillation (See Ch. 15). Thus, a perfect inductor and capacitor connected to each other will continuously transfer electrical energy stored in C to magnetic energy stored in L, and vice versa, with a natural angular frequency

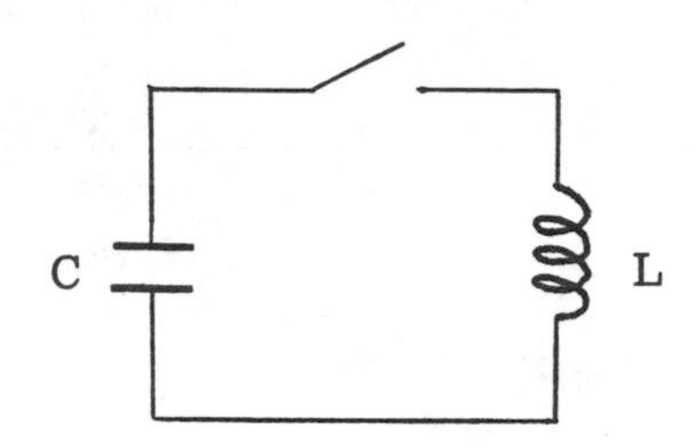

FIGURE 32.2

$$\omega_o = 1/(LC)^{1/2}$$

LC oscillations are analogous to the oscillations of a block-spring system. The energy stored in C is analogous to the potential energy of the spring, whereas the energy stored in L is analogous to the kinetic energy of the block (see Table 32.1 in the text for more details).

In practice, there is resistance in the circuit and electromagnetic radiation. As a consequence, the oscillations are damped.

EXAMPLE 5

A capacitor, $C = 50\ \mu F$, is connected to an inductor, $L = 40$ mH. The initial potential difference across C is 20 V. At $t = 0$ the switch closed. Find: (a) the maximum charge Q_o on C, (b) the maximum energy stored in L; (c) the natural angular frequency; (d) the current when $Q = Q_o/2$. (d) Write Q(t), I(t).

Solution:

(a) The maximum charge on C is simply $Q_o = CV_o = (50\ \mu F)(20\ V) = 1$ mC.

(b) The maximum energies in C and L are equal, so

$$U_{oL} = U_{oC} = Q_o^2/2C = 10 \text{ mJ}$$

(c) The natural angular frequency is

$$\omega_o = (LC)^{-1/2} = 707 \text{ rad/s}$$

(d) The energy of the system is constant, therefore

$$LI^2/2 + Q^2/2C = Q_o^2/2C$$

When $Q = Q_o/2$, we have $LI^2/2 = 0.75(Q_o^2/2C)$, which yields $I = 0.612$ A.

(e) Since $Q = Q_o$ at $t = 0$, we have

$$Q = Q_o \cos(\omega_o t)$$

The current tends to discharge the capacitor, so

$$I = -dQ/dt = I_o \sin(\omega_o t)$$

where $I_o = \omega_o Q_o$.

SOLUTIONS TO SELECTED TEXT EXERCISES AND PROBLEMS

Exercise 13

A toroid with N_1 turns has a rectangular cross-section; see Fig. 32.3. The inner radius is a, the outer radius is b, and the height is h. A coil with N_2 turns is wrapped around the toroid. Find: (a) the flux through the coil; (b) their mutual inductance. (See Example 30.7 in the text.)

Solution:

(a) The magnetic field inside a toroid is given in Example 30.7 of the text. If we divide the cross-sectional area of the toroid into strips of width dr and height h, the infinitesimal flux through such a strip is

$$d\phi = B\,dA = (\mu_o NI/2\pi r)(h\,dr)$$

The total flux is the integral of this from $r = a$ to $r = b$:

$$\phi = \mu_o h N_1 I_1/2\pi \ln(b/a)$$

FIGURE 32.3

(b) The flux we found in part (a) is also ϕ_{21}, the flux in coil 2 due to the field created by coil 1 (the toroid). The mutual inductance is given by

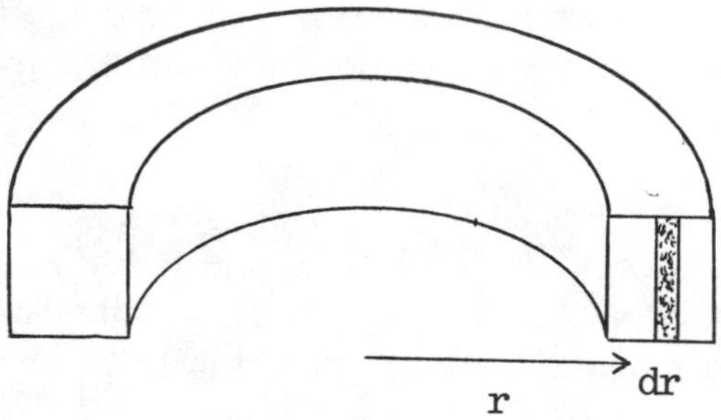

$$M = N_2\phi_{21}/I_1 = (\mu_o h N_1 N_2/2\pi) \ln(b/a)$$

It would not be possible to use the equivalent expression $M = N_1\phi_{12}/I_2$ since ϕ_{12}, the flux created by coil 1, cannot be easily calculated.

Exercise 15
Consider two solenoids with the following specifications: $L_1 = 20$ mH, $N_1 = 80$ turns, $L_2 = 30$ mH, $N_2 = 120$ turns, and $M = 7$ mH. At a certain instant the current in coil 1 is 2.4 A and is increasing at 4 A/s, the current in coil 2 is 4.5 A and increasing at 1.8 A/s. Find the magntiude of: (a) ϕ_{11}; (b) ϕ_{12}; (c) ϕ_{21}; (d) $\mathcal{E}_{11}$; (e) $\mathcal{E}_{12}$; (f) $\mathcal{E}_{21}$.

Solution:
(a) $\phi_{11} = L_1I_1/N_1 = 0.6$ mWb

(b) $\phi_{12} = MI_2/N_1 = 0.39$ mWb

(c) $\phi_{21} = MI_1/N_2 = 0.14$ mWb

(d) $\mathcal{E}_{11} = L_1\, dI_1/dt = 80$ mV

(e) $\mathcal{E}_{12} = M\, dI_2/dt = 12.6$ mV

(f) $\mathcal{E}_{21} = M\, dI_1/dt = 28$ mV

Exercise 25
A solenoid of length 18 cm and radius 2 cm is formed with a tightly wrapped single later of copper wire (diameter 1.0 mm, resistivity 1.7×10^{-8} Ω.m). Estimate its time constant.

Solution:
The resistance of the solenoid of radius a and wrapped with wire of diameter d is

$$R = \rho L/A$$

$$= \rho N(2\pi a)/(\pi d^2/4)$$

where $N = 180$ turns. With $a = 2 \times 10^{-2}$ m and $d = 10^{-3}$ m we find $R = 0.49$ Ω. The field inside an infinite solenoid is $B = \mu_o nI$ and so the flux is $\phi = BA = \mu_o nIA$. If ℓ is the length of the solenoid, the self inductance is

$$L = N\phi/I = \mu_o n^2 A\ell$$

$$= 2.84 \times 10^{-4} \text{ H}$$

Finally, the time constant is $\tau = L/R = 0.58$ ms.

Exercise 31
A resistor $R = 5$ Ω is in series with an inductor $L = 40$ mH and a 20-V battery. The switch is closed at $t = 0$. At what time will the power loss in R equal the rate at which energy is being stored in L? Also express your answer in terms of the number of time constants.

Solution:

The power dissipated in R is $P_R = I^2R$ and the power supplied to L is $P_L = I\mathcal{E}_L = IL\, dI/dt$. When we set $P_R = P_L$ we obtain

$$IR = L\, dI/dt$$

The current rises according to $I = I_o(1 - e^{-t/\tau})$ and so $dI/dt = (I_o/\tau)e^{-t/\tau}$. Using these in the above equation we find

$$(1 - e^{-t/\tau}) = e^{-t/\tau}$$

thus $\exp(-t/\tau) = 0.5$, which leads to $t = \tau \ln 2 = 5.54$ ms

Problem 7

(a) Show that in an underdamped RLC oscillation, the fraction of energy lost per cycle is

$$|\Delta U|/U \approx 2\pi/Q$$

where $Q = \omega L/R$ is called the Q-factor. (Hint: Use Eq. 32.19 to calculate the total energy at two times one period apart. The cosine will have the same value. Then use $e^x \approx 1 + x$ for small x.) (b) If the energy loss per cycle is 2%, what is Q? (c) For the Q found in part (b), suppose R = 0.5 Ω, and L = 18 mH. What is C?

Solution:

(a) The charge on C in a damped oscillation is given by Eq. 32.19. For two times one period apart, the cosine term will have the same value, which we may take to be one. The energies at t and t + T are

$$U_1 = Q_o^2/2 \exp(-Rt/L); \qquad U_2 = Q_o^2/2 \exp[-(t + T)/L]$$

The change in energy is

$$\Delta U = U_1[1 - \exp(-RT/L)]$$

$$\approx U_1 (RT/L) = U_1(2\pi R/\omega L)$$

where we have used $T = 2\pi/\omega$. Thus, $\Delta U/U \approx 2\pi/Q$ where $Q = \omega L/R$.

(b) $Q = 2\pi/0.02 = 314$

(c) With $\omega \approx \omega_o = (LC)^{-1/2}$, we have $Q = (L/C)^{1/2}/R$, so $C = 7.3\ \mu F$.

Problem 12
Given that the charge on the capacitor in an RLC sircuit varies as

$$Q(t) = Q_o \exp(-Rt/2L) \cos(\omega' t)$$

obtain an expression for the current as a function of time. Show that if $R/2L << \omega'$, then the magnitude of the current can be written in the form

$$I(t) \approx A(t) \sin(\omega' t + \delta)$$

where $A(t) = \omega' Q_o \exp(-Rt/2L)$ and $\tan\delta = R/(2L\omega')$.
Solution:
We are given

$$Q = Q_o \exp(-Rt/2L)\cos(\omega' t)$$

The current is $I = dQ/dt$:

$$I = -\omega' Q_o \exp(-Rt/2L)[\sin(\omega' t) + (R/2\omega' L)\cos(\omega' t)]$$

We may express the terms in the square brackets in the form

$$[\sin(\omega' t)\cos\delta + \cos(\omega' t)\sin\delta]$$

Let $\tan\delta = R/2\omega' L$, then $\cos\delta \approx 1$, and $\sin\delta \approx \tan\delta$. Thus

$$I \approx A \sin(\omega' t + \delta)$$

where $A = \omega' Q_o \exp(-Rt/2L)$.

SELF-TEST

1. The current in an inductor is given by $I = 4 + 21t$ A, where t is in seconds. The self-induced emf is 15 mV. Find (a) the self-inductance; (b) the energy stored in this inductor at 0.5 s.

2. The natural frequency of an LC oscillator is 800 kHz. If L = 6 mH, find C.

3. In an RL circuit with L = 50 mH, the current increases from zero to 30% of the final value in 12 ms. (a) What is the resistance? (b) If the source emf is 30 V, at what rate is power supplied to L at 12 ms?

Chapter 33

A C CIRCUITS

MAJOR POINTS

1. The distinction between instantaneous, peak, and root mean square (rms) values of current and potential difference.
2. The use of phasors to determine the phase angle between current and potential difference.
3. RLC series circuit:
   (a) Ohm's law for an ac circuit
   (b) Impedance
4. Resonance in an RLC series circuit.
5. RMS power delivered by an ac source.
6. The operation of a transformer.

CHAPTER REVIEW

INSTANTANEOUS AND RMS VALUES

Instantaneous values of current and potential difference are denoted by lower case letters v and i. It is assumed that the current varies according to

$$i = i_o \sin(\omega t)$$

where i_o is the peak instantaneous value. In a series circuit, the current has the same value at all points in the circuit. The potential difference across each element, and the circuit as a whole, will not in general be in phase with the current, thus we write

$$v = v_o \sin(\omega t + \phi)$$

where ϕ is called the phase angle. For a resistor $\phi = 0$, for an inductor $\phi = +\pi/2$ rad, which means that the potential difference leads the current by one-quarter cycle, and for a capacitor $\phi = -\pi/2$ rad.

The instantaneous potential differences across R, L and C are written as v_R, v_L and v_C and the peak values as v_{oR}, v_{oL}, and v_{oC}. The root mean square (RMS) values of current and potential difference are "average" values that result in the same power dissipation as a direct current of the same magnitude. RMS values are denoted by upper case letters, V and I. The relationship between the RMS and peak values is

$$V = v_o/(2)^{1/2}; \qquad I = i_o/(2)^{1/2}$$

In an RLC series circuit, the sum of the <u>instantaneous</u> potential differences is zero (Kirchhoff's loop rule). However, this does NOT apply to either the peak values or to the RMS values because the peak values for R, L and C occur at different times.

PHASORS

A phasor is a rotating vector used to represent a physical quantity that varies sinusoidally in time. The length of the phasor is the amplitude or peak value of the quantity. The "projection" or component of the phasor onto the vertical axis is the instantaneous value of the quantity.

Through phasor analysis we find that the potential difference across an inductor leads the current by $\pi/2$ radians (one-quarter of a period). The potential difference across a capacitor lags the current by $\pi/2$ radians.

OHM'S LAW

Ohm's law for an RLC series circuit may be written in terms of the <u>peak</u> values or the <u>rms</u> values of current and potential differernce:

$$v_o = i_o Z; \qquad V = I Z$$

where the impedance is

$$Z = [R^2 + (X_L - X_C)^2]^{1/2}$$

and $X_L = \omega L$ and $X_C = 1/\omega C$ are the reactances of L and C. The phasor analysis allows us to depict this relationship as in Fig. 33.1. Impedance plays the same role for an ac circuit that resistance does for a dc circuit: Z tells us the potential difference required to produce unit current.

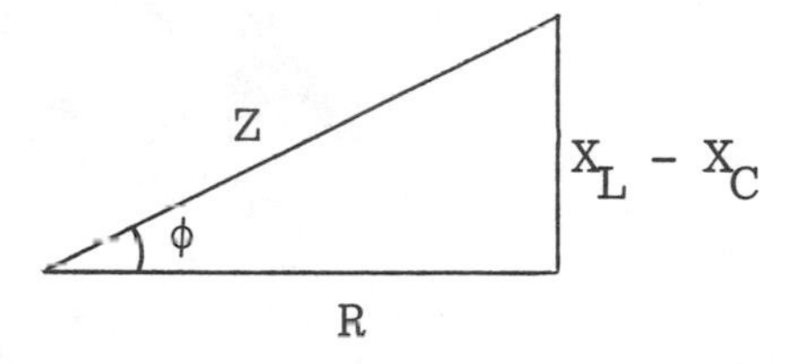

FIGURE 33.1

The phase angle between the instantaneous applied potential difference and the instantaneos current is

$$\tan\phi = (X_L - X_C)/R$$

If ϕ is positive, the v leads i; if ϕ is negative, v lags i.

EXAMPLE 1

In an RLC series circuit $R = 16\ \Omega$ and $L = 250$ mH. For what value of C would the phase angle be (a) $+\pi/4$; (b) $-\pi/4$? The source frequency is 60 Hz.

Solution:

The phase angle is given by

$$\tan\phi = (\omega L - 1/\omega C)/R$$

(a) If $\phi = \pi/4$, then $\tan\phi = 1$, thus

$$\omega L - 1/\omega C = R$$

and, $$1/C = \omega(\omega L - R)$$

With $\omega = 2\pi f = 120\pi$ rad/s, we find $C = 33.9\ \mu F$.
(b) If $\phi = -\pi/4$, then $\tan\phi = -1$, so

$$\omega L - 1/\omega C = -R$$

and

$$1/C = \omega(\omega L + R)$$

This leads to $C = 24.1\ \mu F$.

EXAMPLE 2
A series RLC circuit has the following elements $R = 12\ \Omega$, $L = 30$ mH, and $C = 80\ \mu F$. The rms potential difference of the source is 50 V and its frequency is 60 Hz. Find the rms potential differences V_R, V_L, V_C. What is the relation between these potential differences?

Solution:
First we need to find the impedance and the rms current. The angular frequency is $\omega = 2\pi f = 377$ rad/s. The impedance is

$$Z = [R^2 + (\omega L - 1/\omega C)^2]^{1/2}$$

$$= (144 + 477)^{1/2} = 24.9\ \Omega$$

The rms current through the ciruit is

$$I = V/Z = 2.0\ A$$

The rms potential differences across the elements are

$$V_R = IR = 24\ V$$

$$V_L = IX_L = I\,\omega L = 22.6\ V$$

$$V_C = IX_C = I/\omega C = 66.3\ V$$

Note that $V \neq V_R + V_L + V_C$. This is because the rms values differ by only the factor $(2)^{1/2}$ from the peak values and the peak values occur at different times. However, the phasors, which are related to the peak values, are related by the triangle in Fig. 33.2. Thus,

$$V^2 = V_R^2 + (V_L - V_C)^2$$

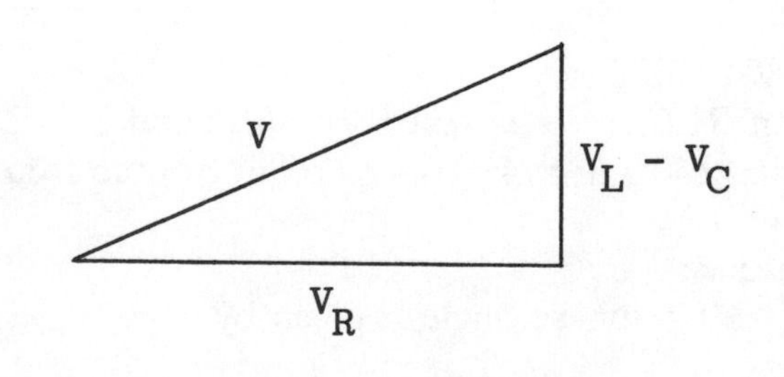

FIGURE 33.2

POWER

The rms power delivered by a source is dissipated only in the resistance in the circuit:

$$P = I^2R = IV\cos\phi$$

where the quantity $\cos\phi$ is called the power factor. Note that V is the rms potential difference across the source, not the resistance.

SERIES RESONANCE

When $X_L = X_C$, that is when $\omega L = 1/\omega C$, the impedance has its minimum value $Z = R$ and the current therefore is a maximum:

$$I_{max} = V/R$$

The (natural) resonance angular frequency at which this occurs is

$$\omega_o = 1/(LC)^{1/2}$$

At resonance, the phase angle is zero; that is, the current is in phase with the potential difference across the circuit. A small R leads to a sharp resonance, whereas a large R results in a broad resonance with a lower height. The power dissipation also displays a resonance at ω_o.

EXAMPLE 3

In an RLC series circuit, $R = 15\ \Omega$, $L = 0.2$ H, and $C = 30\ \mu F$. Find the ratio (a) I/I_{max}, and (b) P/P_{max}, when the frequency is 10% higher than the resonance frequency.

Solution:

(a) The rms current is given $I = V/Z$. At resonance, $Z = R$ and $I_{max} = V/R$, thus

$$I/I_{max} = R/Z$$

With the given values $\omega_o = (LC)^{-1/2} = 408$ rad/s, thus $\omega = 1.1\omega_o = 449$ rad/s. The impedance is

$$Z = [R^2 + (\omega L - 1/\omega C)^2]^{1/2} = 21.6\ \Omega$$

Thus, $I/I_{max} = R/Z = 0.694$.

(b) The power dissipation is $P = I^2R = (V/Z)^2R$, and $P_{max} = (V/R)^2R$, thus

$$P/P_{max} = (R/Z)^2 = 0.482$$

TRANSFORMERS

A transformer consists of a primary coil with N_1 turns and a secondary coil with N_2 turns, usually wound around a common soft-iron core. The ratio of the potential differences across the primary and secondary is equal to the "turns ratio"

$$V_1/V_2 = N_1/N_2$$

This allows one to "step up" or "step down" potential differences. In an ideal transformer there is no energy loss, so the all the power is transferred from the primary to the secondary:

(Ideal transformer) $$I_1 V_1 = I_2 V_2$$

EXAMPLE 4

Tranformer has 140 turns in the primary and 350 turns in the secondary. The power transfer is 90% efficient. Determine the secondary potential difference and current if the primary potential difference is 30 V and the primary current is 4 A.

Solution:

The ratio of the primary and secondary potential differences is

$$V_2/V_1 = N_2/N_1$$

Thus V_2 = (30 V)(350/140) = 75 V. SInce the power transfer is 90% efficient, we have

$$I_2V_2 = 0.9I_1V_1$$

The secondary current is I_2 = 0.9(4 A)(30 V/75 V) = 1.44 A.

SOLUTIONS TO SELECTED TEXT EXERCISES AND PROBLEMS

Exercise 7

A 108 μF capacitor is connected to a source that operates at 80 Hz with a peak potential difference of 24 V. Find: (a) the peak current; (b) the current when the potential difference has its peak value; (c) the current when the potential difference has one-half the (positive) peak value (there are two answers); (d) the instantaneous power delivered to the capacitor at 1 ms.

Solution:

(a) The peak current is $i_o = v_o/X_C = v_o\omega C = 1.3$ A

(b) Since the current lags the potential difference by 90°, the current is zero when v_o has its peak value.

(c) The instantaneous potential difference across C is given by

$$v_C = -v_{oC} \cos(\omega t)$$

If $v_C = 0.5v_{oC}$, we have $-0.5 = \cos(\omega t)$. Thus, $\omega t = 2\pi/3, 4\pi/3$. Using these values, the instantaneous current is

$$i = i_o \sin(\omega t) = \pm 1.13 \text{ A}$$

(d) The instantaneous power delivered to C is

$$p = i\, v_C = -i_o v_{oC} \sin(\omega t)\cos(\omega t)$$

$$= 13.2 \text{ W}$$

Exercise 21
In an RLC series circuit take $R = 40\ \Omega$, $L = 20$ mH, $C = 60\ \mu$F. Find at what frequency the potential difference leads the current by 30°.

Solution:
With a phase angle $\phi = +30°$, we have

$$\tan\phi = (\omega L - 1/\omega C)/R = 0.577$$

From this we obtain

$$\omega^2 L - 23.1\omega - 1/C = 0$$

Solving the quadratic leads to $f = 264$ Hz.

Exercise 37
A step-down transformer has 600 V across the primary and 120 V across the secondary. The secondary has 80 turns. (a) What is the number of turns in the primary? (b) If a load resistance $R_L = 10\ \Omega$ is in the secondary, what is the primary current?

Solution:
(a) The turns ratio is $N_1/N_2 = V_1/V_2$, thus

$$N_1 = (600 \text{ V}/120 \text{ V})80 = 400$$

(b) The secondary current is $I_2 = V_2/R = 12$ A. Assuming an ideal transformer, the power transfer is 100% efficient, thus

$$I_1 V_1 = I_2 V_2$$

Thus, $I_1 = (120 \text{ V}/600 \text{ V})(12 \text{ A}) = 2.4$ A.

Problem 3
For the function shown in Fig. 33.3, determine (a) the average value and (b) the rms value of the potential difference.

Solution:
(a) The area under the function for one 4-s cycle is 8 V.sThus v_{av} = 2 V.

(b) We need consider only one-half cycle. The potential difference function is v = at, where a = 2 V/s. The average value of a function F(t) in the interval t_1 to t_2 is

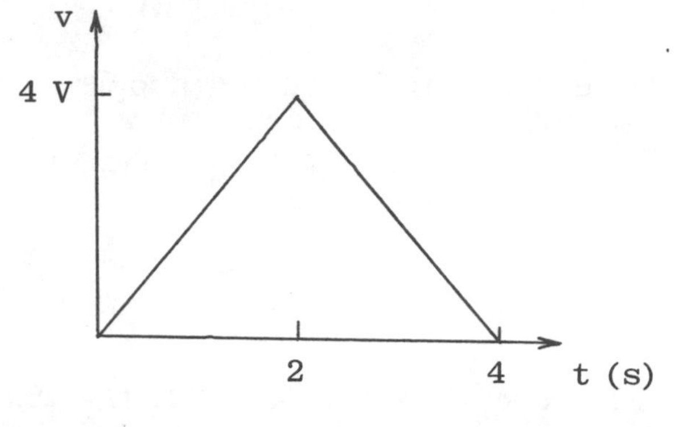

FIGURE 33.3

$$F_{av} = \int_{t_1}^{t_2} F\,dt/(t_2 - t_1)$$

In the present case $F = v^2 = (at)^2$ and $(t_2 - t_1) = T/2$. Thus,

$$(v^2)_{av} = 2/T \int_0^{T/2} (at)^2\,dt$$

$$= (2/T) \left| a^2\, t^3/3 \right|_0^{T/2} = 8/3$$

Thus $v_{rms} = [(v^2)_{av}]^{1/2} = (8/3)^{1/2}$ V.

Problem 13
Figure 33.4 shows a simple filter circuit. The input ac potential difference V_{in}, is across R and C, whereas the output potential difference, V_{out}, is that across C. Show that the ratio V_{out}/V_{in} is

$$V_{out}/V_{in} = (1 + \omega^2 R^2 C^2)^{-1/2}$$

Solution:
The impedance of R and C is $Z = [R^2 + 1/\omega^2 C^2]^{1/2}$. The input potential difference is across Z, that is, $V_{in} = IZ$, whereas the output potential difference is across C, that is, $V_{out} = IX_C = I/\omega C$. Thus,

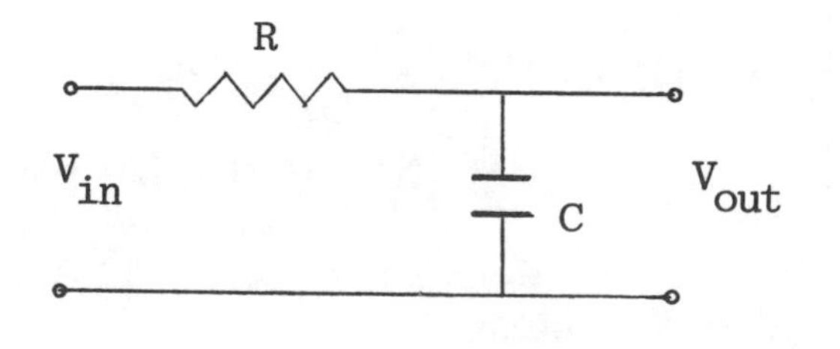

$$V_{out}/V_{in} = (1/\omega C)/Z$$

$$= 1/(\omega C)(R^2 + 1/\omega^2 C^2)^{1/2}$$

FIGURE 33.4

SELF-TEST

1. In an RLC series $R = 80\ \Omega$, $L = 200$ mH, $C = 50\ \mu F$, and the source potential difference is 10^3 V (rms) and its frequency is 120 Hz. Find:
   (a) the impedance;
   (b) the phase angle;
   (c) the rms current through L;
   (d) the peak potential difference across C;
   (e) the rms power supplied by the source;
   (f) the natural frequency f_o;
   (g) the rms current at f_o.

2. In an series RLC circuit, $R = 7\ \Omega$, the rms current is 2 A, and the rms potential difference across the source is 24 V. Find the phase angle given that $\omega L < 1/\omega C$.

Chapter 34

MAXWELL'S EQUATIONS; ELECTROMAGNETIC WAVES

MAJOR POINTS

1. The displacement current
2. (a) Maxwell's wave equation
   (b) Properties of electromagnetic waves
3. The Poynting vector indicates the intensity of an EM wave
4. EM waves transport linear momentum and exert radiation pressure

CHAPTER REVIEW

MAXWELL'S EQUATIONS

All the phenomena of classical electromagnetism are described by Maxwell's Equations. In free space these are stated below. They can be modified to take into account the presence of materials.

Gauss's law $\oint \mathbf{E} \cdot d\mathbf{A} = Q/\epsilon_o$

Gauss's law for the electric field relates the electric flux through a closed surface to the net charge enclosed by that surface. From this we can infer that E α $1/r^2$ for a point charge and that electrostatic field lines begin and end on point charges.

Gauss's law $\oint \mathbf{B} \cdot d\mathbf{A} = 0$

Gauss's law for magnetism tells us that the net magnetic flux through a closed surface is zero. This arises from the fact that there are no magnetic monopoles. It tells us that magnetic field lines a re closed loops.

Faraday's law $\oint \mathbf{E} \cdot d\boldsymbol{\ell} = - d\phi_B/dt$

Faraday's law of electromagnetic induction tells us that there is an induced emf in a closed loop through which the magnetic flux is changing. The lines of the induced electric field are closed loops.

Ampere-Maxwell law $\oint \mathbf{B} \cdot d\boldsymbol{\ell} = \mu_o(I + \epsilon_o\, d\phi_E/dt)$

The Ampere-Maxwell law tells us that a magnetic field can by produced by a conduction current, I, and by displacement current $\epsilon_o\, d\phi_E/dt$, associated with a change in electric flux. These four equations must be supplemented by the Lorentz force,

$$\mathbf{F} = q(\mathbf{E} + \mathbf{v} \times \mathbf{B})$$

and the conservation of charge.

ELECTROMAGNETIC WAVES

Starting with Faraday's law (Eq. 34.5) and the Ampere-Maxwell law (Eq. 34.6) in free space far from the source of the fields (and with $I = 0$), it can be shown that both the electric and magnetic fields satisfy the linear wave equation. For example,

$$\partial^2 E/\partial x^2 = (\epsilon_o \mu_o)\, \partial^2 E/\partial t^2$$

This is a wave equation for electromagnetic waves in free space. A solution to this differential equation is a plane wave in the x direction

$$E_y = E_o \sin(kx - \omega t)$$

$$B_z = B_o \sin(kx - \omega t)$$

The main properties of these waves are:

1. They are transverse waves with E and B perpendicular to each other and to the direction of propagation
2. In vacuum, the speed of the waves is

$$c = (\mu_o \epsilon_o)^{-1/2} \approx 3\text{x}10^8 \text{ m/s}$$

3. E is in phase with B and their magnitudes in vacuum are related by

$$E = cB$$

EM radiation is produced by accelerating charges - often in an antenna connected to an oscillating source of emf.

EXAMPLE 1

A capacitor with circular plates is being charged in an RC circuit. Obtain an expression for the displacement current between the plates.

Solution:

The displacement current is given by

$$I_D = \epsilon_o\, d(EA)/dt$$

Where $E = \sigma/\epsilon_o$ and $Q = \sigma A$, thus $E = Q/\epsilon_o A$, and so $I_D = dQ/dt$. The charge on the capacitor varies according to

$$Q = Q_o(1 - e^{-t/\tau})$$

where τ = RC. Thus,

$$dQ/dt = (Q_o/\tau)\, e^{-t/\tau}$$

The displacement current between the capacitor plates is equal to the conduction current in the circuit.

ENERGY TRANSPORT AND THE POYNTING VECTOR

The instantaneous energy densities of the electric and magnetic fields are equal. The total energy density (J/m^3) can be written in different forms:

$$u = \epsilon_o E^2 = B^2/\mu_o = (\epsilon_o/\mu_o)^{1/2} EB$$

The time-averaged energy density is $u_{av} = u_o/2$:

$$u_{av} = \epsilon_o E_o^2/2 = B_o^2/2\mu_o$$

Note that the peak values E_o and B_o appear in these expressions for u_{av}. The magnitude of the Poynting vector,

$$\mathbf{S} = \mathbf{E} \times \mathbf{B}/\mu_o$$

is the intensity of the wave, that is, the energy that flows per unit time per unit area normal to the direction of propagation of the wave. The SI unit of S is W/m^2. Using the expressions for E_y and Bz given earlier, we see that the instantaneous magnitude of the Poynting vector is $S = E_o B_o \sin^2(kx - \omega t)/\mu_o$ At any given value of x, the average of $\sin^2(\omega t)$ is 1/2, therefore the time-averaged intensity is

$$S_{av} = u_{av} c = E_o^2/2\mu_o c = E_o B_o/2\mu_o$$

EXAMPLE 2

The magnetic field of a plane EM wave is given by

$$B_x = 2x10^{-7} \sin(0.2z + 6x10^7 t)\ T$$

(a) What is the direction of propagation? (b) Write an expression for the electric field; (c) What is the (average) energy density? (d) Find the (average) intensity.

Solution:

(a) The direction of propagation is along the -z axis.

(b) We know that $\mathbf{B} = B\mathbf{i}$ and that the wave is propagating in the -z direction, thus $\mathbf{S} = -S\mathbf{k}$. The fields are related to the direction of **S** by $\mathbf{S} = \mathbf{E} \times \mathbf{B}/\mu_o$. Since $-\mathbf{k} = \mathbf{j} \times \mathbf{i}$, it follows that $\mathbf{E} = +E\mathbf{j}$. The amplitude of the electric field is $E_o = cB_o = 60$ V/m, thus

$$E_y = 60 \sin(0.2z + 6x10^7 t)\ V/m$$

(c) The average energy density is

$$u_{av} = 1/2\ \epsilon_o E_o^2$$

$$= 1/2\ (8.85\text{x}10^{-12}\ \text{F/m})(60\ \text{V/m})^2 = 1.59\text{x}10^{-8}\ \text{J/m}^3$$

(d) The average intensity is

$$S_{av} = u_{av}\, c = E_o B_o / 2\mu_o = 4.77\ \text{W/m}^2$$

EXAMPLE 3
A point source radiates 40 kW. What are the amplitudes of E and B at a distance of 10 km? (b) What is the energy delivered to a disk of area 0.4 m^2 set normal to the waves in 5 min?

Solution:
(a) The energy radiated by a point source spreads uniformly over ever increasing spheres. Thus the intensity is

$$S_{av} = P/A = P/4\pi r^2$$

$$= E_o^2 / 2\mu_o c$$

Thus,

$$E_o^2 = 2\mu_o cP/4\pi r^2$$

$$= 2(4\pi\text{x}10^{-7}\ \text{H/m})(3\text{x}10^8\ \text{m/s})(4\text{x}10^4\ \text{W})/4\pi(10^8\ \text{m}^2)$$

$$= 2.4\text{x}10^{-2}\ \text{V}^2/\text{m}^2$$

Thus $E_o = 0.155$ V/m.
(b) The energy incident in a time t on an area A set normal to the direction of propagation is

$$U = S_{av} A t$$

$$= (P/4\pi r^2)(0.4\ \text{m}^2)(300\ \text{s}) = 3.82\ \text{mJ}$$

MOMENTUM AND RADIATION PRESSURE

The linear momentum p carried by an EM wave is related to the energy U that it transports:

(Momentum) $$p = U/c$$

If the wave is incident normally on a surface that absorbs all the energy, then the momentum transferred to the surface is $p = U/c$. If the surface is perfectly reflecting, the momentum transfer is doubled, $p = 2U/c$.

The radiation pressure exerted by an EM wave incident normally on a surface that completely absorbs the radiation is

(Pressure) $$F/A = S/c = u$$

If the surface is perfectly reflecting, the pressure is doubled.

EXAMPLE 4
A laser has an average 10 W output and a beam of radius 1 mm. Find (a) the average energy in 1 m of the beam; (b) the radiation pressure on a perfectly reflecting surface.

Solution:
(a) The intensity of the beam is

$$S_{av} = P_{av}/A$$

$$= (10\ \text{W})/\pi(10^{-3}\ \text{m})^2 = 3.18\text{x}10^6\ \text{W/m}^2$$

The average energy density is $u_{av} = S_{av}/c$. The energy contained in a length ℓ of the beam is

$$U = u_{av}\ (\text{Volume}) = (S_{av}/c)(\pi r^2 \ell)$$

$$= (1.06\text{x}10^{-2}\ \text{J/m}^3)(\pi 10^{-6}\ \text{m}^3) = 3.33\text{x}10^{-8}\ \text{J}$$

(b) Since the surface is perfectly reflecting the radiation pressure is

$$F/A = 2S/c$$

$$= 2(3.18\text{x}10^6\ \text{W/m}^2)/(3\text{x}10^8\ \text{m/s}) = 2.12\text{x}10^{-2}\ \text{N/m}^2.$$

SOLUTIONS TO SELECTED TEXT EXERCISES AND PROBLEMS

Exercise 7
The circular plates of a parallel plate capacitor have a radius of 2 cm and are separated by 4 mm. They are connected to a 60-Hz ac source with a peak potential difference of 120 V. Find the peak magnetic field halfway from the center to the edge of the plates.

Solution:
Use the expression for B in Example 34.1 with $E = V/d$ and $\epsilon_o\mu_o = 1/c^2$, so

$$B = (dV/dt)r/2dc^2$$

Since $V = V_o \sin(\omega t)$, we have $dV/dt = \omega V_o \cos(\omega t)$. The peak value of $dV/dt = \omega V_o$, thus the peak value of the magnetic field is

$$B_o = \omega V_o R/4dc^2 = 6.28 \text{ x } 10^{-13}\ \text{T}$$

Exercise 11
The magnetic field in a plane electromagnetic wave is given by

$$B_y = 2\text{x}10^{-7} \sin(0.5\text{x}10^3 x + 1.5\text{x}10^{11} t) \text{ T}$$

(a) What is the wavelength and frequency of the wave? (b) Write an expression for the electric field vector.

Solution:
(a) The argument of the sine function is $(kx + \omega t)$, so

$$k = 2\pi/\lambda = 0.5\text{x}10^3 \text{ rad/s, thus } \lambda = 1.26 \text{ cm}$$

$$\omega = 2\pi f = 1.5\text{x}10^{11} \text{ rad/s, thus } f = 23.9 \text{ GHz}$$

(b) The peak electric field is $E_o = cB_o = 60$ V/m. The wave is traveling in the -x direction, that is $\mathbf{S} = -S\mathbf{i}$ and we know that $\mathbf{B} = B\mathbf{j}$. Since $\mathbf{S} = \mathbf{E}\text{x}\mathbf{B}/\mu_o$ and $-\mathbf{i} = \mathbf{k} \text{ x } \mathbf{j}$, we see that $\mathbf{E} = E\mathbf{k}$. Finally,

$$E_z = 60 \sin(500x + 1.5 \text{ x } 10^{11} t) \text{ V/m}$$

Exercise 21
At a distance of 100 m from a point source, the amplitude of the magnetic field is equal in magnitude to 0.1% of the earth's field, that is, about
10^{-7} T. Estimate the power output of the transmitter.

Solution:
The intensity of a point source decreases as the inverse square of the distance. Expressing the intensity in terms of the magnetic field we have

$$S_{av} = B_o^2 c/2\mu_o = P/4\pi r^2$$

thus

$$P = 4\pi r^2 B_o^2 c/2\mu_o$$

$$= 150 \text{ kW}$$

Exercise 25
The intensity of a plane electromagnetic wave is 5 W/m^2. It is incident on a perfectly reflecting surface. Find: (a) the radiation pressure; (b) the force exerted on a panel 60 cm x 40 cm set perpendicular to the direction of wave propagation.

Solution:
(a) Since the surface is perfectly reflecting the radiation pressure is

$$F/A = 2S/c$$

$$= 3.33\text{x}10^{-8}\ \text{N/m}^2$$

(b) The force exerted on the panel is

$$F = 2SA/c = 8\text{x}10^{-9}\ \text{N}$$

Problem 7
A cylindrical coil used as an AM antenna has 250 turns of diameter 1.5 cm. Find the peak emf induced by a 10^4 W station (treated as a point source) broadcasting at 800 kHz at a distance of 2 km. The axis of the coil is parallel to the direction of the magnetic field of the wave.

Solution:
The intensity of a point source decreases as the inverse square of the distance. Expressing S in terms of the magnetic field we have

$$S_{av} = P_{av}/4\pi r^2 = B_o^2 c/2\mu_o$$

$$B_o^2 = 2\mu_o P/4\pi c r^2$$

We find B_o = 1.29 nT. The magnetic flux through the coil (at x = 0 say) is

$$\phi_B = BA = AB_o \sin(\omega t)$$

The induced emf is $\mathscr{E} = -Nd\phi_B/dt = -NAB_o\omega \cos(\omega t)$. Thus the peak emf is

$$\mathscr{E}_o = NAB_o\omega = 2.87\text{x}10^{-4}\ \text{V}$$

SELF-TEST
1. The electric field vector in an electromagnetic wave is given by

$$E_x = 0.12 \sin(1.05\text{x}10^7 y + 3.15\text{x}10^{15} t)\ \text{V/m}$$

(a) Write an expression for the magnetic field.
(b) What is the (average) intensity of the wave?

2. An electromagnetic wave with an intensity of 20 W/m^2 is incident on a surface. Sixty percent of the energy is reflected and 40% absorbed. What is the radiation pressure exerted on the surface?

Chapter 35

LIGHT: REFLECTION AND REFRACTION

MAJOR POINTS

1. (a) The law of reflection
   (b) Definition of refractive index
   (c) Snell's law of refraction
   (d) Total internal reflection
2. Huygen's principle for the construction of wavefronts
3. Dispersion and the spectrum produced by a prism
4. The distinction between real and virtual objects and images
5. (a) Images formed by concave and convex mirrors
   (b) The mirror formula
   (c) Transverse (lateral) magnification
6. Principle ray diagrams

CHAPTER REVIEW

Reflection

The angle of incidence is equal to the angle of reflection. The incident ray, the reflected ray and the normal to the surface all lie in the same "plane of incidence".

Refractive Index

If the speed of light in a medium is v, then the refractive index, n, of the medium is defined as

$$n = c/v$$

where c is the speed light in vacuum. The wavelength of the light is

$$\lambda = \lambda_o/n$$

where λ_o is the wavelength in vacuum.

Refraction

Consider the boundary between two media whose refractive indices are n_1 and n_2 respectively, as in Fig. 35.1. According to Snell's law, the angles of incidence and refraction are related by

$$n_1 \sin\theta_1 = n_2 \sin\theta_2$$

where θ_1 and θ_2 are measured with respect to the normal to the boundary.

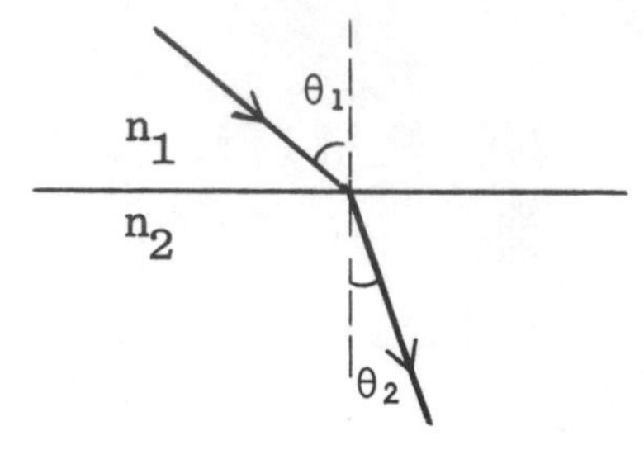

FIGURE 35.1

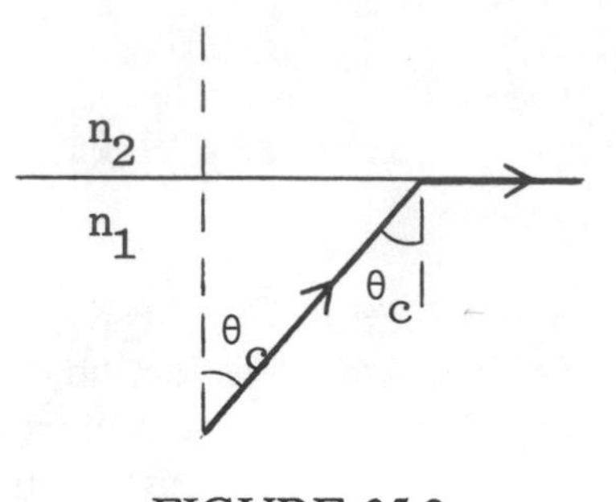

FIGURE 35.2

Total Internal Reflection

Suppose light is incident at a boudary where the refractive index is lower. In general, there will be a refracted ray and a reflected ray. At a critical angle of incidence, θ_c, the angle of refraction is 90°, as shown in Fig. 35.2. For any angle of incidence greater than θ_c, all the light is reflected back into the medium of higher refractive index:

$$n_1 \sin\theta c = n_2$$

where $n_1 > n_2$.

EXAMPLE 1

Light traveling in glass (n = 1.5) is incident at 30° to the normal at the surface on which there is a layer of oil (n = 1.65). Is there total internal reflection at the oil-air interface? See Fig. 35.3.

Solution:

Applying Snell's law at the glass-oil boundary we have

$$1.5 \sin 30^\circ = 1.65 \sin\theta$$

so $\theta = 27.0°$. This is the angle of incidence at the oil-air boundary.

The critical angle for total internal reflection at the oil-air boundary is given by

$$1.65 \sin\theta_c = 1$$

thus $\theta_c = 37.3°$. Since $\theta < \theta_c$, total internal reflection does not occur.

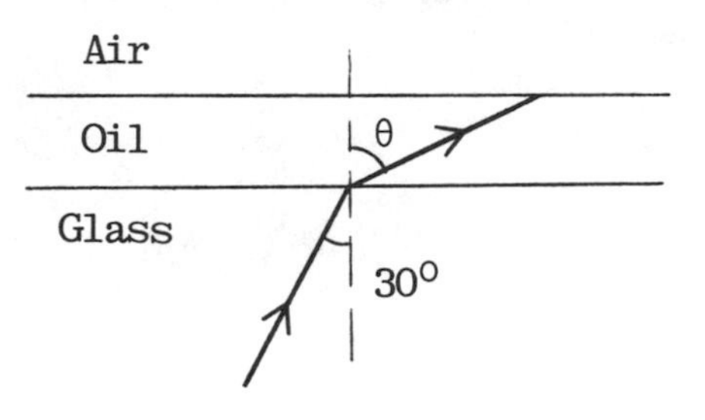

FIGURE 35.3

EXAMPLE 2

A 60° prism is made of glass of refractive index 1.6. Light is incident at on one face at 70° to the normal, see Fig. 35.4. (a) What is the angle of deviation of the ray that emerges from the opposite face? (b) What is the minimum angle of deviation for such a prism?

Solution:

(a) If r_1 is the angle of refraction at the first face, we have

$$\sin 70° = 1.6 \sin r_1$$

thus $r_1 = 36.0°$. The angle between the normals to the faces is the same as the angle between the faces. Thus,

$$60° = r_1 + r_2$$

where $r_2 = 24°$ is the angle at which the light is incident on the second face (see Fig. 35.4). At the second face

$$1.6 \sin r_2 = \sin i_2$$

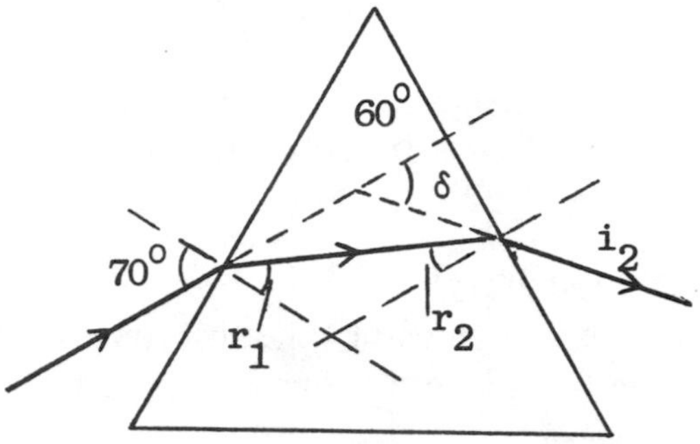

FIGURE 35.4

where i_2 is the angle at which the light emerges. We find $i_2 = 40.7°$. The angle of deviation is

$$\delta = (i_1 - r_1) + (i_2 - r_2)$$

$$= (70° - 36°) + (40.7° - 24°) = 50.7°.$$

(b) The minimum angle of deviation is related to the refractive index by Eq. 35.6 in the text:

$$n = \frac{\sin[(\phi + \delta_{min})/2]}{\sin(\phi/2)}$$

Thus, $\sin[(\phi + \delta_{min})/2] = n \sin(\phi/2)$. With the known values we have

$$\sin(30° + \delta_{min}/2) = 1.6 \sin 30°$$

This yields $\delta_{min} = 46.3°$.

REAL AND VIRTUAL OBJECTS AND IMAGES

The distinction between real and virtual objects and images is made on pp. 707-708 of the text. Here we reproduce Table 35.1 which tells us how to identify each case.

Rays diverge from a real object
Rays converge to a real image
Rays appear to converge to a virtual object
Rays appear to diverge from a virtual image

Note that a virtual image cannot be projected onto a screen. The converging rays associated with a virtual object must be produced by another optical element such as a lens.

SPHERICAL MIRRORS

There are two kinds of spherical mirror, concave and convex, shown in Fig. 35.5. In either case, the focal length is one-half the radius of curvature: f = R/2.

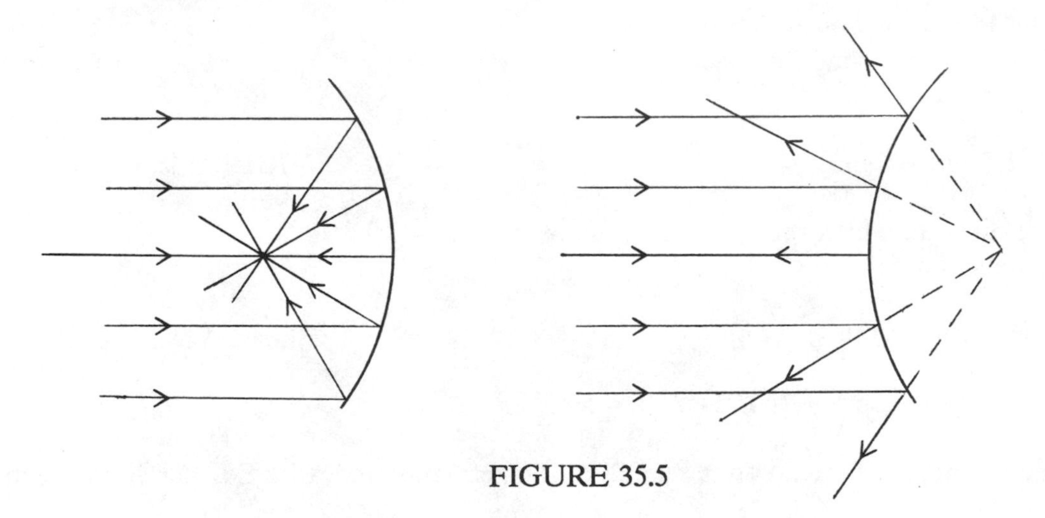

FIGURE 35.5

In the paraxial approximation we consider only rays that are near the central axis and nearly parallel to it (although they are not drawn that way). The mirror formula relates the object distance p, the image distance q and the focal length of the spherical mirror:

$$1/p + 1/q = 1/f$$

SIGN CONVENTION (Mirrors): p, q and f are positve (real) on the left and negative (virtual) on the right. (The light is incident from the left.)

The transverse (lateral) linear magnification of an image is the ratio of the image height to the object height:

$$m = y_I/y_O = -\,q/p$$

If m is positive, the image is erect; if m is negative, the image is inverted.

PRINCIPAL RAY DIAGRAMS

Images in mirrors may be located with any two of the four "principal" rays illustrated in Fig. 35.6.

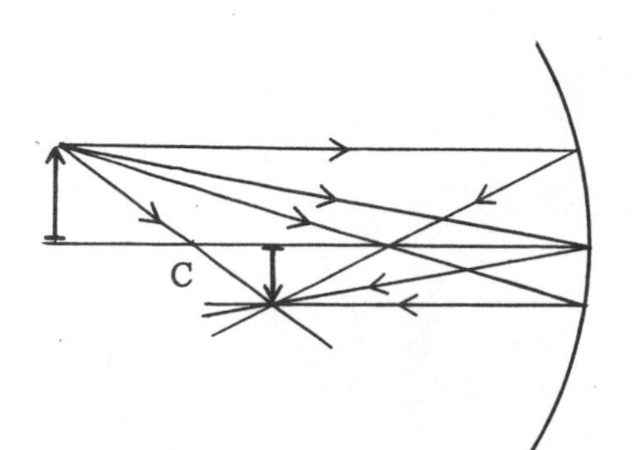

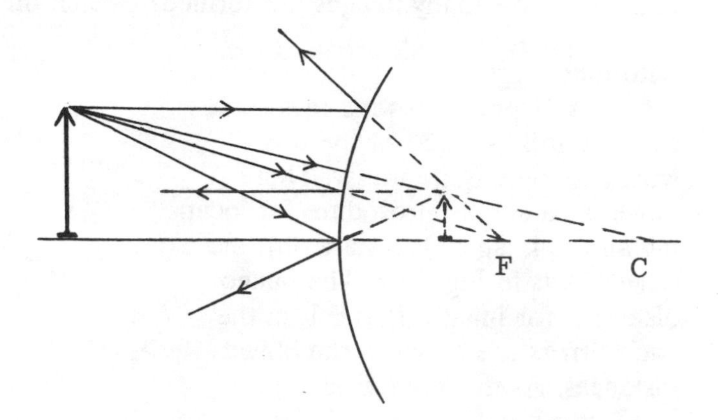

FIGURE 35.6

EXAMPLE 3

An object of height 1 cm is located 24 cm in front of a mirror whose radius of curvature is 36 cm. Locate the image and its size given that the mirror is (a) concave; or (b) convex.

Solution:

(a) The focal length is $f = R/2 = 18$ cm. From the mirror formula

$$1/24 \text{ cm} + 1/q = 1/18 \text{ cm}$$

Thus $q = 72$ cm. The lateral magnitifcation is

$$m = -q/p = -3.$$

The size of the (inverted) image is 3 cm.

(b) The focal length is $f = -R/2 = -18$ cm. From the mirror formula

$$1/24 \text{ cm} + 1/q = -1/18 \text{ cm}$$

Thus $q = -10.3$ cm. The lateral magnification is $m = +0.43$ and the size of the (erect) image is 0.43 cm.

SOLUTIONS TO SELECTED TEXT EXERCISES AND PROBLEMS

Exercise 3

An object is placed between two vertical mirrors that are at 60° to each other as in Fig. 35.7. How many images are formed? Sketch the positions an the orientations of the images.

Solution:

A diagram showing rays such as those in Fig. 35.33 of the text would become quite messy. There is, however, a simple procedure for locating all the images. We start with the object O as in Fig. 35.7. The image distances for images I_1 and I_2 in the two mirrors are equal to the object distances. At this point it is very convenient to "extend" the mirrors backward (the dashed lines). Then I_3 is the image of I_1 in M_2 and I_4 is the image of I_2 in M_1. Finally, the images of I_3 and I_4 in the two mirrors coincide at I_5. If 360° divided by the angle θ between the mirrors is an integer, then the number of images is $(360°/\theta) - 1$. For $\theta = 60°$ we obtain 5 images, as we have just seen.

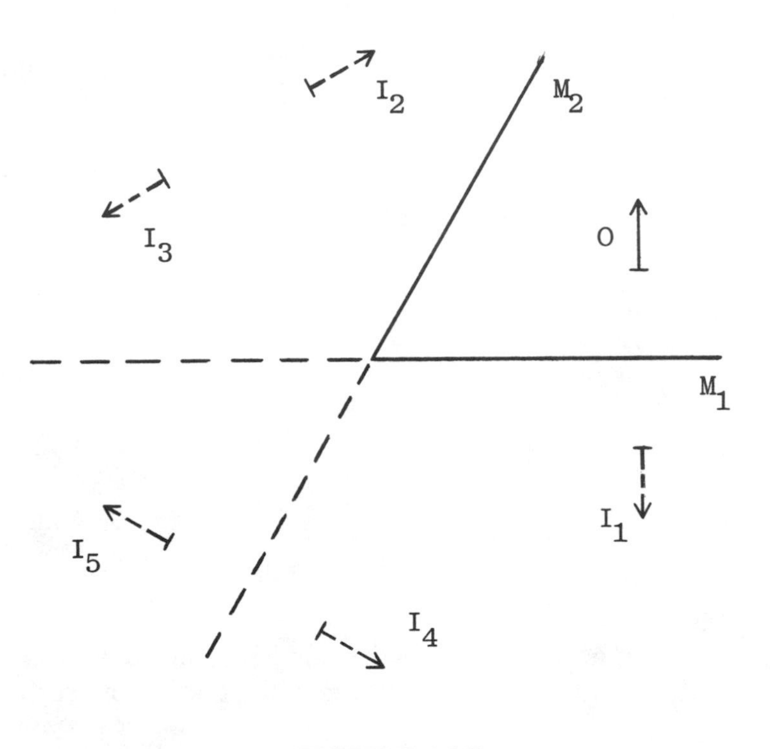

FIGURE 35.7

Exercise 5

Three mirrors are mutually perpendicular. Show that after three reflections, any ray is returned opposite to its original direction. (Hint: Express the direction of the ray in **ijk** notation.)

Solution:

Suppose the original direction is given by $\mathbf{i} + \mathbf{j} + \mathbf{k}$. A reflection in the xy plane wil reverse the direction of the ray along the z axis; that is, it becomes $\mathbf{i} + \mathbf{j} - \mathbf{k}$. Two more reflections in the xz and yz planes results in $-\mathbf{i} - \mathbf{j} - \mathbf{k}$, which is opposite to the original direction.

Exercise 25

A ray is obliquely incident ona prism of refractive index n as shown in Fig. 35.8. Show that the maximum value of α for which there is an emergent ray along face AC is given by $\cos\alpha = n \sin(\phi - \theta_c)$, where θ_c is the critical angle for total internal reflection.

Solution:

Applying Snell's law at the first face we have

$$\cos\alpha = n \sin r$$

The angle between the normals to the faces is equal to the angle between the faces. If the angle of incidence at the second face is equal to the critical angle θ_c, there will be total internal reflection. Since ϕ is the exterior angle in the triangle abc, we have

$$\phi = r + \theta_c$$

Thus Snell's law at the first face takes the form

$$\cos\alpha = n \sin(\phi - \theta_c)$$

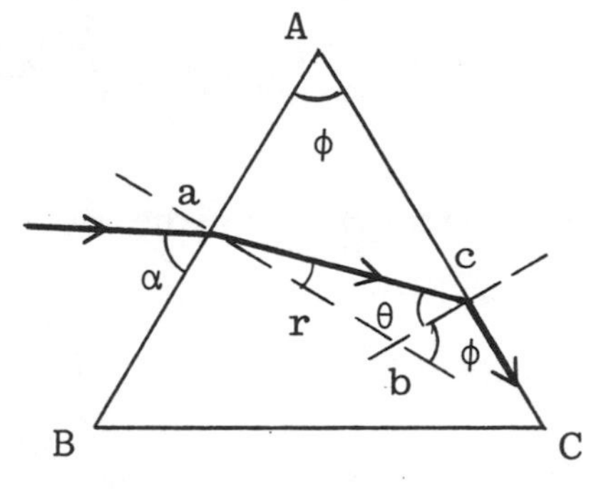

FIGURE 35.8

Exercise 35

A concave mirror produces an image 40% larger when a real object is 20 cm from the mirror. Determine the possible focal lengths of the mirror.

Solution:

We are given that $m = -q/p = \pm 1.4$. If $m = -1.4$, then $q = 1.4p = 28$ cm, and

$$1/20 \text{ cm} + 1/28 \text{ cm} = 1/f$$

which yields $f = 11.7$ cm. If $m = +1.4$, then $q = -1.4p = -28$ cm, and

$$1/20 \text{ cm} - 1/28 \text{ cm} = 1/f$$

which yields $f = 70$ cm.

Problem 3
A beam of light is incident on a flat glass slab of thickness t at an angle of incidence i, as in Fig. 35.9. Show that the lateral displacement, d, of the beam passing through the slab is given by

$$d = t \sin(i - r)/\cos r$$

Solution:

The thickness t can be expressed in terms of the distance AB traveled in the glass, t = AB cosr. Thus,

$$AB = t/\cos r$$

As the figure shows, the lateral displacement

$$d = AB \sin(i - r)$$

Substituting for AB we find

$$d = t \sin(i - r)/\cos r$$

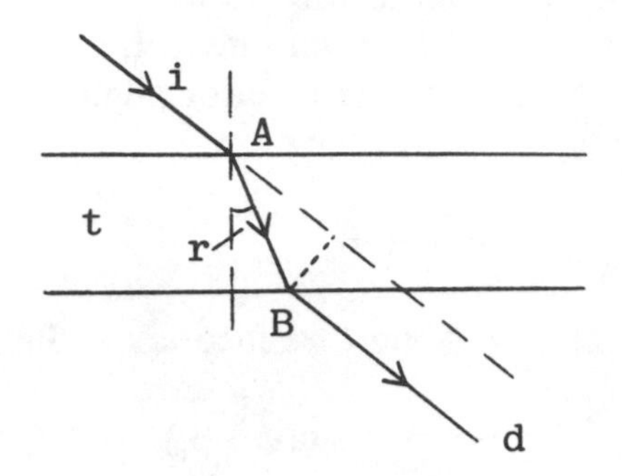

FIGURE 35.9

Problem 9

In Fig. 35.10 light travels from point A to point B after reflection in the mirror. According to Fermat's principle a light ray travels between two points along a path that takes the least time. (a) Show that the time taken is

$$t = (x^2 + a^2)^{1/2}/c + [(L - x)^2 + b^2]^{1/2}/c$$

(b) By setting dt/dx = 0, show that the angle of incidence is equal to the angle of reflection.

Solution:
(a) The distances traveled before and after reflection are given by

$$d_1^2 = a^2 + x^2$$

$$d_2^2 = b^2 + (L - x)^2$$

The time taken to travel from A to B is therefore

$$t = d_1/c + d_2/c$$

$$= (x^2 + a^2)^{1/2}/c + [(L - x)^2 + b^2]^{1/2}/c$$

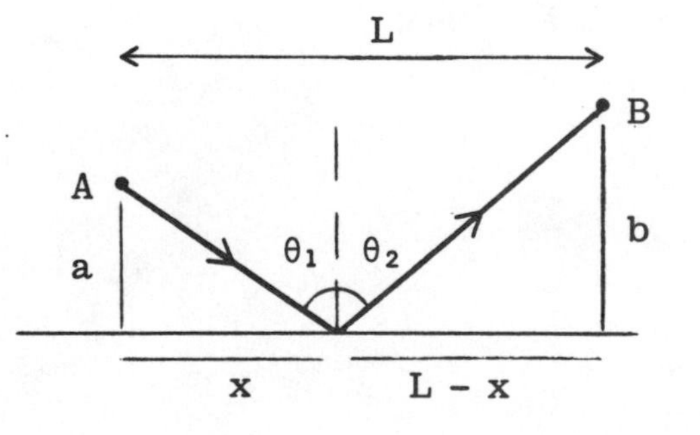

FIGURE 35.10

(b) Taking the derivative of t with respect to x we obtain:

$$c\,dt/dx = x(x^2 + a^2)^{-1/2} - (L - x)[(L - x)^2 + b^2]^{-1/2}$$

At an extremum $dt/dx = 0$. Since $\sin\theta_1 = x/d_1$ and $\sin\theta_2 = (L - x)/d_2$, the condition $dt/dx = 0$ may be written in the form

$$\sin\theta_1 - \sin\theta_2 = 0$$

which means that $\theta_1 = \theta_2$.

SELF-TEST

1. (a) A light ray is incident from air onto a medium of refractive index n at 50° to the normal. The reflected and refracted rays are perpendicular to each other. What is n?

2. A glass block ($n = 1.5$) has a layer of water ($n = 1.33$) on top, as in Fig. 35.11. What is the maximum value of i, the angle of incidence at the vertical face, for the total internal reflection at thc glass-water interface?

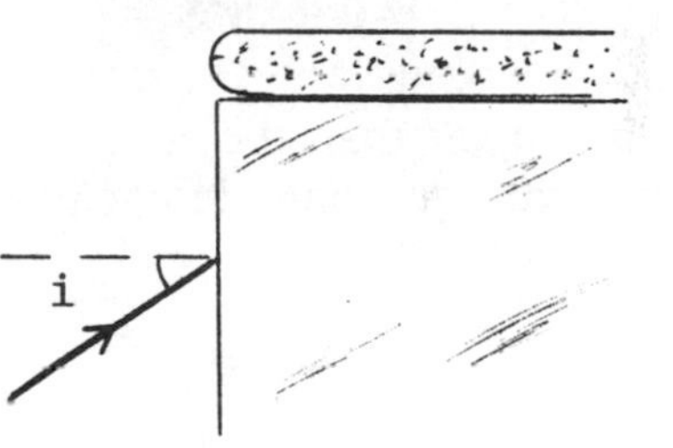

FIGURE 35.11

3. A spherical mirror with a 24 cm radius of curvature produces an image 50% of the size of a real object. Find the position of the object given that the mirror is (a) concave; (b) convex

Chapter 36

LENSES AND OPTICAL INSTRUMENTS

MAJOR POINTS

1. Principal ray diagrams
2. The thin lens formula
3. The angular magnification of a single lens
4. The optics of the compound microscope and telescopes
5. The eye: The correction of near sight and far sight.
6. Spherical boundaries, Lens maker's equation (Optional).

CHAPTER REVIEW

PRINCIPAL RAY DIAGRAMS

In a converging lens, the outer edge is thinner than the center; in a diverging lens, the outer edge is thicker than the center. Any two of the "principal" rays shown in Fig. 36.1 are sufficient to locate an image in each of the lenses.

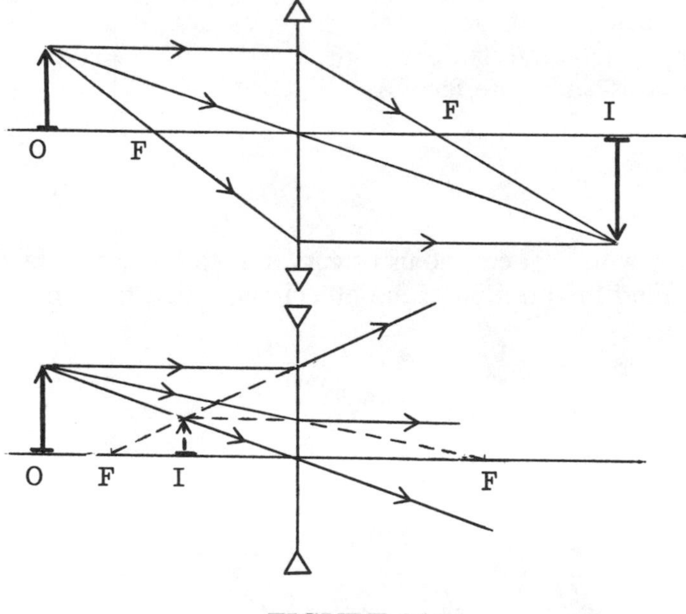

FIGURE 36.1

1. A ray passing through the center of the lens is undeviated
2. A ray directed parallel to the axis passes through a focal point
3. A ray directed twoard, or away from, a focal point, emerges parallel to the axis.

THIN LENS FORMULA

For a thin lens, within the paraxial approximation, the object and image distances, p and q respectively, are related to the focal length f by

$$1/p + 1/q = 1/f$$

SIGN CONVENTION (Lenses): p is positive (real) on the left, negative (virtual) on the right; q is positive (real) on the right, negative (virtual) on the left; f is positive for a converging lens and negative for a diverging lens. (Light is incident from the left.)

The transverse (lateral) linear magnification produced by a lens is defined as the ratio of the image height to the object height

$$m = y_I/y_O = - q/p$$

EXAMPLE 1

A converging lens has a focal length of 12 cm. The image of a real object is enlarged by 50%. Locate the object and the image.

Solution:

We are not told whether the image is real of virtual and thus we must address both possibilities.

(i) If the image is virtual, then $q < 0$ and $m = -q/p = 1.5$, so $q = -1.5p$. From the thin lens formula,

$$1/p - 1/1.5p = 1/f$$

Since $f = 12$ cm, we find $p = 4$ cm and $q = -6$ cm.
(ii) If the image is real, then $q > 0$ and $m = -q/p = -1.5$, so $q = 1.5p$. From the thin lens formula,

$$1/p + 1/1.5p = 1/f$$

Since $f = 12$ cm we find $p = 20$ cm and $q = 30$ cm.

EXAMPLE 2

A converging lens of focal length 9 cm is placed 8 cm behind a diverging lens of focal length -10 cm. A small object is placed 21 cm in front of the diverging lens. (a) Locate the final image; and (b) find its transverse magnification.

Solution:

(a) Applying the thin lens formula to the first lens (L_1) we have

$$1/21 \text{ cm} + 1/q_1 = -1/10 \text{ cm}$$

Thus q_1 = -6.77 cm (to the left of L_1). This image acts as the object for the second lens (L_2). The object distance is p_2 = 14.77 cm, thus

$$1/14.77 \text{ cm} + 1/q_2 = 1/9 \text{ cm}$$

which yields q_2 = 23 cm.
(b) The transverse magnification of the final image is the product of the individual magnifications

$$m_T = m_1 m_2 = (-q_1/p_1)(-q_2/p_2)$$

$$= (6.77/21)(-23/14.77) = 0.50$$

MICROSCOPES, TELESCOPE, THE EYE

Simple Microscope

A simple microscope consists of a single converging lens. The angular magnification is defined as the ratio of the angle subtended by the image produced by the lens to the angle subtended by the object when placed 25 cm from the eye: $M = \beta/\alpha_{25}$. In terms of the object distance (in meters) we find

$$M = 0.25/p$$

If the image is at infinity, then p = f.

Compound Microscope

A compound microscope consists of two converging lenses: An objective with a short focal length f_O and and an eyepiece with a focal length f_E. The angular magnification of a compound microscope when the final image is at infinity is

$$M_\infty = -(\ell/f_O)(0.25/f_E)$$

where the optical tube length ℓ is the distance between the focal points of the objective and the eyepiece.

Telescopes

The astronomical telescope constists of two converging lenses whereas in a terrestrial or Galilean telescope, the eyepiece is a diverging lens (p. 735 in the text). When the final image is at infinity, the angular magnification of a telescope is given by the ratio of the focal lengths of the objective and the eyepiece:

$$M_\infty = -f_O/f_E$$

Eye

The smallest distance from the eye that an object can be brought into focus is called the near point; the farthest distance is called the far point. For a normal eye the near point is at 25 cm and the far point is at infinity. If the near point is greater than 25 cm or the far point is less than infinity, corrective lenses are required.

The power of a lens is simply the inverse of its focal length measured in meters:

$$P = 1/f$$

The unit of (optical) power is the dipotre (D).

EXAMPLE 3

What lens would you prescribe for a person who has (a) a near point at 40 cm; (b) a far point at 250 cm? Draw a simple ray diagram to illustrate each case.

Solution:

(a) The "normal" near point is at 25 cm. For this person, light coming from an object at 25 cm should appear to come from 40 cm, see Fig. 36.2a. The image is virtual:

$$1/25 \text{ cm} - 1/40 \text{ cm} = 1/f$$

thus $f = 66.7$ cm.

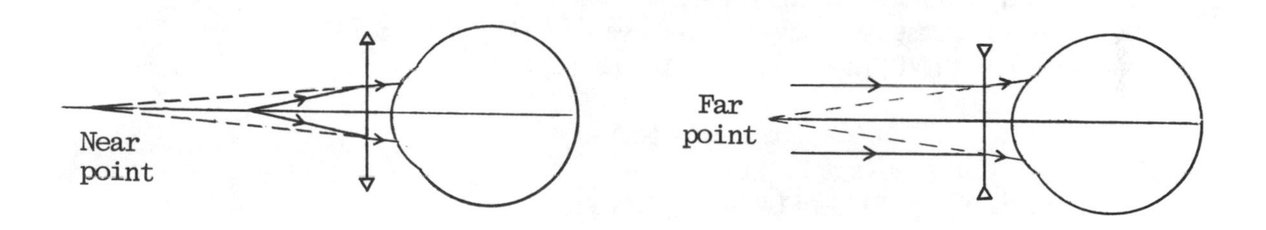

FIGURE 36.2

(b) Light from objects at infinity should appear to come from 250 cm, see Fig. 36.2b. The image is virtual:

$$1/\infty - 1/2.5 \text{ m} = 1/f$$

thus $f = -2.5$ m.

EXAMPLE 4
A person wears a contact lens with a power of -2.5 D. The length of the eyeball is 2 cm. (a) What is the power of the relaxed eye with no glasses?
(b) If the person can add 4 D by using the eye muscles, where is the near point without glasses?

Solution:
(a) First we must locate the far point (FP). Light from infinity must appear to come from the far point, this

$$1/\infty + 1/FP = -2.5,$$

so the FP = -0.4 m. With no glasses, light coming from the far point is focused on the retina. The power of the eye is $P_{eye} = 1/f_{eye}$:

$$1/0.4 \text{ m} + 1/0.02 \text{ m} = P_{eye}$$

Thus $P_{eye} = 52.5$ D. An alternative approach is worth noting. The effective power of the eye-lens combination,

$$P_{eff} = P_{eye} + P_{lens}$$

is such that objects at infinity are focused on the retina, that is, $P_{eff} = 1/0.02 \text{ m} = 50$ D. Since $P_{lens} = -2.5$ D, we see that

$$50 \text{ D} = P_{eye} - 2.5 \text{ D}$$

so $P_{eye} = 52.5$ D, in agreement with the earlier calculation.
(b) When the person increases the power of the eye by 4 D, $P_{eye} = 56.5$ D. Objects at the near point (NP) are focused on the retina, so

$$1/NP + 1/0.02 \text{ m} = 56.5 \text{ D}$$

Thus the near point is NP = 0.154 m (without glasses).

SPHERICAL BOUNDARIES, LENS MAKER'S EQUATION (Optional)

Consider light incident at the spherical boundary between two media with refractive indices n_1 and n_2. The object and image distances are related according to

$$n_1/p + n_2/q = (n_2 - n_1)/R$$

where R is the radius of curvature of the surface. R is positive if the center of curvature is on the right and negative if it is on the left of the spherical boundary.

The lens maker's equation relates the focal length of a lens to the radii of curvature of its two surfaces. For a lens in air,

$$1/f = (n - 1)(1/R_1 - 1/R_2)$$

EXAMPLE 5
A spot 1 cm from the bottom of a hemispherical glass paperweight (n = 1.5) of radius 4 cm. Where is image?

Solution:

The rays approach the boundary from the glass, so $n_1 = 1.5$ and $n_2 = 1$, see Fig. 36.3. The object distance is 3 cm. The center of curvature is on the left of the boundary and so R = -4 cm. Applying the equation for refraction at a spherical boundary we obtain

$$n/R + 1/q = (1 - n)/R$$

$$1.5/3 \text{ cm} + 1/q = (1 - 1.5)/(-4 \text{ cm})$$

which yields q = -2.67 cm. The image is virtual.

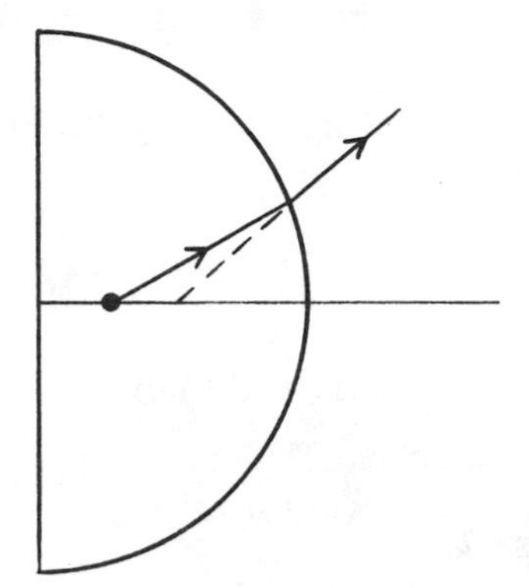

FIGURE 36.3

EXAMPLE 6
A diverging lens has surfaces with radii of curvature 12 cm and 15 cm. What is its focal length? Take n = 1.5.

Solution:

The center of curvature of the left surface is on the left, so $R_1 = -12$ cm; the center of curvature of the right surface is on the right, so $R_2 = +15$ cm, see FIg. 36.4. From the lens'makers' equation:

$$1/f = (n - 1)(1/R_1 - 1/R_2)$$

$$= (1.5 - 1)(-1/12 \text{ cm} - 1/15 \text{ cm})$$

Thus, f = -13.3 cm.

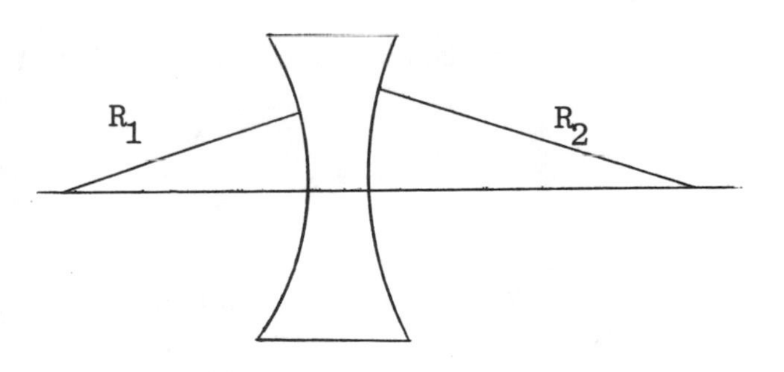

FIGURE 36.4

SOLUTIONS TO SELECTED TEXT EXERCISES AND PROBLEMS

Exercise 11
A converging lens of focal length 20 cm produces a reduced image that is 40% of the size of the object. Locate the object given that the image is (a) real, or (b) virtual.

Solution:
(a) The real image must be inverted so $m = -q/p = -0.4$. Next we use $q = 0.4p$ in the thin lens equation

$$1/p + 1/0.4p = 1/f$$

Using $f = 20$ cm, we find $p = 70$ cm

(b) The virtual image must be erect, thus $m = -q/p = +0.4$ and $q = -0.4p$. From the thin lens equation

$$1/p - 1/0.4p = 1/f$$

which leads to $p = -30$ cm

Exercise 19
A detail on a stamp is 1 mm wide. A converging lens of focal length 4 cm is used to produce a virtual image 40 cm fom the lens (which is close to the eye). Find: (a) the size of the image produced by the lens; (b) the angular magnification.

Solution:
(a) From the thin lens equation we have

$$-1/40 \text{ cm} + 1/p = 1/4 \text{ cm}$$

Thus $p = 3.64$ cm. The lateral magnification is $m = -q/p = y_I/y_O = 11$, so $y_I = my_O = 11$ mm

(b) The angular magnification of a single lens is

$$M = 0.25/p = 6.87$$

Exercise 33
A Galilean telescope consists of a converging lens of focal length 24 cm and a diverging lens of focal length -8 cm separated by 16 cm. The object is 12 m away. (a) Where is the final image? (b) What should be the separation between the lenses for the final image to be at infinity?

Solution:
(a) We first locate the image in the objective:

$$1/12 \text{ m} + 1/q_O = 1/0.24 \text{ m}$$

so $q_O = 24.5$ cm. This image acts as a virtual object for the eyepiece, where the object distance is $p_E = -(24.5 \text{ cm} - 16 \text{ cm}) = -8.5$ cm, thus

$$-1/8.5 \text{ cm} + 1/q_E = -1/8 \text{ cm}$$

We find $q_E = -136$ cm from objective.

(b) For the final image to be at infinity, the focal points of the objective and the eyepiece should coincide. We need $p_E = -8$ cm, so the separation should be 16.5 cm.

Exercise 39

A person requires a lens of +1.5 D in order to read the paper at 25 cm. A few years later, the paper must be held at 40 cm with the same glasses. What is the new prescription required for normal reading?

Solution:

When the paper is at 0.4 m, the light must appear to come from the near point, thus

$$1/0.4 \text{ m} - 1/NP = 1.5$$

The new near point distance is NP = 1 m. When the paper is held at 25 cm, the light should appear to come from the new near point.

$$1/0.25 \text{ m} - 1/1 \text{ m} = 1/f$$

Thus $P = 1/f = +3$ D, is the new prescription. Notice that we did not need to find the original near point when the glasses were new.

Problem 5

A point source and a screen are a fixed distance D apart. There is a convergin lens of focal length f between them. (a) Show that there are two positions of the lens for which a clear image is produced. (b) Show that the distance between the two possible positions of the lens is given by $d = [D(D- 4f)]^{1/2}$.

Solution:

For a given object distance p, the image distance is $q = D - p$, so

$$1/p + 1/(D - p) = 1/f$$

This leads to a quadratic equation in p:

$$p^2 - Dp + fD = 0$$

The solutions of the quadratic are

$$p = D/2 \pm (D^2 - 4fD)^{1/2}/2$$

The difference between the two solutions is $\Delta p = [D(D - 4f)]^{1/2}$.

Problem 11
A thin lens made of a material with a refractive index n_1 is surrounded by a medium of refractive index n_2. Show that the focal length f is given by

$$1/f = (n_1/n_2 - 1)(1/R_1 - 1/R_2)$$

where R_1 and R_2 are the radii of the curvature of the surfaces.

Solution:

If we assume that the image at the first boundary is virtual, as in Fig. 36.5, we have

$$n_2/p_1 - n_1/q_1 = (n_1 - n_2)/R_1$$

This virtual image is on the left of the second surface and so it acts as a real object for this surface:

$$n_1/q_1 + n_2/q_2 = (n_2 - n_1)/R_2$$

On adding these two equations and dividing both sides by n_2 we find

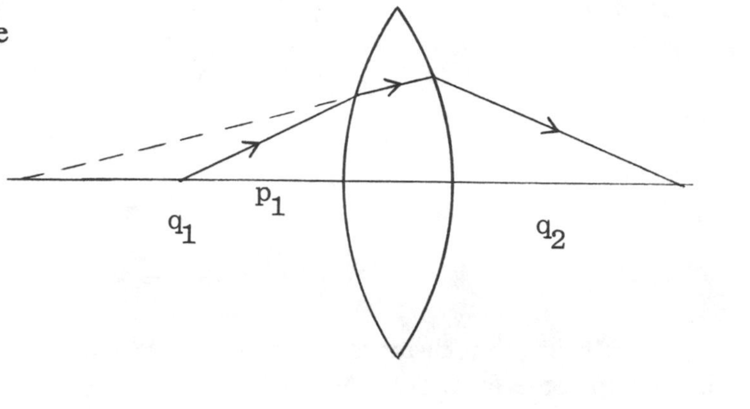

FIGURE 36.5

$$1/f = 1/p_1 + 1/q_2$$

$$= (n_1/n_2 - 1)(1/R_1 - 1/R_2)$$

SELF-TEST

1. An object is placed 60 cm in front of a converging lens (f_1 = 20 cm). A diverging lens (f_2 = -15 cm) is place 18 cm behind the first lens. Locate the final image and its linear magnification.

2. A microscope has a tube length of 15 cm. The focal lengths of the objective and eyepiece are f_O = 1 cm and f_E = 5 cm. The final image is 50 cm from the eyepiece. (a) Locate the object; (b) What is the angular magnification?

3. A nearsighted person wears glasses with a power of -2.5 D. Her near point is at 30 cm when the glasses are worn. When the glasses are taken off, find: (a) far point; (b) near point.

Chapter 37

WAVE OPTICS (I)

MAJOR POINTS

1. The basic features of interference and diffraction
2. Young's double slit experiment demonstrates the wave nature of light
3. To exhibit interference, two sources must be coherent
4. Interference in thin films

CHAPTER REVIEW

INTERFERENCE

Interference arises when two waves are superposed. If the wavefunctions reinforce each other, we have constructive interference; if the wavefunctions tend to cancel each other, we have destructive interference.

In order for two waves to produce a steady interference pattern, the sources must be coherent, that is, they must have a constant phase difference between them. They must be monochromatic, that is, they must emit a single wavelength. (No source is perfectly monochromatic; a narrow range of wavelengths is sufficient in practice.)

YOUNG'S DOUBLE SLIT EXPERIMENT

Figure 37.1 shows two narrow slits, separated by a distance d, illuminated by plane monochromatic wavefronts. If we view the waves on a distant screen, the rays emerging from the slits are nearly parallel. In this case, the path difference between two rays reaching a given point on the screen is $\delta = d \sin\theta$. The conditions for constructive interference (bright fringes) and destructive interference (dark fringes) are

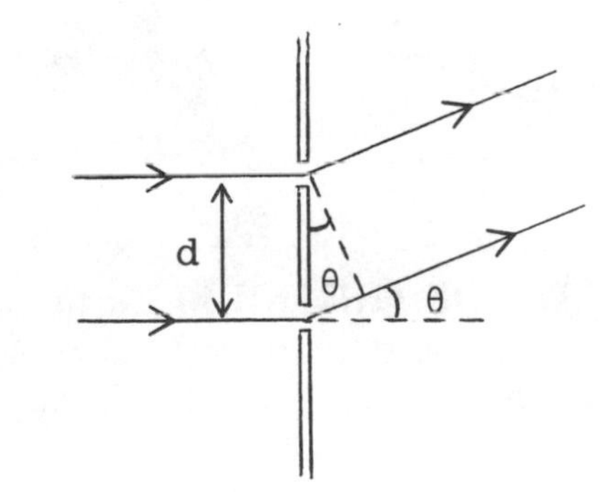

FIGURE 37.1

(Constructive) $\quad d \sin\theta = m\lambda$

(Destructive) $\quad d \sin\theta = (m + 1/2)\lambda$

$m = 0, \pm 1, \pm 2, ...$

The quantity m is called the order of the fringe. Note that if the distance to the screen is not much larger than the wavelength, as is often the case with sound waves, the expression d sinθ is not adequate; the path difference must be calculated properly, see Example 37.1 in the text.

EXAMPLE 1
In a double-slit arrangement, the slits are separated by 0.4 mm and the screen is 1.8 m from the slits. Light containing two wavelengths, λ_1 = 480 nm and λ_2 = 520 nm illuminates the slits. (a) What is the distance from the central peak to the third minimum in the pattern for λ_1? (b) What is the minimum distance from the center of the pattern at which a dark fringe for λ_1 coincides with a bright fringe for λ_2?

Solution:
(a) For the third dark fringe, m = 2, and

$$d \sin\theta = 2.5\lambda_1$$

Thus $\sin\theta = 2.5\lambda_1/d = 3\text{x}10^{-3}$. For such small angles, $\sin\theta \approx \tan\theta$, so the position on the screen is given by

$$y = L \tan\theta \approx L \sin\theta$$

$$= (1.8 \text{ m})(3\text{x}10^{-3}) = 5.4 \text{ mm}.$$

(b) Since $\lambda_1 < \lambda_2$, the condition for the first overlap of a dark fringe for λ_1 and a bright fringe for λ_2 is

$$d \sin\theta = (m + 1/2)\lambda_1 = m\,\lambda_2$$

Thus $(m + 0.5)480 = 520m$ which gives m = 6. The position on the screen is

$$y = L \tan\theta \approx L \sin\theta$$

$$= Lm\lambda_2/d = 1.4 \text{ cm}$$

INTENSITY OF THE DOUBLE-SLIT PATTERN

The phase difference ϕ between the two waves is related to the path difference δ. Since one wavelength corresponds to a change of phase of 2π:

$$\phi/2\pi = \delta/\lambda$$

thus,

$$\phi = 2\pi\delta/\lambda = 2\pi d\sin\theta/\lambda$$

For constructive interference $\phi = 0, 2\pi, 4\pi,..$; and for destructive interference $\phi = \pi, 3\pi, 5\pi, ...$

The intensity distribution for a double slit interference pattern is given by

$$I = 4I_o \cos^2(\phi/2)$$

where I_o is the intensity that would be produced by a single slit. (We assume that the slits are narrow enough that the screen would be uniformly illuminated by a single source.)

EXAMPLE 2
Two slits, separated by d = 0.45 mm, are illuminated with light of wavelength 560 nm. The pattern is viewed on a screen 2.5 m from the slits. What is the intensity at a point 2 mm from the central bright fringe?

Solution:

The phase difference between the waves at the given point on the screen is

$$\phi = 2\pi d \sin\theta/\lambda$$

Since $\sin\theta \approx y/L$,

$$\phi = 2\pi dy/\lambda L$$

$$= 2\pi(4.5\text{x}10^{-4}\text{ m})(2\text{x}10^{-3}\text{ m})/(5.6\text{x}10^{-7}\text{ m})(2.5\text{ m}) = 4.04\text{ rad}$$

The intensity is

$$I = 4I_o \cos^2(\phi/2)$$

$$= 4I_o \cos^2(2.02\text{ rad}) = 0.75I_o$$

THIN FILM INTERFERENCE

When light traveling in one medium meets a boundary where the refractive index is higher, the reflected wave undergoes a 180° phase change, see Fig. 37.2. We assume that the light is incident normally on the film-although for clarity, we draw rays at an angle. Keep in mind that the wavelength in a film of refractive index n is

$$\lambda_F = \lambda/n$$

where λ is the wavelength in vacuum (or, to a good approximation, in air). The conditions for constructive or destructive interference must be worked out for each situation-taking into account phase changes on reflection.

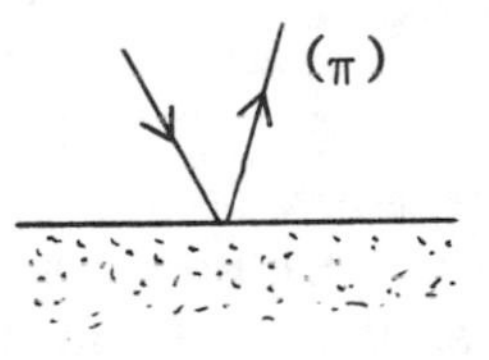

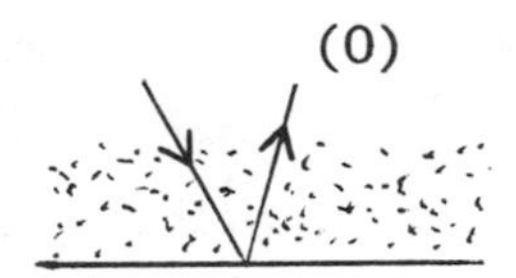

FIGURE 37.2

EXAMPLE 3
An oil film (n = 1.4) on water (n = 1.33) is 750 nm thick and is illuminated with white light. In the reflected light, which visible wavelengths (400-700 nm) are (a) enhanced; (b) missing? (c) How would the answers change if the film were on glass?

Solution:

At the air-oil interface, the reflection occurs at a medium of higher refractive index, so the phase changes by π. At the oil-water interface, the reflection occurs at a medium of lower refractive index and so there is no phase change (Fig. 37.3).

(a) The condition for constructive interference is

$$2t = (m + 1/2)\lambda_F = (m + 1/2)\lambda/n$$

The wavelengths are given by inserting interger values for m into the following equation (with n = 1.4)

$$\lambda = 2nt/(m + 1/2)$$

$$= 2100/(m + 1/2) = 600 \text{ nm}, 467 \text{ nm}$$

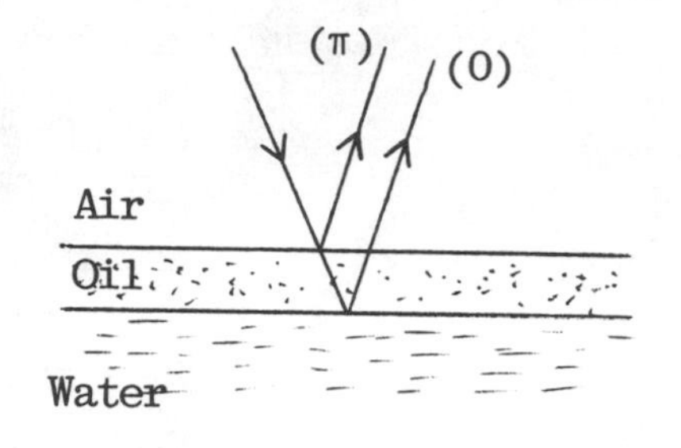

FIGURE 37.3

(b) The condition for destructive interference is

$$2t = m\lambda_F = m\lambda/n$$

The missing wavelengths are found by inserting integer values into

$$\lambda = 2nt/m$$

$$= 2100/m = 700 \text{ nm}, 525 \text{ nm}, 420 \text{ nm}$$

(c) If the oil film were on glass there would be a phase change of π for reflection at this boundary and so the conditions for constructive and destructive interference would be interchanged.

SOLUTIONS TO SELECTED TEXT EXERCISES AND PROBLEMS

Exercise 11

Two line sources transmit microwaves of wavelength 3 cm in phase. How far apart should they be in order to produce a 10° angular separation between the first and second bright fringes on one side of the central peak?

Solution:

$$\sin\theta_2 = \sin(\theta_1 + 10°) = 2\sin\theta_1.$$

Expand $\sin(\theta_1 + 10)$ to find $\tan\theta_1 = \sin 10°/(2 - \cos 10°)$, thus $\theta_1 = 9.71°$

$$d = \lambda/\sin\theta_1 = 17.8 \text{ cm}$$

Exercise 19

Two point sources S_1 and S_2 separated by distance d, as in Fig. 37.4, emit waves at the same wavelength ($\lambda << d$). What is the condition on the distance x such that point P is a point of destructive interference given that (a) S_1 and S_2 are in phase; (b) S_1 and S_2 are π rad out of phase?

Solution:

(a) The path difference between the waves reaching point P is

$$\delta = x - (d - x) = 2x - d$$

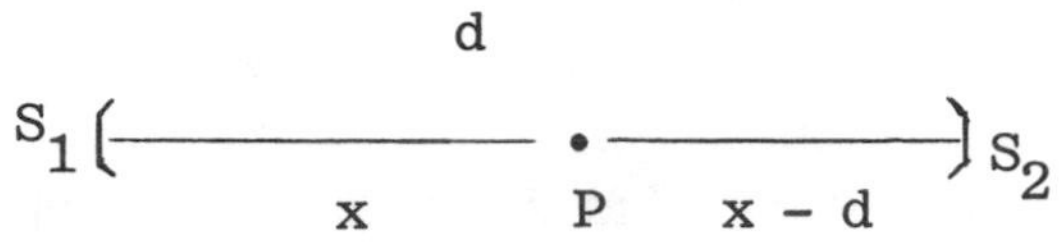

d

S_1 P S_2

x x − d

FIGURE 37.4

For destructive interference we need $\delta = (m + 0.5)\lambda$. Thus we find

$$x = 0.5d - (m + 0.5)\lambda/2$$

(b) The path difference is still $\delta = 2x - d$. Since the sources are π out of phase, the condition for destructive interference is $\delta = m\lambda$. This leads to

$$x = 0.5d - m\lambda/2$$

Exercise 29

Light of wavelength 627 nm illuminates two slits. What is the minimum path difference between the waves from the slits for the resultant intensity to fall to 25% of the central maximum?

Solution:

The intensity distribution for a double-slit pattern is given in Eq. 37.9 in the text:

$$I = 4I_o \cos^2(\phi/2)$$

The intensity of the central maximum ($\phi = 0$) is $4I_o$. We are told $I = I_o$, which means $\cos^2(\phi/2) = 1/4$, thus

$$\cos(\phi/2) = 1/2$$

Thus $\phi/2 = \pi/3$. The relation between the phase difference ϕ and the path difference δ is $\phi/2\pi = \delta/\lambda$, so

$$\phi = 2\pi/3 = 2\pi\delta/\lambda$$

The minimum path difference is therefore $\delta = \lambda/3 = 209$ nm.

Exercise 35
White light is incident normally on a uniform film of water (n = 1.33) lying on a glass plate (n = 1.6). Find the minimum possible thickness of the film given that in the reflected light: (a) 550 nm is enhanced, or (b) 550 nm is missing.

Solution:
(a) The reflections at both boundaries occur at a medium of higher refractive index, therefore both involve a π phase change. The condition for constructive interference is

$$2t = m\lambda_F = m\lambda/n$$

The minimum value of m is 1, thus the minimum thickness is

$$t = \lambda/2n = 207 \text{ nm}$$

(b) In this case, the condition for destructive interference is

$$2t = (m + 1/2)\lambda/n$$

The minimum thickness is $t = \lambda/4n = 103$ nm.

Problem 11
A thin plastic sheet (n = 1.6) placed in front of one slit in a double-slit arrangement results in a shift of the central bright fringe to a location previously occupied by the 12 th bright fringe. Given that the light has a wavelength of 650 nm, what is the minimum thickness of the sheet?

Solution:
When the light travels in air, the change in phase associated with a ditsnace t is $\phi_1 = 2\pi t/\lambda$. For a film of thickness t, the change in phase is $\phi_2 = 2\pi t/\lambda_F = 2\pi nt/\lambda$. Therefore, the phase difference associated with the introduction of the film is

$$\Delta\phi = \phi_2 - \phi_1 = 2\pi t(n - 1)/\lambda$$

Since there is a shift of 12 fringes, we have that $\Delta\phi = 12(2\pi)$, so

$$t = \lambda\ \Delta\phi/2\pi(n - 1) = 12\lambda/(n - 1) = 13\ \mu\text{m}$$

Here is an alternative approach.
Suppose that within a distance t in air, there are m wavelengths, that is, $t = m\lambda$. Since there is a shift of 12 fringes, there must be an extra 12 wavelengths introduced by the film, that is, $t = (m + 12)\lambda_F = (m + 12)\lambda/n$.

$$t = m\lambda = (m + 12)\lambda/n$$

From this we obtain m = 20 and then t = 20x650 nm = 13 μm.

Problem 15.

When white light is incident normally on a thin film in air, 550 nm is missing from the reflected light. Assuming the film has the minimum possible thickness, find the phase differences between the two interfering beams for (a) 400 nm, and (b) 700 nm. Estimate the factors by which the reflected intensities of (c) 400 nm and (d) 700 nm are reduced relative to that for constructive interference.

Solution:

Since there is a π phase change on reflection at one surface, the condition for destructive interference is $2t = m\lambda_F$. If the film has the minimum thickness ($m = 1$) then $2t = \lambda/n$, or $2nt = 550$ nm.

The phase difference between the waves reflected at the two surfaces associated with the path difference is

$$\phi = 2\pi\delta/\lambda_F = 2\pi(2nt)/\lambda$$

In addition one must include the π phase change on reflection at the bottom surface. The total phase difference is therefore

$$\phi = 2\pi(550 \text{ nm})/\lambda - \pi$$

(a) The phase difference for 400 nm is $\phi = 2\pi(550 \text{ nm})/400 \text{ nm} - \pi = 5.50$ rad

(b) The phase difference for 700 nm is $\phi = 2\pi(550 \text{ nm})/700 \text{ nm} - \pi = 1.80$ rad

(c) The factor by which the intensity of 400 nm is reduced is

$$I/4I_o = \cos^2(\phi/2) = 0.85$$

(d) For 700 nm, we find $I/4I_o = 0.39$.

SELF-TEST

1. In a double slit experiment the 7 th order bright fringe is 5.4 mm from the central peak on a screen 1.8 m from the slits. What is the slit separation if the wavelength used is 560 nm?

2. A film of oil ($n = 1.4$) is on glass ($n = 1.6$). When white light is incident normally, the only visible wavelengths missing in the reflected light are 420 nm and 588 nm. (a) What is the condition for destructive interference in the reflected light? (b) What is the minimum thickness of the film?

Chapter 38

WAVE OPTICS (II)

MAJOR POINTS

1. The location of the minima in a single slit diffraction pattern
2. Rayleigh's criterion for the resolution of images
3. The location of the principal maxima of grating spectrum
4. The use of phasors to determine the intensity distribution for multiple slits and a single slit

CHAPTER REVIEW

SINGLE SLIT DIFFRACTION

The angular positions of the minima in the diffraction pattern produced by a single slit of width a are given by

(Minima) $$a \sin\theta = m\lambda \qquad m = 1, 2, 3,...$$

Note that $m \neq 0$. The secondary maxima lie approximately midway between the minima.

EXAMPLE 1

A single slit of width 0.1 mm is illuminated by light of wavelength 550 nm. The diffraction pattern is observed on a screen 80 cm from the slit. What is the width of the central peak?

Solution:

The first minimum occurs where

$$\sin\theta = \lambda/a$$

Since θ is small, the position of the first minimum is

$$y = L \tan\theta \approx L \sin\theta$$

The width of the central maximum is $2y = 2L\lambda/a = 8$ mm.

RAYLEIGH'S CRITERION

Consider light from a distant point source incident on a circular aperture of diameter a. The position of the first minimum in the Fraunhofer diffraction pattern is given by

$$a \sin\theta = 1.22\ \lambda$$

According to Rayleigh's criterion, the images of the two point sources are resolved if the first minimum of one diffraction pattern coincides with the central maximum of the other. Thus the critical angular separation between the sources is (Fig. 38.1)

$$\theta_c = 1.22\lambda/a$$

where we have used the approximation $\sin\theta \approx \theta$ for small angles.

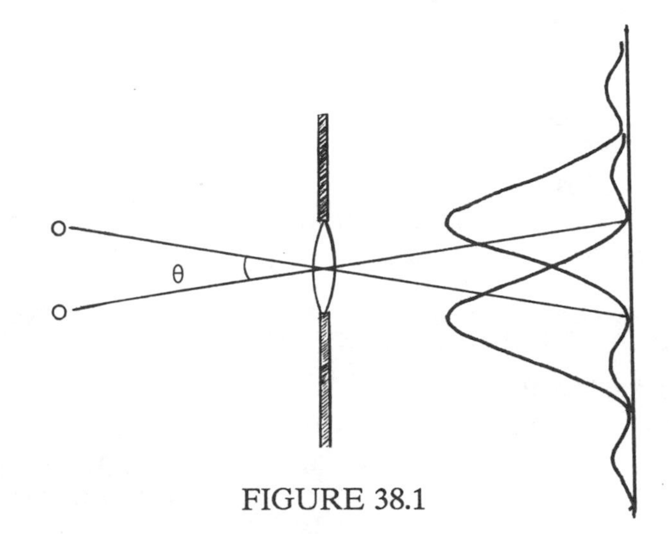

FIGURE 38.1

EXAMPLE 2

An astronaut is in a space shuttle at an altitude of 160 km. What is the size of the smallest object that can be resolved by (a) an eye with a pupil diameter of 5 mm; (b) a telescope with a 1.4 m diameter mirror? Take the wavelength to be 450 nm.

Solution:

The angle (in radians) subtended by an object of size x at a distance D is

$$\theta = x/D$$

where in the present example D = 160 km. At the critical angular

$$\theta_c = 1.22\lambda/a = x/D$$

so $x = 1.22\lambda D/a$
(a) For $a = 5\text{x}10^{-3}$ m, we find x = 17.6 m.

(b) For a = 1.4 m, we find x = 6.3 cm.

These values assume ideal conditions - for example we have ignored atmospheric turbulence. Nonetheless, unconfirmed reports indicate that a reconnaisance satellite can record objects as small as 15 cm.

GRATINGS

A grating consists of many fine slits (or rulings on a glass plate). The interference pattern consists of bright sharp fringes. The positions of these principal maxima, are given by

$$d \sin\theta = m\lambda \quad (m = 0, 1, 2,..)$$

where d is the distance between adjacent slits. The integer m is called the order of the fringe (Fig. 38.2).

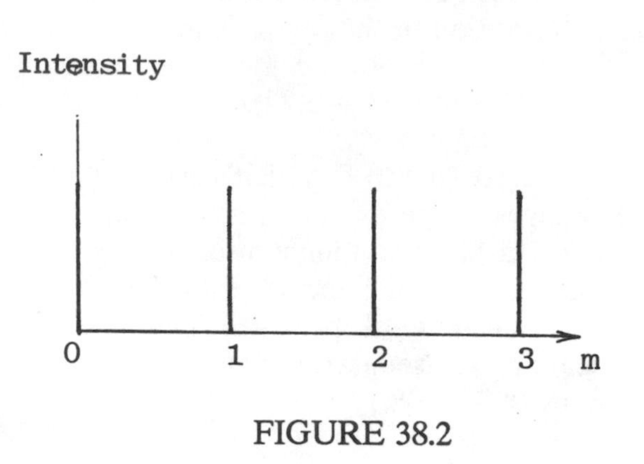

FIGURE 38.2

EXAMPLE 3

When light of wavelength 550 nm is incident normally on a grating, the second order principal maximum is observed at an angle of 21°. (a) What is the separation between adjacent lines? (b) What is the total number of principal maxima that can be observed?

Solution:

(a) We are given $\theta = 21^{\circ}$ for $m = 2$, thus

$$d \sin 21^{\circ} = 2\lambda$$

We find $d = 2(550 \text{ nm})/\sin 21^{\circ} = 3.06 \times 10^{-6}$ m

(b) The maximum possible value of θ is 90°, so the maximum value of m is

$$m = d \sin 90^{\circ}/\lambda = 5.6$$

Thus a total of five principal maxima can be observed.

PHASORS

When two or more coherent waves are superimposed, the phasor technique allows to determine the amplitude and phase of the resultant. The angle between adjacent phasors is the phase difference, ϕ, between the the waves that arrive at a particular point on the screen. Note that

$$\phi = 2\pi\delta/\lambda$$

where δ is the path difference between the waves. The resultant amplitude at any point on the screen may be found by drawing the phasors as shown, for example, in Fig. 38.14 of the text.

INTENSITY OF THE SINGLE SLIT DIFFRACTION PATTERN

The intensity distribution produced by a single slit of width a is

$$I = I_o \sin^2(\alpha/2)/(\alpha/2)^2$$

where

$$\alpha = 2\pi a \sin\theta/\lambda$$

and I_o is the intensity at $\theta = 0$.

EXAMPLE 4

A slit of width 0.15 mm is illuminated with light of wavelength 600 nm.
The screen is 3.2 m from the slit. What is the intensity at a point 2.5 mm from the center of the pattern?

Solution:

For such small angles we take

$$\sin\theta \approx \tan\theta = y/L = 2.5 \text{ mm}/2.2 \text{ m} = 7.81\text{x}10^{-4}$$

Thus

$$\alpha = 2\pi a \sin\theta/\lambda$$

$$= 2\pi(1.5\text{x}10^{-4} \text{ m})(7.81\text{x}10^{-4})/(6\text{x}10^{-7} \text{ m}) = 1.227 \text{ rad}$$

The intensity at this point is

$$I = I_o \sin^2(\alpha/2)/(\alpha/2)^2$$

$$= 0.90\, I_o$$

SOLUTIONS TO SELECTED TEXT EXERCISES AND PROBLEMS

Exercise 5

In a single-slit diffraction pattern, the distance between the first and second minima is 3 cm on an screen 2.80 m from the slit. Find the slit width given that the wavelength of the light is 480 nm.

Solution:

The minima are given by a $\sin\theta = m\lambda$. Let $\sin\theta_1 = \lambda/a$ and $\sin\theta_2 = 2\lambda/a$. For small angles, the position on the screen located at a distance L is

$$y = L \tan\theta \approx L \sin\theta$$

The distance between the first and second minima is

$$\Delta y = L(\sin\theta_2 - \sin\theta_1) = L\lambda/a$$

thus, the slit width is $a = L\lambda/\Delta y = 0.045$ mm

Exercise 13

Two tiny objects are 25 cm from an eye. What is the smallest separation between the objects if they are to be resolved when the pupil diameter is 3mm? The wavelength of the light is 500 nm.

Solution:

If x is the distance between the objects and L is the distance from the eye, the critical angle is

$$\theta_c = 1.22\lambda/a = x/L$$

thus,

$$x = 1.22\lambda L/a$$

$$= 1.22(5\text{x}10^{-7}\text{ m})(0.25\text{ m})/(3\text{x}10^{-3}\text{ m}) = 5.08\text{x}10^{-5}\text{ m}.$$

Exercise 19

How many complete orders are formed for the visible range 400-700 nm by a grating that has 6000 lines/cm?

Solution:

The line spacing is $d = (1/6000)$ cm $= 1.67\text{x}10^{-6}$ m. The maximum possible value for θ is 90° and this will occur first for the longest wavelength, that is, 700 nm. For $\theta = 90°$, $d \sin\theta = m\lambda$ gives us

$$m = d/\lambda$$

$$= 1.67\text{x}10^{-6}\ \text{m}/7\text{x}10^{-7}\ \text{m} = 2.4$$

Thus, there are two complete orders.

Exercise 31
A grating of width 2.8 cm has 4200 lines/cm. What is the minimum difference in wavelength that can be resolved in second order at 550 nm?

Solution:

The resolving power of a grating is given by Eq. 38.14 in the text:

$$R = \lambda/\Delta\lambda = Nm$$

where N is the number of slits (lines) and m is the order. In the present example $N = 1.176\text{x}10^4$ and $m = 2$, therefore

$$\Delta\lambda = \lambda/Nm$$

$$= (550\ \text{nm})/2(1.176\text{x}10^4) = 0.023\ \text{nm}$$

Exercise 41
Sunlight is reflected off the calm surface of a pond. At what angle above the horizon should the sun be located for the reflected light to be linearly polarized?

Solution:

According to Brewster's law, Eq. 38.16 in the text, the polarizing angle, at which the reflected light is linearly polarized, is given by

$$\tan\theta_p = n_2/n_1$$

where $n_2 > n_1$ and light travels initially in the medium of refractive index n_2. In the present example, $n_1 = 1$ and $n_2 = 1.33$, so $\tan\theta_p = 1.33$, therefore $\theta_p = 53.1°$, which is measured with respect to the normal to the surface. The question asks for the angle of the sum above the horizon, so the answer is 36.9°.

Problem 11
In order to find an expression for the intensity distribution due to N sources, use Fig. 38.16 (in the text) to obtain expressions for the phasors E_o and E_{oT} in terms of ϕ and R. Express E_{oT} in terms of E_o; then use the fact that the intensity is proportional to the square of the field to show that

$$I = I_o \sin^2(N\phi/2)/\sin^2(\phi/2)$$

where $I_o \propto E_o^2$ is the intensity due to one source.

Solution:

From Fig. 38.3 we see that the amplitude of a single source is

$$E_o = 2\,R\,\sin(\phi/2)$$

where ϕ is the angle between adjacent phasors. The magnitude of the resultant phasor is

$$E_{oT} = 2\,R\,\sin(N\phi/2)$$

Taking the ratio of these, we have

$$E_{oT}/E_o = \sin(N\phi/2)/\sin(\phi/2)$$

Since the intensity is prportional to the square of the field strength:

$$I = I_o\,\sin^2(N\phi/2)/\sin^2(\phi/2)$$

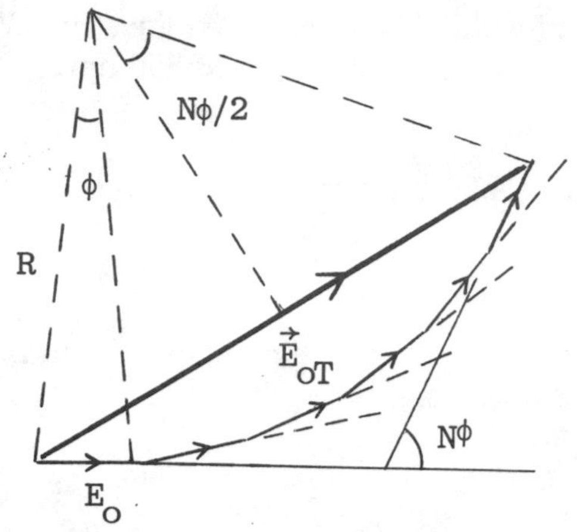

where $I_o \propto E_o^2$ is the intensity due to a single source.

FIGURE 38.3

SELF-TEST

1. (a) In a double-slit experiment, the slit separation is 0.42 mm. If there are 11 bright fringes within the central diffraction peak, what is the slit width?
   (b) A diffraction grating has a 1 st order blue (500 nm) line at 11°. At what angle is the first order red (700 nm) line?

2. (a) Two point sources are 2 m apart. At what distance would they be resolvable by the eye according to the Rayleigh criterion? Take the diameter of the pupil to be 5 mm and the wavelength to be 650 nm.

3. Light of wavelength 600 nm illuminates a slit of width 0.2 mm. What is the intensity, relative to the central peak, of the diffraction pattern at a point 3 mm from the central peak on a screen located 2.8 m from the slit?

Chapter 39

SPECIAL RELATIVITY

MAJOR POINTS

1. The two postulates of special relativity
2. The relativity of simultaneity
3. The phenomena of time dilation and length contraction
4. The relativistic Doppler effect for light
5. (a) The Lorentz transformation of coordinates
   (b) The addition of velocities
6. Mass-energy equivalence
7. The relativistic definitions of linear momentum and kinetic energy

CHAPTER REVIEW

THE TWO POSULATES

The two postulates of special relativity are

1. All physical laws have the same form in all inertial reference frames

2. The speed of light in free space is the same in all inertial frames. It does not depend on the motion of the source or the observer.

Recall that in an inertial frame a body free of external influence will stay at rest or move at constant velocity. A frame moving at constant velocity relative to a known inertial frame is also an inertial frame. Note that an "event" occurs at a specific position x and time, t. A length is $L = \Delta x$ and a time interval, or period, is $T = \Delta t$.

The Relativity of Simultaneity

Spatially separated events that are simultaneous in one frame are not simultaneous in another moving relative to the first.

Time Dilation

A proper time, T_o, is the time interval between two events as measured in the rest frame of the clock. In this frame the events occur at the same position.

Two spatially separated clocks record a greater time interval between two events than the proper time recorded by a single clock that is present at both events.

$$T = \gamma T_o$$

where

$$\gamma = 1/(1 - v^2/c^2)^{1/2}$$

Length contraction

The length, L_o, of a rod measured in its rest frame is its proper length. The length measured in a frame relative to which the rod moves, is smaller:

$$L = L_o/\gamma$$

On pages 802 and 803 you will find examples of how a moving object appears to a single observer.

EXAMPLE 1

An electron moves at 0.6c along a 3.2 km linear accelerator. (a) What is the length of the accelerator in the electron's frame? How long does it take to electron to travel the length of the accelerator according to observers in (b) the laboratory frame and (c) the electron's frame?

Solution:

(a) The proper length is $L_o = 3.2$ km and the quantity $\gamma = 5/4$. In the electron's frame, the length is

$$L = L_o/\gamma = 0.8L_o = 2.56 \text{ km.}$$

(b) In the laboratory frame, the time to travel L_o is

$$\Delta t = L_o/v = 1.78\text{x}10^{-5} \text{ s}$$

(c) The electron measures the proper time, $T_o = \Delta t' = \Delta t/\gamma$ or

$$\Delta t' = L/v = 1.42\text{x}10^{-5} \text{ s}$$

THE DOPPLER EFFECT

Suppose that a source with a natural frequency f_o, measured in its rest frame, is moving away from a single observer at a relative speed v. The frequency measured by the single observer is

$$f = [(c - v)/(c + v)]^{1/2} f_o$$

If the source and observer approach each other, the sign of v changes.
Note that this effect depends only on the relative velocity of the source and observer.

EXAMPLE 2
A radar source emits waves of frequency f_o toward a plane that is moving away. What is the speed of the plane if the fractional frequency shift measured at the airport is $\Delta f/f_o = 10^{-6}$?

Solution:

The relative motion of source and observer is such as to decrease the observed frequency. Thus, the frequency of the signal reaching observers on the plane is

$$f_1 = [(1 - \beta)/(1 + \beta)]^{1/2} f_o$$

where $\beta = v/c$. The signal is reflected at f_1, and then picked up at the source at a frequency

$$f_2 = [(1 - \beta)/(1 + \beta)]^{1/2} f_1$$

$$= [(1 - \beta)/(1 + \beta)] f_o$$

Thus,

$$\Delta f = f_2 - f_o = -2\beta f_o/(1 + \beta)$$

$$\approx -2\beta f_o$$

Thus, $v \approx c\Delta f/2f_o = (3\text{x}10^8 \text{ m/s})(5\text{x}10^{-7}) = 150 \text{ m/s}$.

THE LORENTZ TRANSFORMATION

Let us say that frame S' moves in the +x direction relative to frame S and that the space and time coordinates of an event in frame S are (x, t) and those in frame S' are (x', t'). According to the Lorentz transformation of coordinates we have:

$$x' = \gamma(x - vt); \qquad t' = \gamma(t - vx/c^2)$$

At low speeds $\gamma \approx 1$, so these equations reduce to the classical (Galilean) equations: $x' = x - vt$; $t = t'$. The inverse transformation involves just a change in sign for v: $x = \gamma(x' + vt')$ and $t = \gamma(t' + vx'/c^2)$. The y and z coordinates are the same in both frames.

The phenomena of time dilation, length contraction and the lack of synchronization of clocks can be easily derived from the Lorentz transformation equations. See Examples 39.3 and 39.4 in the text.

The Addition of Velocities

Using the notation that v_{AB} is the velocity of A relative to B along the x axis, we have

$$v_{AB} = (v_{AC} + v_{CB})/(1 + v_{AC}v_{CB}/c^2)$$

EXAMPLE 3
Rocket A, moving at 0.8c relative to earth, is chasing a spaceship B moving at 0.6c relative to earth (Fig. 39.1). What is the velocity of the rocket relative to the spaceship?

Solution:
The velocity of B relative to A is

$$v_{BA} = (v_{BE} + v_{EA})/(1 + v_{BE}v_{EA}/c^2)$$

$$= (0.8c - 0.6c)/(1 - 0.48)$$

$$= 0.37c$$

This relative velocity is nearly double the classically expected value of 0.2c.

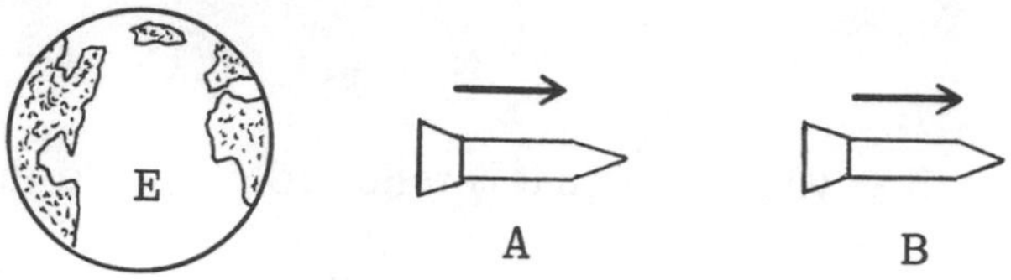

E A B

FIGURE 39.1

ENERGY AND MOMENTUM

Linear Momentum
The relativistic definition of the linear momentum of a particle is

$$p = \gamma m_o v = mv$$

The quantity $m = \gamma m_o$ is called the relativistic mass of the particle.

Mass-Energy equivalence
The total energy of a particle is

$$E = mc^2$$

This equation expresses the equivalence between mass and energy.

Kinetic Energy
The kinetic energy of a particle the difference between its total energy and its rest energy m_oc^2:

$$K = (m - m_o)c^2 = (\gamma - 1)m_oc^2$$

The total energy and the linear momentum are related by

$$E^2 = m_o^2c^4 + p^2c^2$$

EXAMPLE 4
The kinetic energy of an electron is one half of its rest energy. Find:
(a) its speed; (b) its linear momentum.

Solution:
(a) We are given

$$K = m_oc^2/2 = (\gamma - 1)m_oc^2$$

Thus,

$$\gamma = 1.5 = (1 - v^2/c^2)^{-1/2}$$

From which we find v = 0.745c = 2.24×10^8 m/s.
(b) The linear momentum is

$$p = mv = \gamma m_o v$$

$$= (1.5)(9.11 \times 10^{-31}\ kg)(0.745c)$$

$$= 3.05 \times 10^{-22}\ kg.m/s$$

EXAMPLE 5
What is the percentage error in using the classical expression for kinetic energy, $K_C = 1/2\ m_o v^2$ at (a) 0.1c; (b) 0.2c; (c) 0.5c, and (d) 0.7c?

Solution:
The relativeistic kinetic energy is $K_R = (\gamma - 1)m_oc^2$ and the classical kinetic energy is $K_C = 1/2\ m_o v^2$. The fractional error is

$$\Delta K/K_R = 1 - K_C/K_R$$

$$= 1 - \beta^2/2(\gamma - 1)$$

where $\beta = v/c$.
(a) For $\beta = 0.1$, $\gamma = 1.0050$, thus $\Delta K/K_R = 0.75$ %
(b) For $\beta = 0.2$, $\gamma = 1.0206$, thus $\Delta K/K_R = 3.0$ %
(c) For $\beta = 0.5$, $\gamma = 1.1547$, thus $\Delta K/K_R = 19.2$ %
(d) For $\beta = 0.7$, $\gamma = 1.4003$, thus $\Delta K/K_R = 38.8$ %

SOLUTIONS TO SELECTED TEXT EXERCISES AND PROBLEMS

Exercise 5
Use the results in Exercise 2 for $v << c$ to obtain an expression for (a) $(T - T_o)/T_o$, where $T = \gamma T_o$; and (b) $(L - L_o)/L_o$, where $L = L_o/\gamma$.

Solution:

We use the binomial expansion $(1 + z)^n \approx 1 + z$ for $z << 1$. With $\beta = v/c$, we have

$$\gamma = (1 - \beta^2)^{-1/2} \approx 1 + \beta^2/2$$

$$1/\gamma = (1 - \beta^2)^{1/2} \approx 1 - \beta^2/2$$

(a) $\Delta T/T_o = \gamma - 1 \approx +v^2/2c$

(b) $\Delta L/L_o = 1 - 1/\gamma \approx -v^2/2c$

Exercise 13
Spaceship B overtakes spaceship A at a relative speed of 0.2c. Observers in A measure the length of B to be 150 m. (a) What is the proper length of B? How long does it take B to pass a given point on A as measured by observers (b) in A, and (c) in B?

Solution:
(a) The quantity $\gamma = 1.02$. The proper length of B is

$$L_o = \gamma L = (1.02)(150) = 153 \text{ m}$$

(b) Since one observer in A makes both the initial and final time measurements, the proper time is

$$T_o = L/v = 150 \text{ m}/0.2c = 2.5\ \mu s$$

(c) Observers in B measure a dilated time interval

$$T = \gamma T_o = 2.55\ \mu s$$

Exercise 23
A speed trap uses 3-cm wavelength radar waves. What is the Doppler shift in frequency for a car moving at 108 km/h toward the source as measured by the policeman?

Solution:
The natural frequency of the waves is $f_o = 10^{10}$ Hz, and the speed of the car is $v = 30$ m/s. The frequency received by observers in the car is

$$f_1 = [(c + v)/(c - v)]^{1/2} f_o$$

The frequency of the reflected signal received by the police is

$$f_2 = [(c + v)/(c - v)]^{1/2} f_1$$

The frequency shift is

$$\Delta f = f_2 - f_o = 2vf_o/(c - v)$$

Using the given values we find $\Delta f = 2$ kHz.

Exercise 33
When it is 10^8 m away in the earth's frame (S), a rocket (frame S') traveling toward the earth emits a flash. On receipt, it is immediately transmitted back to the rocket. How long does it take the flash to return to the rocket according to the earth frame?

Solution:
The time taken for the flash to reach the earth, as measured in the earth's frame, is

$$\Delta t_1 = L/c = 333 \text{ ms}$$

In this time the rocket travels $v\Delta t_1$ and so its distance from earth when the signal is received is $(L - v\Delta t_1)$. If it takes a time Δt_2, according to the the earth frame, for the returning signal to reach the rocket, the distance the returning signal has to travel is $(L - v\Delta t_1) - v\Delta t_2$. Therefore,

$$(L - v\Delta t_1) - v\Delta t_2 = c\Delta t_2$$

From this we obtain

$$\Delta t_2 = (L - v\Delta t_1)/(c + v) = 37 \text{ ms}$$

Exercise 59
A proton has a kinetic energy of 40 GeV. What is (a) its speed, and (b) its linear momentum?

Solution:
(a) With $m_o = 1.67\text{x}10^{-27}$ kg, we have $m_oc^2 = 1.50\text{x}10^{-10}$ J = 938 MeV.
We are given $K = (\gamma - 1)m_oc^2 = 40$ GeV, thus,

$$\gamma = 43.6 = (1 - v^2/c^2)^{-1/2}$$

which leads to $v = 0.9997c$.
(b) The linear momentum is

$$p = \gamma m_o v$$

$$= 43.6(1.67\text{x}10^{-27} \text{ kg})(0.9997c) = 2.2 \times 10^{-17} \text{ kg.m/s}$$

Problem 7
A light beam travel at θ' to the x' axis of frame S', which moves at velocity v in the +x direction of frame S. If θ is the angle measured to the x axis, show that

$$\cos\theta = (\cos\theta' + \beta)/(1 + \beta\cos\theta')$$

where $\beta = v/c$. [Hint: Use the Lorentz transformation and note that $\cos\theta = dx/(cdt)$]. (b) Given $\beta = 0.9$, evaluate θ for $\theta' = 30°, 60°$, and $90°$. Why is this called the headlight effect?

Solution:
(a) The cosine of the angles is given by the ratio u_x/c or u_x'/c, thus, $\cos\theta = (dx/dt)/c$; $\cos\theta' = (dx'/dt')/c$. We use the transformation of velocities to write an expression for $\cos\theta$ in terms of u_x':

$$\cos\theta = u_x/c = (u_x' + v)/(1 + vu_x'/c^2)c =$$

$$= (\cos\theta' + \beta)/(1 + \beta\cos\theta')$$

(b) With the given values we find $\theta = 7.04°, 15.1°, 25.8°$. The energy upto $\theta = 90°$ is confined to angles upto only $\theta' = 25.8°$.

Problem 17
A particle of rest mass m_o moves at speed u along the +x axis in frame S. The linear momentum of the particle is $p_x = \gamma(u)m_ou$ and its energy is $E = \gamma(u)m_oc^2$, where $\gamma(u) = (1 - u^2/c^2)^{1/2}$. Frame S' moves at velocity v along the +x axis of frame S. Show that the linear momentum and energy in frame S' are related to the values in frame S' according to

$$E' = \gamma(E - vp_x); \qquad p_x' = \gamma(p_x - vE/c^2)$$

where $\gamma = (1 - v^2/c^2)^{-1/2}$. Thus E and p_x transform in exactly the same way as x and t. [Hint: The speed of the particle in frame S' is
$u' = (u - v)/(1 - uv/c^2)$.]

Solution:
The linear momenta and energies in the two frames are

$$p_x = \gamma(u)m_ou; \qquad p_x' = \gamma(u')m_ou'$$

$$E = \gamma(u)m_oc^2; \qquad E' = \gamma(u')m_oc^2$$

To find $\gamma(u')$ we use the velocity transformation equation

$$\gamma(u') = (1 - u'^2/c^2)^{-1/2} = c(c^2 - u'^2)^{-1/2}$$

$$= [1 - c^2(u - v)^2/(c^2 - uv)^2]^{-1/2}$$

$$= (c^2 - uv)/[c^2(c^2 - v^2) \qquad - u^2(c^2 - v^2)]^{1/2}$$

$$= \gamma(1 - uv/c^2)\gamma(u), \text{ where } \gamma = \gamma(v)$$

Substitute $\gamma(u')$ into E' and p_x' (which also requires u'):

$$E' = \gamma m_o c^2(1 - uv/c^2)\gamma(u) = \gamma(E - vp_x)$$

$$p_x' = \gamma m_o \gamma(u)(u - v) = \gamma(p_x - vE/c^2)$$

SELF-TEST

1. A train moves along a platform at speed v = 0.8c. A single observer on the platform measures an interval of 5 μs between the times the front and the rear go past him. (a) What is the time interval measured by observers on the train? (b) What is the length of the train in the frame of the platform?
   (c) What is the length of the train according to observers on the train?

2. A spaceship B is moving away from the earth E with a velocity of 0.8c relative to earth. A rocket A is fired from earth toward the spaceship at 0.6c relative to earth. What is the velocity of the rocket relative to the spaceship?

3. The total energy of a particle is four times its rest energy. Find (a) its kinctic energy; (b) its speed.

4. The wavelength of a certain spectral line from a distance galaxy is observed to have a 10% shift towards the red end of the spectrum. What is the speed of the source?

Chapter 40

EARLY QUANTUM THEORY

MAJOR POINTS

1. (a) Characteristics of blackbody radiation
   (b) The failure of classical physics to explain the spectrum of blackbody radiation
2. Planck's radiation law and Einstein's quantum hypothesis
3. The photoelectric effect:
   (a) The failure of classical physics to explain several aspects of the photoelectric effect;
   (b) Einstein's equation; the photon concept
4. The Compton effect involves the scattering of X rays by electrons
5. Balmer's formula for the hydrogen spectrum
6. Rutherford's nuclear model of the atom
7. Bohr's theory of the hydrogen atom
8. Bohr's correspondence principle

CHAPTER REVIEW

BLACKBODY RADIATION

A blackbody is an ideal system that absorbs all the radiation incident on it. A small opening to a cavity is approximately a blackbody. The radiation emitted by a blackbody depends only on the temperature of the walls, not on the material of which the walls are made. The spectral distribution of blackbody radiation has a peak that shifts toward shorter wavelengths as the temperature increases, see Fig. 40.2 in the text. The wavelength at which the maximum radiation occurs is related to the temperature by Wein's displacement law:

$$\lambda_{max}T = 2.898 \times 10^{-3} \text{ m.K}$$

Classical physics, which led to the Rayleigh-Jeans law, completely failed to account for the shape of the experimentally obtained curve. Max Planck was able to theoretically obtain the curve by assuming that the energy of the vibrating atoms in the walls was quantized: Energy could assume only certain discrete values. The exact nature of the assumptions were clarified by Einstein:

The energy of an oscillator, whose frequency is f, is quantized in steps of hf. The energy of an oscillator in the n th. level is

$$E_n = nhf \qquad\qquad n = 0, 1, 2, \ldots$$

where $h = 6.626 \times 10^{-34}$ J.s is Planck's constant. Besides leading to the correct spectral distribution for blackbody radiation, further justification for the quantum hypothesis came from the explanation of the photoelectric effect and the Compton effect.

EXAMPLE 1
A bulb radiates 5 W as a point source of green light (550 nm). Find: (a) the intensity at 2 m; (b) the number of photons incident on a 0.3 m^2 area in 8 s at this distance.

Solution:
(a) The energy emitted by a point source spreads over ever-increasing spheres, therefore the intensity at a distance r is

$$I = P/A = P/4\pi r^2$$

$$= (5\ W)/4\pi(2\ m)^2 = 9.95\times10^{-2}\ W/m^2.$$

(b) The energy transported by n photons is $\Delta E = nhf$, so the intensity is

$$I = P/A = \Delta E/A\Delta t = nhf/A\Delta t$$

The number of photons incident is therefore

$$n = I\ A\ \Delta t/hf$$

$$= (9.95\times10^{-2}\ W/m^2)(0.3\ m^2)(8\ s)/(6.63\times10^{-34}\ J.s)(3\times10^8\ m/s)$$

$$= 1.20\times10^{24}\ \text{photons}$$

THE PHOTOELECTRIC EFFECT

In the photoelectric effect, electrons are emitted from a surface when radiation is incident on it. Several features of this effect cannot be explained classically:

1. There is a threshold frequency, f_o, below which photoemission does not occur.
2. The maximum kinetic energy of the electrons does not depend on the intensity of the radiation. (The number electrons does depend on the intensity, as one would expect classically.)
3. Photoemission occurs almost instantly, even for very weak light. Classically, one should have to wait for a long time in this case.

Einstein's Equation

Einstein proposed that light consists of quanta of energy, now called photons, whose energy is given by

$$E = hf$$

In the photoelectric effect a single photon gives its energy to a single electron. The maximum kinetic energy of the ejected electrons is given by

$$K_{max} = hf - \phi$$

where ϕ, the work function, is the minimum energy required to remove an electron from the surface. The threshold frequency is given by

$$hf_o = \phi$$

The maximum kinetic energy can be determined by measuring the stopping potential difference V_o, required to bring the emitted electrons to rest:

$$K_{max} = eV_o.$$

Combining the above we find

$$eV_o = h(f - f_o)$$

A plot of V_o versus f is a straight line with a slope of h/e, which is independent of the material.

EXAMPLE 2

The threshold wavelength for a metal is 270 nm. (a) What is the work function? (b) What is the stopping potential when the wavelength of the incident light is 220 nm?

Solution:

(a) The work function is

$$\phi = hf_o = hc/\lambda_o$$

$$= (6.63\text{x}10^{-34}\ \text{J.s})(3\text{x}10^{8}\ \text{m/s})/(2.7\text{x}10^{-7}\ \text{m}) = 7.0\text{x}10^{-19}\ \text{J}$$

or $\phi = 4.38$ eV.

(b) Since $f = c/\lambda$, we have

$$V_o = (hc/e)(1/\lambda - 1/\lambda_o)$$

$$= 1.04\ \text{V}$$

THE COMPTON EFFECT

In the Compton effect an X-ray photon is scattered by an electron. The energy of the outgoing X-ray is lower than the initial value, thus the scattered wavelength, λ', is longer. The change in wavelength $\Delta\lambda = \lambda' - \lambda$, is given by

$$\Delta\lambda = \lambda_c (1 - \cos\theta)$$

where θ is the angle between the incident and scattered X-rays and $\lambda_c = h/m_o e = 2.43$ pm is called the Compton wavelength (Fig. 40.1). Note that $\Delta\lambda$ is independent of λ.

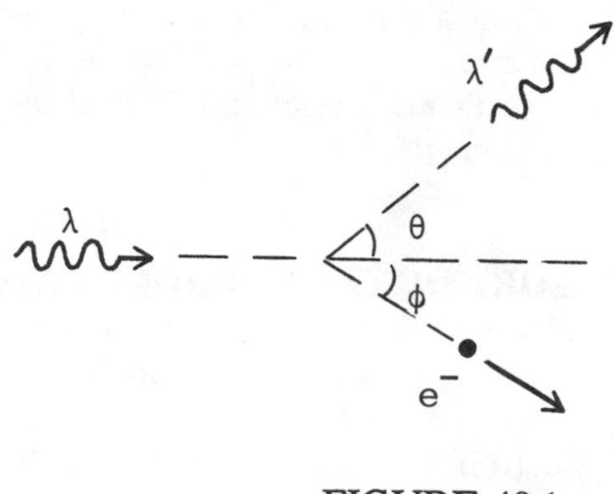

FIGURE 40.1

EXAMPLE 3

An X-ray with a wavelength of 0.650 nm is scattered by an electron. The electron's recoil speed is $1.2\text{x}10^6$ m/s. Find: (a) The wavelength of the scattered X-ray; (b) the energy of the scattered X ray; (c) the angle between the incident and scattered beam.

Solution:

(a) From the conservation of energy the kinetic energy of the scattered electron is

$$K = hc(1/\lambda - 1/\lambda')$$

where $K = mv^2/2 = 6.56\text{x}10^{-19}$ J (we may use the classical expression for K since the electron's speed is much less than c). Thus,

$$1/\lambda' = 1/\lambda - K/hc$$

$$= 1/0.65 \text{ nm} - 6.56\text{x}10^{-19} \text{ J}/hc$$

This leads to $\lambda' = 0.6514$ nm

(b) The X-ray energy is $E' = hc/\lambda' = 3.05\text{x}10^{-16}$ J = 1.91 keV.

(c) Since $\Delta\lambda = \lambda' - \lambda = \lambda_c(1 - \cos\theta)$, we have

$$\cos\theta = 1 - (\lambda' - \lambda)/\lambda_c$$

$$= 1 - 1.4 \text{ pm}/2.43 \text{ pm}$$

Thus, $\theta = 64.9°$.

RUTHERFORD'S NUCLEAR MODEL OF THE ATOM

As a result of experiments involving the scattering of α particles (these are nuclei of helium atoms that are emitted by some radioactive elements), Rutherford proposed in 1911 that an atom consists of a tiny (radius $\approx 10^{-14}$ m) positively charged nucleus that is surrounded by electrons at a distance of about 10^{-10} m. In this model, most of the atom consists of "empty space". Rutherford's nuclear model formed the basis of Bohr's theory of the visible spectrum of the hydrogen atom.

BOHR'S THEORY OF THE HYDROGEN ATOM

Bohr assumed that the hydrogen atom consists of a tiny positive nucleus (which we now know is a proton) and a single electron. The theory is based on three postulates:

1. The electron moves only in certain circular orbits called stationary states. The motion can be described classically.

$$mv^2/r = ke^2/r^2$$

2. Radiation occurs only when an electron goes from one allowed orbit to another with a lower energy. The radiation frequency is given by

$$hf = E_m - E_n$$

where E_m and E_n are the energies of the two states.

3. The angular momentum of the electron is restricted to integer multiples of $h/2\pi$:

$$mvr = nh/2\pi$$

From these postulates one finds that the energy of the n th state, $E_n = -ke^2/2r_n$, is

$$E_n = -mk^2e^4/2\hbar^2n^2$$

The radii of the orbits are $r_n = n^2r_B$, where $r_B = \hbar^2/mke^2 = 5.29\text{x}10^{-11}$ m. The Bohr theory may be applied to other "one-electron" systems such as He^+, or Li^{2+}. If Z is the atomic number, then

$$E_n = -13.6Z^2/n^2 \qquad \text{eV}$$

EXAMPLE 4
An electron in the hydrogen atom makes a transition from the n = 5 level to the n = 2 level. Find: (a) the wavelength of the emitted photon; (b) the angular momentum of the electron in the lower state; (c) the radius in the lower state; (d) the speed in the lower state.

Solution:
(a) We first need to find the difference in energy between the two levels.

$$\Delta E = 13.6\ \text{eV}(1/2^2 - 1/5^2)$$

$$= 2.86\ \text{eV} = 4.57\text{x}10^{-19}\ \text{J}$$

Since $\Delta E = hc/\lambda$, we have

$$\lambda = hc/\Delta E = 435\ \text{nm}$$

(b) The angular momentum for n = 2 is

$$L = 2h/2\pi = h/\pi = 2.4\text{x}10^{-34}\ \text{J.s}$$

(c) The radius of the n = 2 orbit is

$$r = n^2 r_B = 4r_B = 2.08\text{x}10^{-10}\ \text{m}$$

(d) Since L = mvr,

$$v = L/mr = 1.1\text{x}10^6\ \text{m/s}$$

BOHR'S CORRESPONDENCE PRINCIPLE

According to Bohr's correspondence principle the results of a new theory, such as special relativity or quantum theory, should reduce, in the appropriate limit to the classical results. Thus when the speed of a particle is much less than the speed of light, the predictions of special relativity should be the same as those of classical physics which has been successful in this domain. Similarly, for large values of a quantum number, the predictions of quantum theory should agree with classical expectations.

SOLUTIONS TO SELECTED TEXT EXERCISES AND PROBLEMS

Exercise 5
Given that the sun's surface temperature is 5760 K, find the total power radiated into space (taken to be at 0 K). The sun's radius is $6.96\text{x}10^8$ m. (See Exercise 4).

Solution:
The intensity of radiation from a blackbody at temperature T is given by Eq. 40 in the text

$$R = \sigma T^4$$

where $\sigma = 5.67\text{x}10^{-8}$ W/m^2.K^4. The total power radiated from a spherical surface of area $4\pi r^2$ is

$$\text{Power} = RA = \sigma T^4(4\pi r^2)$$

$$= (5.67\text{x}10^{-8}\ \text{W/m}^2.\text{K}^4)(5760\ \text{K})^4(4\pi)(6.96\text{X}10^8\ \text{m})^2$$

$$= 3.8 \times 10^{26}\ \text{W}$$

Exercise 21

When radiation of wavelength 350 nm is incident on a surface, the maximum kinetic energy of the photoelectrons is 1.2 eV. What is the stopping potential for a wavelength of 230 nm?

Solution:

The maximum kinetic energy of the photoelectrons is

$$K_{max} = eV_o = hf - \phi$$

so,

$$\phi = hc/\lambda - eV_o$$

$$= 3.55\ \text{eV} - 1.2\ \text{eV} = 2.35\ \text{eV}$$

The new stopping potential is given by

$$eV_o = hc/\lambda - \phi = 5.4\ \text{eV} - 2.35\ \text{eV} = 3.05\ \text{eV}$$

Thus $V_o = 3.05$ V.

Exercise 35

X rays with an energy of 50 keV are scattered by 45°. Find the frequency of the scattered photons.

Solution:

Since $E = hc/\lambda$, the original wavelength of the X rays is

$$\lambda = hc/E = 0.0248\ \text{nm}$$

The shift in wavelength is

$$\Delta\lambda = \lambda_c(1 - \cos\theta)$$

$$= (2.43\text{x}10^{-12}\ \text{m})(1 - \cos 45°) = 7.1\text{x}10^{-13}\ \text{m}$$

Since $\Delta\lambda = \lambda' - \lambda$, we have $\lambda' = 2.48\text{x}10^{-11}$ m $+ 0.07\text{x}10^{-11}$ m $= 2.55\text{x}10^{-11}$ m and then $f' = 1.2 \times 10^{19}$ Hz

Exercise 47
An electron orbits a nucleus with a charge Ze. Show that the speed for the n th level is given by $v_n = 2.2\text{x}10^6 Z/n$.

Solution:

Newton's second law for an electron orbiting a nucleus with charge Ze, is

$$kZe^2/r = mv^2/r$$

thus

$$v = (kZe^2/mr)^{1/2} \qquad \text{(i)}$$

From the quantization of angular momentum, $mvr = n\hbar$, we have $v = n\hbar/mr$. Equating these two expressions for v we find the radius of the n th orbit is

$$r = n^2\hbar^2/Zmke^2$$

Substitute this r into (i) to find

$$v = Zke^2/n\hbar \ = 2.2\text{x}10^6 Z/n$$

Problem 7
Derive Eq. 40.16 for the Compton effect. (Hint: First use Eq. 40.14 and 40.15 to eliminate ϕ and obtain an expression for p^2. Second, use $E^2 = p^2c^2 + M_o^2c^4 = (K + m_oc^2)^2$ and Eq. 40.13 to obtain another expression for p^2.)

Solution:

From Eqs. 40.14 and 40.15 we obtain expressions for $p\cos\phi$ and $p\sin\phi$, see Fig. 40.1 above:

$$p^2\cos^2\phi = (h/\lambda - h\cos\theta/\lambda')^2$$

$$p^2\sin^2\phi = (h\sin\theta/\lambda')^2$$

When these are added we find

$$p^2 = h^2[1/\lambda^2 - 2\cos\theta/\lambda\lambda' + 1/\lambda'^2] \qquad \text{(i)}$$

For the electron we have $p^2c^2 = E^2 - m_o^2c^4 = (K + m_oc^2)^2 - m_oc^4$, where from the conservation of energy (Eq. 40.13): $K = hc(1/\lambda - 1/\lambda')$. Thus,

$$p^2 = K^2/c^2 + 2Km_o = h^2(1/\lambda - 1/\lambda')^2 + 2hm_oc(1/\lambda - 1/\lambda') \qquad \text{(ii)}$$

Equate (i) and (ii) to find $\Delta\lambda = \lambda' - \lambda = (h/m_oc)(1 - \cos\theta)$.

Problem 9
The total intensity, R, radiated from the surface of a blackbody is found by multiplying the integral of the energy density over all wavelengths, $U = \int u_\lambda \, d\lambda$, by c/4, that is, $R = Uc/4$. Derive the Stefan-Boltzmann law, $R = \sigma T^4$, where the constant $\sigma = 2\pi^5 k^4/15c^2h^3$. Set $x = hc/\lambda kT$ and note that

$$\int_0^\infty x^3 \, dx/(e^x - 1) = \pi^4/15$$

Solution:

If we let $x = hc/\lambda kT$, then

$$dx = -hc d\lambda/\lambda^2 kT$$

and Eq. 40.5 becomes $u = 8\pi hc/\lambda^5(e^x - 1)$.

$$R = (c/4) \int u \, d\lambda = [2\pi(kT)^4/h^3c^2] \int_0^\infty x^3 \, dx/(e^x - 1)$$

Using the given integral we find $R = \sigma T^4$.

SELF-TEST

1. When light of wavelength 214 nm is incident on a material, the stopping potential is 2 V. Find (a) the work function; (b) the maximum speed of the photoelectrons.

2. An X-ray with a wavelength of 0.20 nm is scattered by an electron. At what scattering angle θ for the photon is the kinetic energy of the electron equal to 12 eV?

3. The energy levels of the He^+ ion are given by $E_n = -54.4/n^2$ eV.
   Electrons with energy 50 eV are used to excite the atomic electrons, which are initially in the ground state. What frequencies will be observed in the emitted radiation?

4. A laser emits 1 mW at a wavelength of 632.8 nm. The beam diameter is 1mm. Find (a) the number of photons emitted per m^2 per second; (b) the linear momentum carried by the beam per m^2 per second.

Chapter 41

WAVE MECHANICS

MAJOR POINTS

1. De Broglie's hypothesis that particles have wave properties
2. Schrodinger's wave equation is used to predict the behavior of matter waves
3. The wave function tells us the probability of finding a particle in a given region
4. The phenomenon of tunneling through a potential barrier
5. The Heisenberg uncertainty principle: One cannot measure certain pairs of quantities simultaneously to arbitrary precision.

CHAPTER REVIEW

DE BROGLIE'S HYPOTHESIS

De Broglie suggested that since light has both wave and particle characteristics, perhaps particles also had wave properties. The wavelength associated with a particle is

$$\lambda = h/p$$

where p is the linear momentum. This relationship is also valid for the photon: $p = E/c = hf/c = h/\lambda$. If a particle is accelerated from rest by a potential difference V, then $mv^2/2 = qV$, so $p = mv = (2mqV)^{1/2}$ and

$$\lambda = h/(2mqV)^{1/2}$$

We have used the classical expressions for kinetic energy and linear momentum. This is valid for speeds upto about 0.1c. This would correspond to an energy of about 3 keV for an electron and 5 MeV for a proton.

De Broglie's hypothesis helped "explain" Bohr's third postulate regarding the quantization of angular momentum:

$$L = mvr = nh/2\pi$$

Since $mv = h/\lambda$, we find

$$n\lambda = 2\pi r$$

This looks like a condition for standing waves: Only those orbits that can fit an integral number of wavelengths are allowed.

EXAMPLE 1
A particle is accelerated from rest by a potential difference of 1.5 kV. What is the de Broglie wavelength if the particle is (a) an electron; (b) a proton?

Solution:
At the given accelerating potential difference we may use the classical expression for kinetic energy, thus

$$\lambda = h/(2mqV)^{1/2}$$

where $q = e$.
(a) For an electron $\lambda = 31.7$ pm; (b) For a proton $\lambda = 0.74$ pm

SCHRODINGER'S WAVE EQUATION

The time-independent Schrodinger wave equation for a particle in a stationary state ($E =$ constant) with a potential energy U is

$$d^2\psi/dx^2 + 2m(E - U)/\hbar^2 = 0$$

The square of the wavefunction ψ tells us the probability of finding the particle in a given region:

$$\psi^2\, dV = \text{Probability of finding the particle within volume } dV$$

In one dimension, the total probability of finding the particle anywhere along the x axis must be one:

$$\int_{-\infty}^{\infty} \psi^2\, dx = 1$$

When the wavefunction satisfies this condition it is said to be normalized.

The wavefunction for a given system depends on the boundary conditions that must be satisfied. (This is analogous to the conditions for standing waves on a string on in the air column in a pipe.) Once the allowed wavefunction have been found, the allowed energies can also be determined.

Particle in a box

For a particle confined to a box of side L the allowed wavefunctions are

$$\psi(x) = A\sin(n\pi x/L) \qquad n = 1, 2, 3, ...$$

For a normalized wavefunction, $A = (2/L)^{1/2}$. From thus wavefunction we infer that there are some points at which the particle will not be found. Since $\lambda_n = 2L/n$ and $p = h/\lambda$, we find the allowed energies, $p^2/2m$, are

$$E_n = n^2h^2/8mL^2 \qquad n = 1, 2, 3, ...$$

Note, in particular, the $n \neq 0$. The minimum energy the particle can have is called the zero-point energy.

Tunneling Through a Barrier

A particle with an energy less that the height of a barrier can penetrate into the classically forbidden region. If the barrier is not too wide, there is a small, but finite probability that the particle can "tunnel" through the barrier and emerge on the other side (Fig. 41.1). This phenomenon occurs, for example, in α particle (radioactive) decay of nuclei.

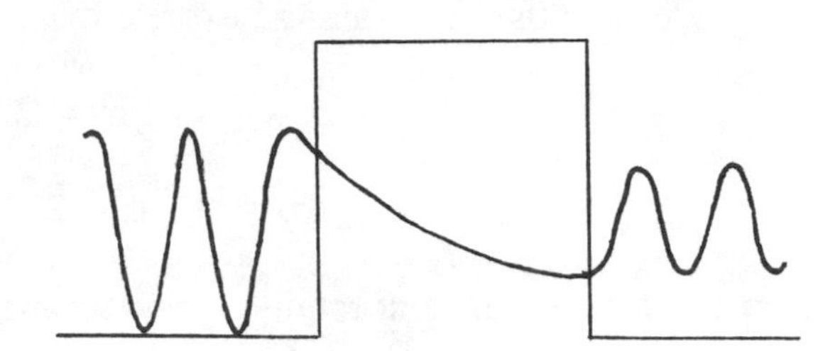

FIGURE 41.1

EXAMPLE 2

A proton ($1.67\text{x}10^{-27}$ kg) is confined to an impenetrable one-dimensional box of length $4\text{x}10^{-15}$ m. Find (a) the two lowest energy levels; (b) the wavelength of the photons that would be emitted when the proton returns to the ground state.

Solution:

(a) The energy levels are

$$E_n = n^2h^2/8mL^2 = 2.06\text{x}10^{-12}\ n^2\ \text{J} = 12.85n^2\ \text{MeV}$$

The first two values are 12.9 MeV, 51.4 MeV

(b) The wavelength is determined by $\Delta E = hc/\lambda$, so

$$\lambda = hc/\Delta E = 3.22\text{x}10^{-23}\ \text{m}$$

HEISENBERG UNCERTAINTY PRINCIPLE

According to the Heisenberg uncertainty principle it is not possible to make simultaneously measurements of certain pairs of quantities to arbitrary precision. For example, the uncertainties in a particle's position and linear momentum are related by

$$\Delta x\ \Delta p_x \approx h$$

Similarly, the uncertainty in a particle's energy is related to the uncertainty in the lifetime of the state:

$$\Delta E\ \Delta t \approx h$$

These are fundamental restrictions associated with the wave-particle duality of matter and light. They do not arise from experimental difficulties.

EXAMPLE 3
An electron has a kinetic energy of 1.8 keV. If its momentum is measured with an uncertainty of 0.1%, what is the minimum uncertainty in is position?

Solution:

We may use the classical expressions for kinetic energy (in joules) and momentum, so

$$p = (2mK)^{1/2} = 2.29\text{x}10^{-23} \text{ kg.m/s}$$

$$\Delta p = 10^{-3}p = 2.29\text{x}10^{-26} \text{ kg.m/s}$$

From the Heisenberg uncertainty principle $\Delta p \, \Delta x \approx h$, so

$$\Delta x \approx h/\Delta p = 2.9\text{x}10^{-8} \text{ m}$$

The position of the electron cannot be determined more precisely than this.

SOLUTIONS TO SELECTED TEXT EXERCISES AND PROBLEMS

Exercise 7
A 1 g pellet moves at 10 m/s. For what slit width would the first diffraction minimum be at 0.5°? Is this a practical experiment?

Solution:

The de Broglie wavelength of the pellet is

$$\lambda = h/mv = 6\text{x}10^{-32} \text{ m}$$

The position of the first diffraction minimum is given by a $\sin\theta = \lambda$, so

$$a = \lambda/\sin 5° = 7.6\text{x}10^{-30} \text{ m}$$

This is smaller than a nucleus; the experimental is not practical!

Exercise 23
The ground state energy of an electron in an infinite well is 20 eV. (a) What is the energy of the first excited level? (b) What is the length of the well?

Solution:
(a) Since $E_n = n^2E_1$, we have $E_2 = 4E_1 = 80$ eV
(b) Since $E_n = n^2h^2/8mL^2$, we find

$$L = h/(8mE_1)^{1/2}$$

Using $E_1 = 3.2\text{x}10^{-18}$ J, we find L = 0.137 nm

Problem 1
Use the relativistic expressions for kinetic energy and linear momentum to show the following: (a) $\lambda \approx h/(2m_oK)^{1/2}$, if $K << m_oc^2$, and (b) $\lambda = hc/K$, if $K >> m_oc^2$.

Solution:
(a) From $K = (\tau - 1)m_oc^2$, we have $K/m_oc^2 = (\tau - 1) << 1$. This means $\tau \approx 1$ or $v <<$ c. Therefore we may use the classical expression, $K = p^2/2m$, so $p = (2m_oK)^{1/2}$ and

$$\lambda = h/p = h/(2m_oK)^{1/2}$$

(b) From Eq. 39.27, $E^2 = p^2c^2 + m_o^2c^4$. At high energies, $E \approx pc$.
Also, at high energy, $E \approx K$, so $p \approx K/c$ and

$$\lambda = h/p = hc/K$$

Problem 5
Consider the ground state wavefunction for a particle in an infinite potential well that extensd from $x = 0$ to $x = L$. What is the probability of finding the particle from $x = L/4$ to $3L/4$?

Solution:
In the ground state the wave function is $\psi = (2/L)^{1/2} \sin(\pi x/L)$.
The probability of finding the particle is given by

$$\text{Probability} = \int \psi^2\, dx = \int (2/L) \sin^2(\pi x/L)\, dx$$

$$= (1/L) \int (1 - \cos(2\pi x/L)\, dx$$

$$= (1/L)[x - (L/2\pi)\sin(2\pi x/L)]$$

$$= 1/2 + 1/\pi = 0.818$$

SELF-TEST

1. An electron is accelerated from rest by a potential difference of 900 V. What is its de Broglie wavelength?

2. (a) The uncertainty in the velocity of a proton is $3\text{x}10^6$ m/s. Determine the uncertainty in its position. Take $m = 1.67\text{x}10^{-27}$ kg.
(b) The uncertainty in the lifetime of an excited state in an atom is $5\text{x}10^{-8}$ s. What is the intrinsic linewidth (in Hz) of the line?

3. An electron is trapped in an infinite well of width 0.15 nm. What are the first three energy levels?

Chapter 42

ATOMS AND SOLIDS

MAJOR POINTS

1. Four quantum numbers are needed to specify the state of an electron in an atom
2. According to the Pauli exclusion principle, no two electrons in an atom can have the same four quantum numbers
3. The periodic table can be formed by using the four quantum numbers and the Pauli exclusion principle
4. The band theory of solids explains the differences in conductivity between conductors, insulators and semiconductors.

CHAPTER REVIEW

THE HYDROGEN ATOM

Energy

The energy of an electron in a hydrogen atom is determined by the principal quantum number, n:

$$E_n = -13.6 \text{ eV}/n^2$$

Orbital angular momentum

The orbital angular momentum, L, is determined by the orbital quantum number ℓ:

$$L = [\ell(\ell + 1)]^{1/2}\ h/2\pi$$

where ℓ is restricted to the values

$$\ell = 0, 1, ..., (n - 1)$$

The component of L along a given axis, for example the direction of a magnetic field, is also quantized:

$$L_z = m_\ell\ h/2\pi$$

where the orbital magnetic quantum number, m_ℓ, can take on the values

$$m_\ell = 0, \pm 1; \pm 2; ..., \pm \ell$$

In an external magnetic field, the orientation of L is restricted to certain discrete angles relative to B: $\cos\theta = m_\ell/L$

Spin

The electron has an intrinsic spin angular momentum, S:

$$S = [s(s + 1)]^{1/2} h/2\pi$$

where the spin quantum number s = 1/2. In the presence of a magnetic field, the component of S along the field can assume only one of two values

$$S_z = m_s h/2\pi$$

where the spin magnetic quantum number, $m_s = \pm 1/2$.

The orbital magnetic quantum number m_ℓ helps to account for the fact that in the presence of a magnetic field, single spectral lines split into a group of finer lines. Even in the absense of an external field, some lines are actually two finer lines close together. The spin magnetic quantum number, which can assume only two values, helps to account for this doublet structure.

To sum up, the four quantum numbers are n, ℓ, m_ℓ and m_s.
The energy states with a given value of n are said to form a shell; the states with a given value of ℓ are said to form a subshell. For historical reasons (arising from spectroscopy) the shells and subshells are given letter designations:

$$n = 1, 2, 3, \ldots \text{ are called K, L, M, } \ldots$$

$$\ell = 0, 1, 2, 3, 4, \ldots \text{ are called s, p, d, f, } ..$$

If a there are two electrons (corresponding to $m_s = \pm 1/2$) in a state with n = 2 and ℓ = 3, the configuration is designatied as $2f^2$. One electron in the n = 3 and ℓ = 1 would be written as $3p^1$.

Since the angular momentum of an atom changes when it makes a transition from a higher to a lower state, and angular momentum must be conserved, it follows that the emitted photon must carry away angular momentum.

EXAMPLE 1

Enumerate the possible quantum numbers for a state with n = 2. What is the total number of possible states?

Solution:

The possible states are listed in the table below.

n	ℓ	m_ℓ	m_s
2	0	0	$\pm 1/2$
	1	0	$\pm 1/2$
	1	1	$\pm 1/2$
	1	-1	$\pm 1/2$

The total number of states in $2n^2 = 8$.

WAVEFUNCTIONS FOR THE HYDROGEN ATOM

The wavefunction for an electron in the lowest 1s state is

$$\psi_{1s} = (1/\pi r_o^3)^{1/2} \exp(-r/r_o)$$

where $r_o = 5.29 \times 10^{-11}$ m is the Bohr radius.

The radial probability density function P(r) is defined such that the probability of finding the electron in the spherical shell from r to r + dr is P(r)dr, where

$$P(r) = 4\pi r^2 \psi^2$$

For example, for the 1s state,

$$P_{1s} = (4r^2/r_o^3)\, e^{-2r/ro}$$

EXAMPLE 2

Find the value of the radial propability density function for an electron in the 2s state at the point $r = 5r_o$.

Solution:

From Eq. 42.9 in the text we have

$$P_{2s}(r) = (r^2/8r_o^3)(2 - r/r_o)^2 \exp(-r/r_o)$$

At $r = 5r_o$,

$$P_{2s}(5r_o) = (25/8r_o)(81)e^{-5} = 3.22 \times 10^{10}\ m^{-1}$$

In Fig. 42.5 in the text, note that this is close to the maximum value of P_{2s}. The Bohr picture of well-defined orbits with most of the atom being empty space is clearly not correct.

EXAMPLE 3

An electron in a hydrogen atom is in the 1s state. What is the probability of finding it in the region from $r = r_o$ to $r = 2r_o$? Note from the table of integrals in Appendix C that:

$$\int x^2 \exp(-ax) = -(1/a^3)(2 + 2ax + a^2x^2)\exp(-ax)$$

Solution:

The expression for P_{1s} is given in the section above.

$$\int P(r)\, dr = (4/r_o^3) \int_{r_o}^{2r_o} r^2 \exp(-2r/r_o)\, dr$$

$$= -(4/r_o^3)(r_o^3/8)(2 + 4r/r_o + 4r^2/r_o^2) \exp(-2r/r_o)$$

$$= -(1 + 2r/r_o + 2r^2/r_o^2) \exp(-2r/r_o)$$

$$= -13e^{-4} + 5e^{-2} = 0.439$$

X-RAYS

When an electron makes a transition between one of the inner shells in an atom, the photon emitted is in the X-ray region. The transitions are often to a vacant state in the K or L shells produce a line spectrum. A continuous X-ray spectrum is produced as high-energy electrons are declerated when they enter a heavy metal target. The shortest wavelength is produced when an electron loses all its kinetic energy in one step:

$$\lambda_o = hc/eV$$

The frequency of the K_α line is given by Moseley's law (see Fig. 42.9 in the text):

$$(f)^{1/2} = a(Z - 1)$$

where a is a constant and Z is the atomic number of the element.

SOLUTIONS TO SELECTED TEXT EXERCISES AND PROBLEMS

Exercise 17

For an electron in the 1s state, calaculate the radial probability density, P(r), at (a) $r_o/2$; (b) r_o, and (c) $2r_o$, where r_o is the Bohr radius.

Solution:

The radial probability density for the 1s state is

$$P_{1s} = (4r^2/r_o^3) \exp(-2r/r_o)$$

(a) At $r = r_o/2$, $P = e^{-1}/r_o = 0.37/r_0$

(b) At $r = r_o$, $P = 4e^{-2}/r_o = 0.54/r_0$

(c) At $r = 2r_o$, $P = 16e^{-1}/r_o = 0.29/r_0$

Exercise 27
The energy of the n = 2 state in molybdenum is E_2 = -2870 eV. Given that the wavelengths of the K_α and K_β lines are 0.71 nm and 0.63nm respectively, determine the energies E_1 and E_3.

Solution:
The energy of a photon (in eV) is related to its wavelength (in nm) by

$$E = hc/\lambda = 1240/\lambda$$

Thus the energy differences associated with the K_α and K_β lines are

$$\Delta E(K_\alpha) = 1240/0.71 \text{ nm} = 1746 \text{ eV}$$

$$\Delta E(K_\beta) = 1240/0.63 \text{ nm} = 1968 \text{ eV}$$

Since we are given E_2, we can find the the other two levels:

$$E_1 = -2870 - 1746 = -4616 \text{ eV}$$

$$E_3 = -4616 + 1968 = -2648 \text{ eV}$$

Exercise 33
Sodium, which as a single electron in the 3s state, emits doublet lines at 589.0 nm and 589.6 nm in making transitions to the ground state (which consists of a single level). (a) What is the difference in energy between the excited states? (b) What is the effective magnetic field experienced by the electron?

Solution:
(a) The difference in the energy levels is

$$\Delta E = hc(1/\lambda_1 - 1/\lambda_2) = 2.14 \text{ meV}$$

(b) The difference in potential energy between the parallel and antiparallel orientations is

$$\Delta U = 2\mu_B B$$

$$= 2.14 \text{ meV} = 3.42\text{x}10^{-22} \text{ J}$$

Since $\mu_B = 9.27\text{x}10^{-27}$ J/T, we find B = 18.4 T.

Problem 1
Show that the most probable value of r for an electron in the 2s state of hydrogen is at $r \approx 5.2r_o$.

Solution:

From Eq. 42.9, we have, using $x = r/r_o$,

$$8r_o\, P = x^2(2 - x)^2\, e^{-x}$$

At the minima and maxima $dP/dx = 0$, so

$$(8r_o)\, dP/dx = x(2 - x)(x^2 - 6x + 4)e^{-x} = 0$$

We know that $P = 0$ at $x = 0$ and $x = 2$. The maxima are given by the solutions of the quadratic: $x = 3 \pm (5)^{1/2} = 5.24;\ 0.764$.

SELF-TEST

1. The obital quantum number for an electron is 3 and the orbital magnetic quantum number is 2. Find: (a) The angular momentum; (b) The component of the angular momentum along an external magnetic field; (c) The angle between the angular momentum and the field. (d) The component of the orbital magnetic moment along the field.

2. The radial probability density for the 2s state of hydrogen is

$$P_{2s}(r) = (r^2/8r_o^3)(2 - r/r_o)^2 \exp(-r/r_o)$$

Evaluate this function at the maximum points (a) $r = 0.764r_o$; (b) $r = 5.24r_o$. (State your answer in terms of r_o.)

Chapter 43

NUCLEAR PHYSICS

MAJOR POINTS

1. The binding energy of a nucleus
2. Radioactivity: α, β and γ decay
3. The radioactive decay law
4. In the process of fission a heavy nucleus into two lighter fragments
5. In the process of fusion two light nuclei combine to form a heavier nucleus

CHAPTER REVIEW

PROPERTIES OF NUCLEI

A nucleus consists of protons and neutrons, both of which are called nucleons. The number of protons is called the atomic number (Z). The atomic mass number A is the total number of nucleons in a nucleus, A = Z + N, where N is the number of neutrons. A particular nuclear species is called a nuclide and is designated as

$$^{A}_{Z}X$$

where X is the chemical symbol for the element. Isotopes of a given element have the same number of protons (Z) in the nuclei but different numbers of neutrons (N). Their chemical properties are identical.

The unified mass unit (u) is defined such that the mass of a neutral atom of the isotope ^{12}C is exactly 12 u. Using the equation $E = mc^2$, we find

$$1 \text{ u} = 1.6606\text{x}10^{-27} \text{ kg} = 931.5 \text{ MeV/c}^2$$

Nuclear Radius

Nucleons are tightly packed together and nuclei are roughly spherical. The radius depends on the mass number according to

$$R = 1.2A^{1/3} \text{ fm}$$

where 1 fm = 10^{-15} m. Thus the radius of a nucleus is typically about 10^{-14} m, which should be compared with the typical size of an atom which is about 10^{-10} m.

Binding Energy

The mass of a nucleus is less that the sum of the masses of its nucleons. The binding energy is the mass difference multiplied by c^2:
$BE = \Delta m.c^2$. In MeV's:

$$BE = [Zm_H + Nm_n - m_X](931.5\ MeV/u)$$

where m_H and m_X are the masses (in u) of the neutral atoms.

The binding energy per nucleon rises sharply for light nuclei, reaches a peak of about 8.8 MeV/nucleon at Fe, and then gradually decreases until uranium, see Fig. 43.2 in the text.

EXAMPLE 1

Calcuate the binding energy per nucleon in ^{15}O and ^{15}N. These two nuclides are examples of "mirror nuclei" in which the numbers of protons and neutrons are interchanged. Use mass values from Appendix E in the text.

Solution:

$$BE(^{15}O) = [8x1.007825\ u + 7x1.008665\ u - 15.003066\ u](931.5\ MeV/u)$$

$$= 112.0\ MeV$$

$$BE(^{15}N) = [7x1.007825\ u + 8x1.008665\ u - 15.000109\ u](931.5\ MeV/u)$$

$$= 115.5\ MeV$$

The fact that the binding energies of mirror nuclei are nearly equal implies that attractive the nuclear force is not affected by the presence or absence of electrical charge.

RADIOACTIVITY

A nucleus is theoretically unstable with respect to a given type of decay if its mass is greater than the sum of the masses of the decay products. The "disintegration energy", Δmc^2, appears as kinetic of the decay particles.

α decay

In α decay, a nucleus X emits an α particle (helium nucleus) ^4He and is transformed to a nuclide Y:

$$X \rightarrow Y + \alpha$$

For example, $^{238}U \rightarrow {}^{234}Th + {}^4He$. The disintegration energy is

(α decay) $$Q = (m_X - m_Y - m_\alpha)c^2$$

ß decay

In $ß^-$ decay, the emitted particle is an electron. The mass number is unchanged but the charge or atomic number of the daughter nucleus increases by one. An massless antineutrino is also emitted. For example,

$$^{14}C \rightarrow {}^{14}N + e^- + \overline{\nu}$$

The disintegration energy is

($ß^-$ decay) $$Q = (m_X - m_Y)c^2$$

In $ß^+$ decay, the emitted particle is a positron. In this case, the atomic number of the daughter nucleus decreases by one. A massless neutrino is also emitted. For example,

$$^{13}N \rightarrow {}^{13}C + e^+ + \nu$$

The distintegration energy is (see Problem 6 in the text)

($ß^+$ decay) $$Q = (m_X - m_Y - 2m_e)c^2$$

γ decay

Following α or ß decay, nuclei are often left in an excited state. The nucleus then returns to the ground state by emitting one or more high-energy photons called γ rays. The mass number and atomic number are unchanged.

EXAMPLE 2

For the α decay

$$^{232}U \rightarrow {}^{228}Th + \alpha$$

(a) Find the disintegration energy. (b) What are the kinetic energies of the throium nucleus and the alpha particle?

Solution:

(a) Using values in Appendix E of the text, we find the disintegration energy is

$$Q = (232.03713\text{ u} - 228.02872\text{ u} - 4.002603\text{ u})(931.5\text{ MeV/u})$$

$$= 5.41\text{ MeV}$$

(b) In order to find the kinetic energy of the nucleus, we note that linear momentum must be conserved. Assuming that the original nucleus is at rest,

$$p_\alpha + p_T = 0$$

The disintegration energy goes into kinetic energy of the products, so

$$Q = p_\alpha^2/2m_\alpha + p_T^2/2m_T$$

$$= (1/m_\alpha + 1/m_T)p_T^2/2$$

$$= [(m_T + m_\alpha)/m_\alpha]K_T$$

Thus,

$$K_T = m_\alpha Q/(m_T + m_\alpha)$$

$$= (4 \text{ u})(5.41 \text{ MeV})/(232 \text{ u}) = 9.32\text{x}10^{-2} \text{ MeV}$$

Finally, $K_\alpha = Q - K_T = 5.32$ MeV.

RADIOACTIVE DECAY LAW

The process of radioactive decay is random: One cannot know which nuclei will decay or when they will do so. When there is a large number of nuclei present in a sample, the rate of decay is proportional to the number of nuclei that are present:

$$dN/dt = -\lambda N$$

where λ is called the decay constant. From this relation we find that

$$N = N_o \exp(-\lambda t)$$

where N_o is the number of nuclei at $t = 0$ (Fig. 43.1). The half-life, $T_{1/2}$, is the time at which the number of nuclei decreases to 50% of the initial number. It can be shown that

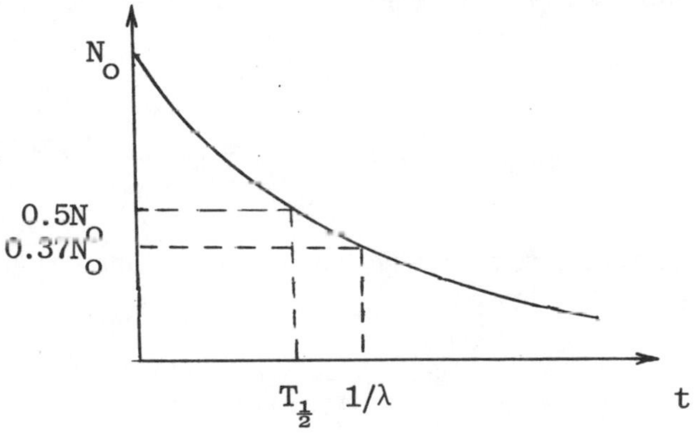

FIGURE 43.1

$$T_{1/2} = 0.693/\lambda$$

The decay rate, $R = -dN/dt = \lambda N$, also decreases exponentially:

$$R = R_o \exp(-\lambda t)$$

EXAMPLE 3
The initial decay rate of a radioactive sample is 16 mCi. After 2 h, the rate drops to 12 mCi. Find: (a) the half-life; (b) the initial number of nuclei; (c) the number of nuclei that decay in the first 4 h.

Solution:
(a) We are given

$$R/R_o = 0.75 = e^{-\lambda t}$$

Thus,

$$-\lambda t = \ln(0.75)$$

The half-life is

$$T = 0.693/\lambda = -0.693t/\ln(0.75)$$

$$= 4.82 \text{ h} = 1.73\text{x}10^4 \text{ s}.$$

(b) Since 1 Ci = $3.7\text{x}10^{10}$ decays/s, the initial decay rate is $R_o = 5.92\text{x}10^8$ decays/s. The decay constant is $\lambda = -\ln(0.75)/t = 4\text{x}10^{-5}\ \text{s}^{-1}$. Since $R_o = \lambda N_o$, we have

$$N_o = R_o/\lambda$$

$$= (5.92\text{x}10^8\ \text{s}^{-1})/(4\text{x}10^{-5}\ \text{s}^{-1}) = 1.48\text{x}10^{13}$$

(c) The number of nuclei that remain at time t is $N = N_o \exp(-\lambda t)$, therefore the number that decay in this time is

$$N_o - N = N_o(1 - e^{-\lambda t})$$

$$= (1.48\text{x}10^{13})[(1 - \exp(-0.576)] = 6.48\text{x}10^{12}$$

NUCLEAR REACTIONS

In a nuclear reaction a target nucleus X is bombarded by a particle a. The result is a nucleus Y and another particle b: a + X → Y + b, which can also be written as a(X, Y)b. The reaction energy is

$$Q = (m_a + m_X - m_Y - m_b)c^2$$

If Q is is negative, there is a threshold energy for the incoming particle below which the reaction will not occur. One can also express Q in terms of the initial and final kinetic energies:

$$Q = K_Y + K_b - K_X - K_a$$

FISSION

In the process of fission, a heavy nucleus such as ^{235}U, splits into two smaller fragments. The process is usually stimulated by the absorption of a neutron which then leaves the nucleus in an excited state. The excited nucleus oscillates violently and finally it undergoes fission, releasing a few neutrons in addition to the two larger fragments. Energy is released because the binding energy per nucleon for the uranium is less than that of the fragments (see Fig. 43.2 in the text).

FUSION

In the process of fusion two light nuclei coalesce to form a larger nucleus. Energy is released because the binding energy per nucleon of the final nucleus is larger than that of the initial nuclei (see Fig. 43.2 in the text). In order for two nuclei to undergo fusion, they must have enough initial kinetic energy to overcome the large Coulomb repulsion between them. One method of doing this to raise the temperature of a plasma (completely ionized gas) to about 10^8 K.

SOLUTIONS TO SELECTED TEXT EXERCISES AND PROBLEMS

Exercise 9

What is the radius of the gold isotope ^{197}Au? If the radius of an α particle 1.8 fm, what initial kinetic energy, in eV, must it have to "touch" the surface of the gold nucleus? Assume that the gold stays at rest.

Solution:

The radius of the nucleus is

$$R = 1.2A^{1/3} = 6.98 \text{ fm}$$

When the α particle is closest to the nucleus it is momentarily at rest so all its energy is potential energy. From the conservation of mechanical energy we have $E_i = E_f$:

$$K_\alpha = kq_1q_2/(R_1 + R_2)$$

$$= (9\text{x}10^9 \text{ N.m}^2/\text{C}^2)(2e)(79e)/(8.8\text{x}10^{-15} \text{ m})$$

$$= 4.14\text{x}10^{-12} \text{ J} = 25.9 \text{ MeV}$$

Exercise 15

(a) What is the energy required to remove one neutron from ^7Li? (b) Compare the result of part (a) with the average binding energy per nucleon for this nuclide.

Solution:
(a) If a neutron is removed from ^{7}Li it becomes ^{6}Li. Let us calculate the binding energy for both nuclei.

$$BE(^7Li) = [3(1.007825\ u) + 4(1.008665\ u) - 7.016005\ u](031.5\ MeV/u)$$

$$= 39.24\ MeV$$

Similarly, BE(^{6}Li) = 33.18 MeV. The energy required to remove one neutron is the difference in the binding energies, ΔBE = 6.06 MeV.

(b) For ^{7}Li, the binding energy per nucleon is BE/A = 5.61 MeV.

Exercise 17
The radioactive isotope ^{60}Co is used in the treatment of tumors. It undergoes β^- decay with a half-life of 5.25 y. What is the initial decay rate of a 0.01 g sample?

Solution:
The number of nuclei (or atoms) initially present is given by
The molecular mass M (in kg/mol) is the mass of N_A (Avogadro's number) atoms, so the mass of N_o atoms is $m = (N_o/N_A)M$. Thus, the initial number of nuclei in the 0.01 g sample is

$$N_o = mN_A/M = (10^{-5}\ kg)(6.022x10^{23}\ mol^{-1})/(0.06\ kg/mol)$$

$$= 1.0037x10^{20}\ \text{nulcei}$$

The initial decay rate is

$$R_o = \lambda N_o = 0.693N_o/T_{1/2}$$

$$= 4.2x10^{11}\ Bq$$

Exercise 33
The isotopes of uranium, ^{235}U and ^{238}U, are radioactive with half-lives of 7.13x10^8 y and 4.47x10^9 y, respectively. The present ratio of ^{235}U to ^{238}U is about 0.007. What was the ratio 10^9 y ago?

Solution:
Since $N = N_o \exp(-\lambda t)$ and $\lambda = 0.693/T$, where T is the half-life, we have $N_o = N \exp(0.693/T)$. The ratio of the initial numbers of the two nuclides is

$$N_{o1}/N_{o2} = (N_1/N_2) \exp[0.693(1/T_1 - 1/T_2)t]$$

where "1" refers to ^{235}U. Since t = 10^9 y, we find

$$N_{o1}/N_{o2} = (0.007)\exp[0.693(1.403 - 0.224)] = 0.0158.$$

Exercise 53
A "prompt" neutron released in a fission reaction has a kinetic energy of 1 MeV. It passes through a moderator which reduces its kinetic energy to 0.025 eV. If the neutron loses 50% of its kinetic energy in each collision, how many collisions are needed?

Solution:

After n collisions the energy is reduced by a factor $(1/2)^n$, thus

$$0.025 \text{ eV}/10^6 \text{ eV} = (1/2)^n$$

which can be rewritten as $4\text{x}10^7 = 2^n$. Taking logaritms we find

$$n\log 2 = 7.602$$

which leads to n = 25.3. So the number of collisions needed is 26.

Exercise 55
Show that the energy released in the D-D reaction $^2H(d, n)^3He$ is 3.27 MeV.

Solution:

The reaction energy is

$$Q = (2\text{x}2.014102 \text{ u} - 1.008665 - 3.016029 \text{ u})(931.5 \text{ MeV/u})$$

$$= 3.27 \text{ MeV}$$

SELF-TEST

1. Identify the two missing particles in the following decays:

   (a) $^{19}Ne \rightarrow ? + e^+ + ?$; (b) 64Cu $\rightarrow ? + e^- + ?$

2. The nuclide 60Co decays with a half-life of 5.3 y. (a) What is the initial activity of a 0.02 g sample. (b) At what time does the activity fall to 30% of the initial rate?

3. What is the energy released in the fusion reaction $n(^6Li, ^3H)\alpha$. Use the table of atomic masses in Appendix E of the text.

ANSWERS TO SELF-TESTS

CHAPTER 1
Question 1.
We find $(674.9)(5.429)/(0.58) = (6317) = 6.3\text{x}10^3$.

Question 2.
1 inch = 2.54 cm and 1 d = $8.64\text{x}10^4$ s, thus

$$0.02 \text{ inch/d} = (0.02 \text{ inch/d})(25.4 \text{ mm/1 inch})(1 \text{ d/86400 s})$$

$$= 5.88\text{x}10^{-6} \text{ mm/s.}$$

Question 3.
The dimension of x is length: [x] = L. Thus each term on the right side must also have the same dimension. Clearly, [A] = L, and

$$[B\ t] = L; \quad \text{so} \quad [B] = LT^{-1}$$

$$[C\ t^2] = L; \quad \text{so} \quad [C] = LT^{-2}$$

CHAPTER 2
Question 1.
(a) We must first find the components of the vectors. The angle **B** makes with the y axis is 35°. Thus,

$\mathbf{A} = A\cos 25°\mathbf{i} + A\sin 25°\mathbf{j}$
$= 1.81\mathbf{i} + 0.85\mathbf{j}$ m

$\mathbf{B} = -B\sin 35°\mathbf{i} + B\cos 35°\mathbf{j}$
$= -1.71\mathbf{i} + 2.46\mathbf{j}$ m

(b) $\mathbf{C} = \mathbf{B} - \mathbf{A} = -3.52\mathbf{i} + 1.61\mathbf{j}$ m, or
C = 3.87 m at 24.6° above the -x axis.
(Note C_x is negative and C_y is positive, thus the angle is in the second quadrant, that is, 155.4°
(c) The triangle is illustrated in Fig. 2.12.

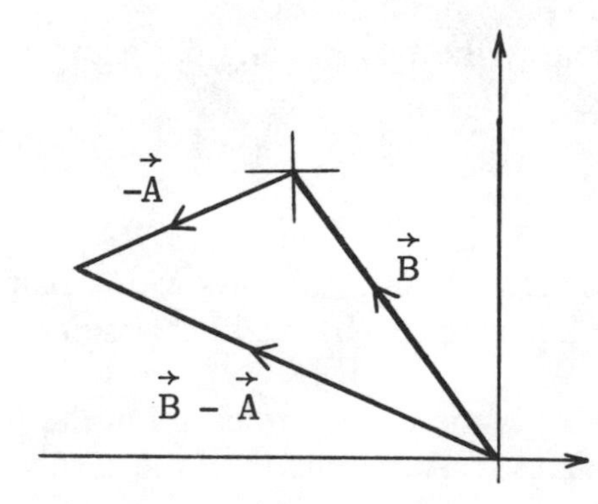

FIGURE 2.12

Question 2.
Note that we are given only the plane of the vectors and the angle between them, not their directions. Thus, we cannot determine the components.

(a) **A.B** = AB cosθ = (3)(2) cos40° = 4.60 m^2

(b) **A x B** = AB sinθ **n̂** = (3)(2) sin40° **n̂** = 3.86 **n̂** m^2
To find **n̂** use the right-hand rule, rotating **A** toward **B**. This gives the -x direction, thus **A x B** = -3.86**i** m^2.

Question 3.
(a) **A.B** = -2 - 6 + 4 = -4 m^2, therefore (**A.B**)**C** = -4**C** = -20**k** m^3.

(b) **A x B** = (4**k** - 8**j**) + (-3**k** - 12**i**) + (-**j** - 2**i**) = -14**i** - 9**j** + **k** m^2. Thus, (**A x B**).**C** = 5 m^3

(c) C(**A x B**) = -70**i** - 45**j** + 5**k** m^3;

(d) Not allowed.

CHAPTER 3

Question 1.
(a) a = Δv/Δt = -2 m/s^2. Note that the acceleration is not zero even though the instantaneous velocity is zero.

(b) The sum of the areas tells us the displacement. Areas below the t axis are negative:
Δx = 4 m + 1 m - 1 m - 2 m = +2 m

(c) To find the average speed we need the distance traveled. All areas are taken as positive, thus the distance is (1 + 1 + 2) m = 4 m and the average speed is 4 m/4 s = 1 m/s.

(d) The body speeds up between 3 s and 4 s when v and a are negative.

Question 2.
(a) First write the positions as functions of time. If the car travels for a time of t second, the truck travels for time (t - 1). We choose the origin at the initial position of the car.

$$x_C = 2t^2; \qquad x_T = 14 + (t - 1)^2$$

Set $x_C = x_T$ to find t = 3 s. Substitute 3 s into either x_C or x_T to find x = 18 m.
(b) The velocities on meeting are

$$v_C = 4t = 12 \text{ m/s}; \quad v_T = 2(t - 1) = 4 \text{ m/s}.$$

Question 3.
A sketch with axes is shown in Fig. 3.15. We first need to find the initial velocity. From $y = y_o + v_o t + 1/2\ at^2$,

$$0 = 40 + v_o(4) - 5(4)^2$$

so, $v_o = 10$ m/s. Use the same equation to find when y = 25 m:

$$25 = 40 + 10t - 5t^2$$

which yields t = 3 s.

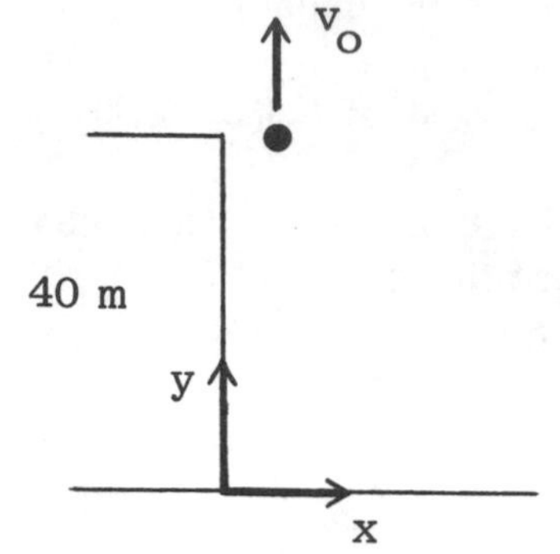

FIGURE 3.15

(b) $v^2 = v_o^2 + 2a(y - y_o)$, becomes $0 = 10^2 - 20(y - 40)$, and so y = 45 m.

(c) $v = v_o + at = 10 - 10(4) = -30$ m/s.
Note the negative sign.

CHAPTER 4
Question 1.
(a) With the origin at the base of the building, $x = v_o \cos\theta\ t$ becomes

$$16 = v_o \cos\theta \qquad \text{(i)}$$

With the given numbers $y = y_o - v_o \sin\theta t - 5t^2$ takes the form $0 = 44 - v_o \sin\theta(2) - 5(2)^2$, which can be simplified to

$$12 = v_o \sin\theta \qquad \text{(ii)}$$

Square and add (i) and (ii) to find $v_o = 20$ m/s. Take the ratio (ii) over (i) to find $\tan\theta = 3/4$, thus $\theta = 37°$.
(b) The angle at which it lands is given by

$$\tan\theta = v_y/v_x = (-v_o \sin\theta - 10t)/(v_o \cos\theta)$$

$$= -32/16 = -2$$

Thus, $\theta = 63.4°$ below the horizontal.

Question 2.
Since $a = v^2/r$, we have $r = v^2/a = 0.5$ m. The period is

$$T = 2\pi r/v = 1.57 \text{ s.}$$

The number of revolutions in 5 s is (5 s/1.57 s) = 3.18 rev.

Question 3.
The vector triangle in Fig. 4.9 shows

$$\mathbf{v}_{PG} = \mathbf{v}_{PA} + \mathbf{v}_{AG}.$$

We use the simpler notation:

$$\mathbf{R} = \mathbf{v}_{PG};\ \mathbf{P} = \mathbf{v}_{PA}, \text{ and } \mathbf{W} = \mathbf{v}_{AG},$$

thus, $\mathbf{R} = \mathbf{P} + \mathbf{W}$, so $\mathbf{W} = \mathbf{R} - \mathbf{P}$, and,

$$W_x = R_x - P_x = 150 \cos 50° - 160 \cos 30° = -42.1 \text{ km/h}$$

$$W_y = R_y - P_y = 150 \sin 50° - 160 \sin 30° = 34.9 \text{ km/h}$$

The velocity of the wind relative to land is
$\mathbf{W} = -42.1\mathbf{i} + 34.9\mathbf{j}$ km/h or, 54.7 km/h at 39.7° N of W.

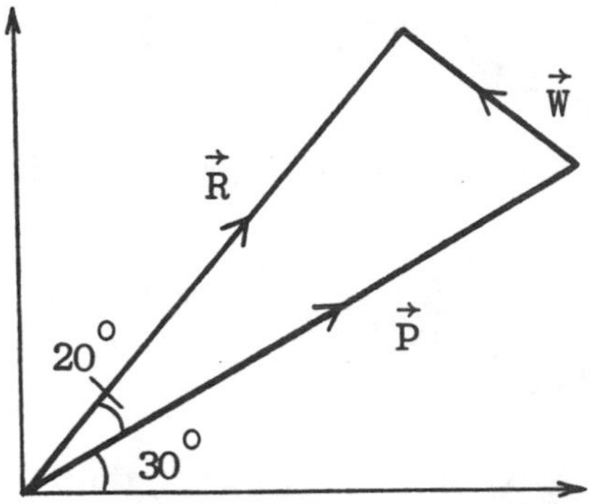

FIGURE 4.9

CHAPTER 5
Question 1.
The forces acting on the blocks are shown and the FBD for m_2 is drawn in Fig. 5.15. Since m_1 hangs vertically and it is greater than m_2 it is clear that the acceleration of m_1 will be downward. The axes for the blocks are chosen such that the positive x (or y axis) lies along a. (In the absence of friction we could choose the direction of the acceleration arbitrarily. However, this cannot be done when friction is present. See Ch. 6)

The component form of Newton' second law is

$$\Sigma F_y = m_1 g - T = m_1 a \qquad \text{(i)}$$

$$\Sigma F_x = T - m_2 g \sin\theta = m_2 a \qquad \text{(ii)}$$

Adding these two we find

$$m_1 g - m_2 g \sin\theta = (m_1 + m_2)a$$

Thus $a = 46.1 \text{ N}/11 \text{ kg} = 4.19 \text{ m/s}^2$. From (i) we find $T = 39.3$ N

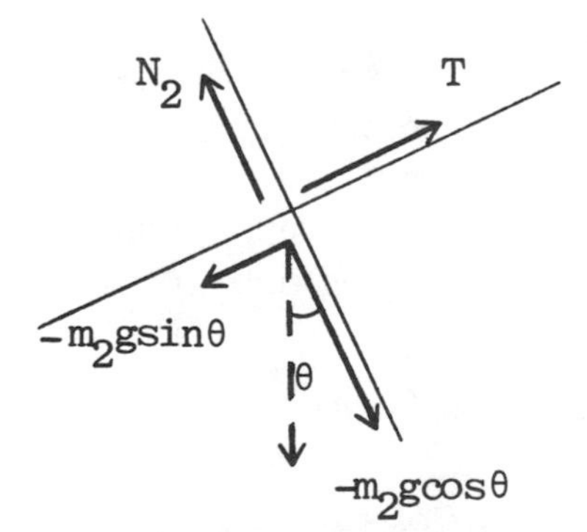

FIGURE 5.15

Question 2.
To find the acceleration we can treat the blocks as a single system. We assume the acceleration is up the incline and choose the axes accordingly. The FBD is shown in Fig. 5.16a.

$$\Sigma F_x = F\cos\theta - (m_1 + m_2)g\sin\theta = (m_1 + m_2)a$$

$$(60\text{ N})\cos 30° - (49\text{ N})\sin 30° = 5\,a$$

so, $a = 5.49\text{ m/s}^2$.

(b) The force exerted on m_1 by m_2, that is F_{12}, is directed up along the incline. Considering only the forces acting on m_1, the FBD in Fig. 5 16b shows that

$$\Sigma F_x = F_{12} - m_1 g\sin\theta = m_1 a$$

Thus $F_{12} = m_1(g\sin\theta + a) = 15.3$ N.

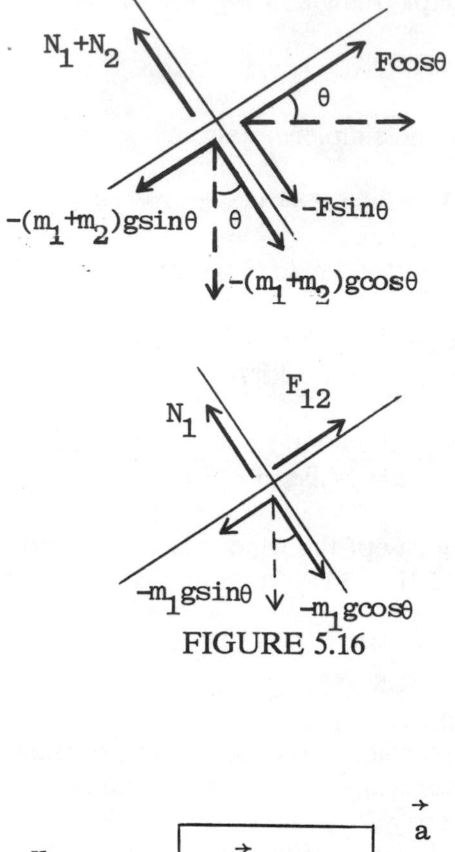

FIGURE 5.16

Question 3.
(a) The acceleration of the elevator is upward. The forces acting on the woman are shown in Fig. 5.17 with the axes indicated.

$$\Sigma F_y = N - mg = ma$$

Her apparent weight is $N = m(g + a) = 678$ N.

(b) Her true weight is $mg = 588$ N and so the magnitude of the reaction to this is also 588 N. The "reaction" force acts on the earth - not on the elevator. (The force "equal and opposite" to N acts downward on the floor of the elevator.)

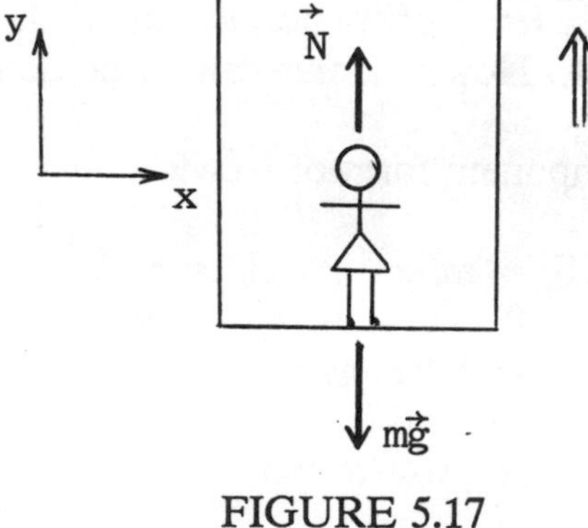

FIGURE 5.17

CHAPTER 6
Question 1.
It can be shown (see Problem 11) that the maximum angle of the incline such that the block (without the external F) does not start to slide is given by $\tan\theta = \mu_s$. For this reason we state that θ is larger than this critical value. If F is large enough, it will tend to push the block up the incline and the friction will be directed down the incline, as in Fig. 6.14a. Thus,

$$\Sigma F_x = F - f - mg\sin\theta = 0; \qquad \Sigma F_y = N - mg\cos\theta = 0$$

where $f = \mu N$. Thus $F = mg\sin\theta + f = mg(\sin\theta + \mu \cos\theta) = 21.6$ N

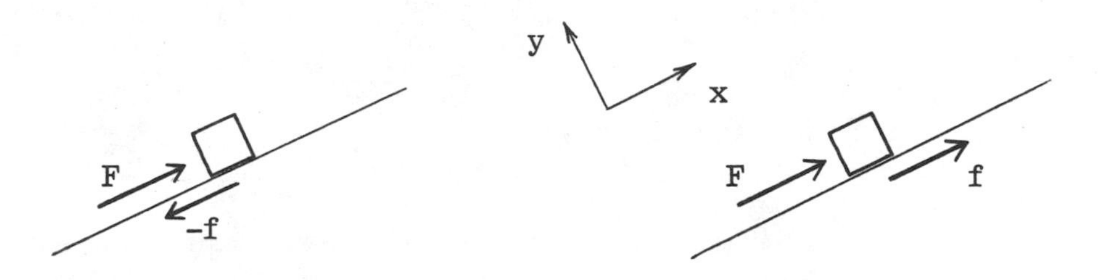

FIGURE 6.14

If F is too small, the block will tend to slide downward and so the friction will be directed up the incline, as in Fig. 6.14b, so

$$\Sigma F_x = F + f - mg\sin\theta = 0$$

Thus $F = mg\sin\theta - f = mg(\sin\theta - \mu \cos\theta) = 3.5$ N

Question 2.
We choose the +x axis to point along the acceleration which is horizontal and directed toward the center of the circle, as shown in Fig. 6.15.

$$\Sigma F_x = T \sin\theta = mv^2/r \qquad \text{(i)}$$

$$\Sigma F_y = T \cos\theta - mg = 0 \qquad \text{(ii)}$$

The radius of the circular motion is $r = L \sin\theta = 1.2$ m. From (ii) we have $T = mg/\cos\theta = 36.8$ N. We could substitute this into (i) to find v, or we could take the ratio (i)/(ii) to find $\tan\theta = v^2/rg$, so

$$v^2 = rg \tan\theta = 2.98 \text{ m/s.}$$

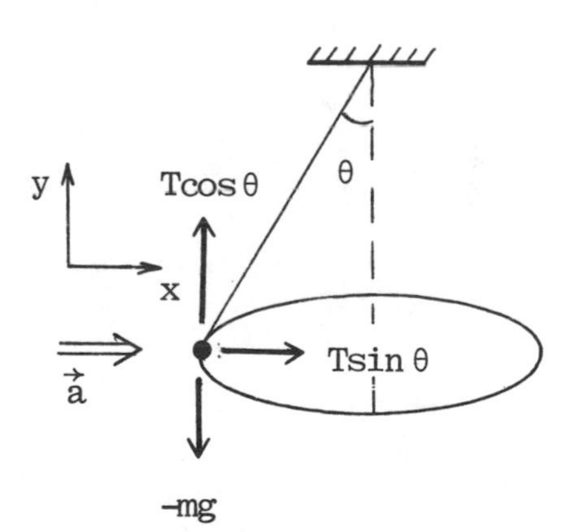

FIGURE 6.15

Question 3.
(a) First note that the radius r of an orbit is $r = R + h$, thus $r_A = 2R$ and $r_B = 3R$. The orbital speed was obtained in the review section: $v = (GM/r)^{1/2}$. Thus,

$$v_B/v_A = (r_A/r_B)^{1/2}$$

$$= (2/3)^{1/2} = 0.816$$

(b) From Kepler's third law, $T^2 = \kappa r^3$, thus,

$$T_B/T_A = (r_B/r_A)^{3/2}$$

$$= (3/2)^{3/2} = 1.84$$

CHAPTER 7

Question 1.

(a) The force F is at an angle θ to the incline, so

$$W_F = Fd\cos\theta = 41.6\ \text{J}$$

$$W_g = -\, mg\,\Delta y = -mgd\sin\theta = -9.8\ \text{J}$$

$$W_{sp} = -1/2\,k(x_f^2 - x_i^2) = -\,1/2\,kd^2 = -4.8\ \text{J}$$

Taking the components of the forces perpendicular to the incline:

$$\Sigma F_y = N - mg\cos\theta - F\sin\theta = 0$$

so, $N = mg\cos\theta + F\sin\theta = 51.2$ N. The work done by friction is

$$W_f = -fd = -\mu\,N\,d = -\,4.92\ \text{J}$$

(b) From the work-energy theorem

$$\Delta K = W_{NET} = W_F + W_g + W_{sp} + W_f = 22.1\ \text{J}.$$

(c) From $K = 1/2\,mv^2$, we find v = 4.2 m/s. The instantaneous power delivered by F is

$$P = \mathbf{F.v} = F\,v\cos\theta = 218\ \text{W}$$

Question 2.

(a) $K_A = K_B$ means $1/2\,m_A\,v_A^2 = 1/2\,m_B\,v_B^2$. Since $m_A = 2m_B$, we see that

$$v_B = (2)^{1/2}\,v_A = 1.414v_A.$$

(b) When they slow down $K_A = 0.5K_B$, thus removing the common factor of 1/2:

$$m_A\,(v_A - 4)^2 = 0.5\,m_B\,(v_B - 4)^2$$

Since $m_A = 2m_B$, we find, after taking the square root, $2(v_A - 4) = (v_B - 4)$. Next we use $v_A = 1.414v_B$ from part (a) to find

$$v_A^2 - 8v_A + 8 = 0$$

So $v_A = 6.83$ m/s, and $v_B = 4.83$ m/s.

Question 3.

$$W = \int_2^5 F_x\, dx = \left| -5/x \right|_2^5 = 5(1/2 - 1/5) = 1.5 \text{ J.}$$

CHAPTER 8

Question 1.
If we choose $U_g = 0$ at the lowest point,

$$E_i = mgD\sin\theta$$

$$E_f = 1/2\ mv^2 + 1/2\ kD^2$$

$$W_{NC} = -fD = -\ \mu ND = -\mu(mg\cos\theta)D$$

From $\Delta E = E_f - E_i = W_{NC}$,

$$1/2\ mv^2 + 1/2\ kD^2 - mgD\sin\theta = -\mu(mg\ \cos\theta)D$$

$$0.25v^2 + 0.12 - 0.8 = -0.06$$

This leads to $v = 0.79$ m/s

Question 2.
First we must find the speed at the lowest point by using the conservation of energy. The vertical position of the bob is $L(1 - \cos\theta)$

$$E_i = mgy = mgL(1 - \cos\theta)$$

$$E_f = 1/2\ mv^2$$

From $E_f = E_i$ we find $v = 4$ m/s (with $g = 10$ N/kg). Next we apply Newton's second for the circular motion:

$$T - mg = mv^2/L$$

Thus $T = m(g + v_2/L) = 27$ N.

Question 3.
The initial energy is

$$E_i = 1/2\ m\ (0.85v_{esc})^2 - GmM/R$$

where $v_{esc} = (2GM/R)^{1/2}$. The final energy is purely potential energy:

$$E_f = -GmM/(R + h)$$

From the conservation of energy

$$(0.85)^2/R - 1/R = -1/(R + h)$$

which yields $h = 2.6R$.

Question 4.
(a) $F_x = -\partial U/\partial x = aU_o \exp(-ax)$
(b) $F_x = -C/x$ find $U(x)$ given $U = 0$ at x_o.

$$U(x) - U(x_o) = -\int (-C/x)\,dx = C\int dx/x$$

$$= C\ln(x/x_o)$$

Since $U(x_o) = 0$, $U(x) = C\ln(x/x_o)$

CHAPTER 9
Question 1.
(a) We apply the conservation of linear momentum to the x and y directions

$$\Sigma p_x: \quad 0 = -m_1v_1\sin 30^\circ + m_2v_{2x}$$

$$\Sigma p_y: \quad -m_1u_1 + m_2u_2 = -m_1v_1\cos 30^\circ + m_2v_{2y}$$

We find $v_2 = 1.50\mathbf{i} + 1.58\mathbf{j}$ m/s, and the speed is $v_2 = 2.18$ m/s.
(b) The initial and final kinetic energies are

$$K_i = 1/2\, m_1u_1^2 + 1/2\, m_2u_2^2 = 12.9 \text{ J}$$

$$K_f = 1/2\, m_1v_1^2 + 1/2\, m_2v_2^2 = 9.45 \text{ J}$$

The collision is not elastic.

Question 2.
For the collision we must apply the conservation of momentum. (Since the collision is completely inelastic, kinetic energy is not conserved.)

$$mu = (m + M)V$$

After the collision we apply the conservation on energy.

$$E_i = 1/2\,(m + M)V^2 = (mu)^2/2(m + M)$$

$$E_f = 1/2\, kA^2$$

On setting $E_f = E_i$, we find $A^2 = (mu)^2/k(m + M)$, which yields $A = 0.522$ m.

Question 3.

(a) The impulse is the change in linear momentum

$$\mathbf{I} = \Delta\mathbf{p} = m(\mathbf{v}_f - \mathbf{v}_i)$$

$$= (0.15)(40\mathbf{j} - 30\mathbf{i}) = 6\mathbf{j} - 4.5\mathbf{i} \text{ kg.m/s.}$$

(b) The average force is

$$\mathbf{F}_{av} = \Delta\mathbf{p}/\Delta t = 2000\mathbf{j} - 1500\mathbf{i} \text{ N}$$

Question 4.

In an elastic collision both momemtum and kinetic energy are conserved. One can avoid having to deal with squared speeds by using Huygen's relation for the relative velocity. Since we do not know the directions of the final velocity we simply the (x) components as v_1 and v_2.

Σp_x:	$m_1 u_1 - m_2 u_2 = m_1 v_1 + m_2 v_2$	(i)
Relative Velocity	$u_2 - u_1 = -(v_2 - v_1)$	(ii)

Substituting values and switching the left and right sides of (ii) we find

$$4 = 2v_1 + 3v_2 \qquad \text{(iii)}$$

$$-7 = v_1 - v_2 \qquad \text{(iv)}$$

We multiply (iv) by 3 and add (iii) to obtain $\mathbf{v}_1 = -17/5\ \mathbf{i}$ m/s. From (iv) we find $\mathbf{v}_2 = 18/5\ \mathbf{i}$ m/s. You should confirm that both the initial and final kinetic energies are 31 J.

CHAPTER 10

Question 1.

The object is placed in the first quadrant of the coordinate system. We treat the missing triangle as an object of negative mass. Its CM is located at (10 m, 5 m). The CM of the rectangle is at (6 m, 3 m). If σ is the areal mass density (kg/m^2) the mass of the rectangle is $m_1 = 72\sigma$ and of the triangle $m_2 = -9\sigma$

$$x_{CM} = [(72\sigma)(6 \text{ m}) + (-9\sigma)(10 \text{ m})]/(63\sigma) = 5.43 \text{ m}$$

$$y_{CM} = [(72\sigma)(3 \text{ m}) + (-9\sigma)(5 \text{ m})]/(63\sigma) = 2.71 \text{ m}$$

Question 2.
We take the origin to be at Jack's original position.

$$x_{CM} = [(70 \text{ kg})(0 \text{ m}) + (50 \text{ kg})(3 \text{ m})]/120 \text{ kg}$$

$$= 1.25 \text{ m}$$

Jack is initially 1.25 m to the left of the CM, as in Fig. 10.9a. After they exchange positions, Jack is 1.25 m to the right of the CM, Fig. 10.9b. The CM itself does not move. Since Jill was initially 1.75 m to the right of the CM, the boat has moved (1.75 - 1.25) = 0.5 m to the left.

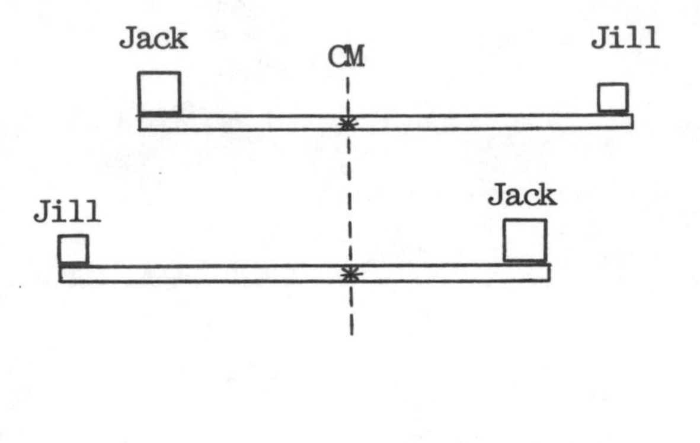

FIGURE 10.9

CHAPTER 11
Question 1.
We apply the equation F = ma to the block and $\gamma = I\alpha$ to the pulley.

$$mg\sin\theta - T = ma \qquad \text{(i)}$$

$$TR = I\alpha \qquad \text{(ii)}$$

Since $I = 0.5MR^2$ and $\alpha = a/R$, from (ii) we find $T = Ma/2$. Using this in (i) we find

$$a = mg\sin\theta/(m + M/2) = 2.5 \text{ m/s}^2$$

Using this in (i) gives T = 5 N.
(b) The angular acceleration is $\alpha = a/R = 5 \text{ rad/s}^2$. The angular displacement is $\Delta\theta = 1/2\ \alpha t^2 = 0.5(5 \text{ rad.s}^2)(9 \text{ s}^2) = 22.5$ rad. Since 1 rev corresponds to 2π rad, the number of revolutions is 3.58 rev.

Question 2.
Let us take the counterclockwise sense as positive. The lever arm for the weight is a horizontal line of length (0.3 m)cos20°, thus the torque due to the weight is

$$\tau_1 = -r_1 F_1 = -(0.3 \text{ m})(\cos 20°)(49 \text{ N}) = -13.8 \text{ N.m}$$

Since the rope is at 60° to the vertical it is at 30° to the horizontal. Since the rod is at 20° (below) the horizontal where the rope is attached, the angle between the rod and the rope is 50°. The torque due to the rope is

$$\tau_2 = +r_2 F_2 = (0.6 \text{ m})(30 \text{ N})(\sin 50°) = +13.8 \text{ N.m}$$

Since the net torque is zero, the rod is in equilibrium.

Question 3.
We divide the triangle into infinitesimal strips of width dx, as in Fig. 11.14. The area of such an element is $dA = y\,dx = (hx/b)dx$ and its mass is

$$dm = \sigma y dx = \sigma(xh/b)dx$$

The contribution of this element to the moment of inertia is

$$dI = x^2\,dm = (\sigma h/b)\,x^3\,dx$$

On integrating this from $x = 0$ to b, we find

$$I = (\sigma h/b)(b^4/4)$$

Since the total mass is $M = \sigma bh/2$, we find $I = Mb^2/2$.

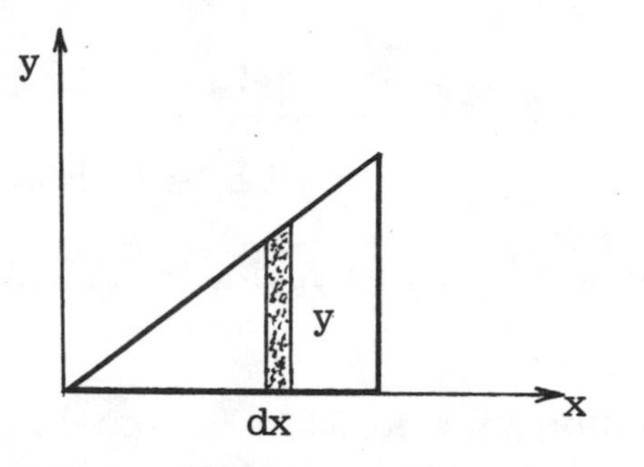

FIGURE 11.14

Question 4.
From the conservation on energy we have $\Delta K + \Delta U_g = 0$:

$$1/2\,(m_1 + m_2)v^2 + 1/2\,I\omega^2 + (m_1gd - m_2gd) = 0$$

If there is no slipping of the rope, $v = \omega R$. Using this and $I = 0.5MR^2$

$$0.5(m_1 + m_2 + M/2)v^2 = (m_2 - m_1)gd$$

Thus $v = 2.1$ m/s.

CHAPTER 12

Question 1.
The position is $\mathbf{r} = 2\mathbf{i} + 3\mathbf{j}$ m and its velocity is $\mathbf{v} = 7\cos25°\mathbf{i} + 7\sin25°\mathbf{j}$ m/s. The angular momentum is

$$\ell = m\mathbf{r} \times \mathbf{v} = (1.5\text{ kg})(2\mathbf{i} + 3\mathbf{j}\text{ m}) \times (6.34\mathbf{i} + 2.96\mathbf{j}\text{ m/s})$$

$$= -19.7\mathbf{k}\text{ kg.m/s}$$

Question 2.
(a) The angular velocity changes because the moment of inertia changes and angular momentum is conserved. The initial and final angular momenta are

$$L_1 = I_1\omega_1 = (0.5MR^2)\,\omega_1$$

$$L_2 = I_2\omega_2 = (0.5MR^2 + mR_2)\,\omega_2$$

Note that we treat the child as a point particle; we do not simply write $I_2 = 0.5(m + M)R^2$. (This implies that the mass of the child becomes uniformly spread over the platform!) On setting $L_1 = L_2$ we find $\omega_2 = 1.5$ rad/s.

(b) Again, angular momentum is conserved. The angular velocity changes because the moment of inertia changes. The child goes to $r = R/2$, thus

$$L_2 = (MR^2/2 + mR^2)\ \omega_2$$

$$L_3 = (MR^2/2 + mR^2/4)\ \omega_3$$

On setting $L_3 = L_2$, we find $\omega_3 = 1.85$ rad/s.

Question 3.

The net force is zero, $\Sigma\mathbf{F} = 0$. We take the horizontal component of the force at the pivot to be $+H\mathbf{i}$ and the vertical component to be $+V\mathbf{i}$. The components of $\Sigma\mathbf{F} = 0$ are

$$\Sigma F_x = H - T\cos\theta = 0 \qquad \text{(i)}$$

$$\Sigma F_y = V + T\sin\theta - W_1 - W_2 = 0 \qquad \text{(ii)}$$

The net torque is zero. If we take torques about the pivot, H and V make no contribution.

$$\Sigma\tau = L\,(T\sin\theta) - (L/2)W_1 - (L)W_2 = 0 \qquad \text{(iii)}$$

From (iii), $T = 54$ N. Using this in (i) we find $H = 46.8$ N and from (ii) we find $V = 15$ N.

CHAPTER 13

Question 1.

Since $mg = GmM/r^2$, we have $g = GM/r^2$. We are told that $g = 0.2g_o$, where $g_o = GM/R^2$ is the value at the earth's surface. If h is the altitude then $r = R + h$, so

$$1/(R + h)^2 = 1/5R^2$$

This leads to $(R + h) = (5)^{1/2}R$, and $h = 1.24R$.

Question 2.

From $GmM/r^2 = mv^2/r$, we find $v = (GM/r)^{1/2}$, and since $T = 2\pi r/v$, we obtain Kepler's third law $T^2 = (4\pi^2/GM)r^3$. The mass $M = \rho(4\pi R^3/3)$, thus

$$T^2 = 3\pi r^3/G\rho R^3$$

Since $r = R + h$, we find

$$\rho = (3\pi/GT^2)[(R + h)/R]^3$$

With the given values, $\rho = 5460$ kg/m^3.

Question 3.
First draw the directions of the forces exerted on m_2 by m_1 and m_3, as shown in Fig. 13.6. Next, find the magnitudes of the forces.

$$F_{21} = Gm_2m_1/r^2 = G(5\ kg^2)/25\ m^2$$

$$= 0.20G$$

$$F_{23} = Gm_2m_3/r^2 = G(3\ kg^2)/9\ m^2$$

$$= 0.33G$$

The components of the resultant force are

$$F_{2x} = -F_{21}\sin 37° = -8.06\text{x}10^{-12}\ N$$

$$F_{2y} = -F_{21}\cos 37° - F_{23} = -3.27\text{x}10^{-11}\ N$$

Thus, $\mathbf{F}_2 = (-0.81\mathbf{i} - 3.27\mathbf{j})\text{x}10^{-11}$ N

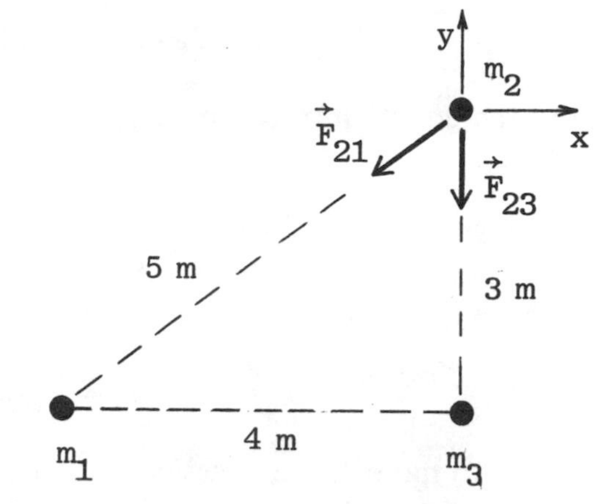

FIGURE 13.6

Question 4.
(a) From Kepler's third law

$$T^2 = (4\pi^2/GM)a^3$$

where M is the mass of the sun. The semimajor axis is $a = (r_A + r_P)/2 = 2.28\text{x}10^{11}$ m. On inserting the values we find $T = 5.92\text{x}10^7$ s = 1.88 y.

(b) The energy for an elliptical orbit is

$$E = -\ GmM/2a = 1.88\text{x}10^{32}\ J$$

(c) From the conservation of angular momentum we have

$$r_Av_A = r_Pv_P$$

thus v_A = 21.9 km/s.

CHAPTER 14

Question 1.
From Archimedes principle, the weight of the water displaced is equal to the weight of the log. If V is the volume of the log, then

$$\rho_1 g(0.8V) = \rho_2 g(0.75V)$$

Thus $\rho_2 = \rho_1(0.8/0.75) = 1067$ kg/m^3.

Question 2.
Since the heights of the tubes are the same, $y_1 = y_2$, Bernoulli's equation takes the form $P_1 + \rho v_1^2 = P_2 + \rho v_2^2$, so the difference in pressure is

$$P_1 - P_2 = 1/2\ \rho(v_2^2 - v_1^2)$$

According to the equation of continuity,

$$A_1 v_1 = A_2 v_2$$

thus $v_2 = 0.8$ m/s. The pressure at the bottom of a manometer tube is $P = P_o + \rho gh$, where h is the height of the liquid, so

$$P_1 - P_2 = \rho g(h_1 - h_2)$$

Comparing this to the ealier expression, we have

$$h_1 - h_2 = (v_2^2 - v_1^2)/2g = 0.024 \text{ m.}$$

CHAPTER 15

Question 1.
(a) The angular frequency is $\omega = (k/m)^{1/2} = 6$ rad/s. Since $x = A \sin(\omega t + \phi)$, it follows that $v = \omega A\cos(\omega t + \phi)$. With the given values:

$$-2 = A \sin(6t + \phi)$$

$$2 = A \cos(6t + \phi)$$

Square and add these to find A = 2.83 cm. Taking the ratio we find $\tan(6t + \phi) = -1$. Since the sine is negative and the cosine is positive, we see that $(6t + \phi)$ is in the fourth quadrant. Thus $(6t + \phi) = (\pi/4 + \phi) = 7\pi/4$, so $\phi = 3\pi/2$ rad.
(b) The total energy is

$$E = 1/2\ mv^2 + 1/2\ kx^2 = 1/2\ kA^2$$

Using the given x and v or the A found in part (a) we find E = 7.21 mJ. Note that this expression could have been used to find A.

Question 2.
The angular frequency of a physical pendulum is given by $\omega = 2\pi f = (mgd/I)^{1/2}$ where d is the distance between the center of gravity and the pivot. Thus, the moment of inertia is given by

$$I = mgd/4\pi^2 f^2$$

$$= (1.8 \text{ kg})(9.8 \text{ N/kg})(0.2)/4\pi^2(4 \text{ Hz}^2) = 2.23\text{x}10^{-2} \text{ kg.m}^2$$

Question 3.
The angular displacement is given by $\theta = \theta_o \sin(\omega t + \phi)$. The speed of the bob is given by

$$v = Ld\theta/dt = \omega L\theta_o \cos(\omega t + \phi)$$

The maximum value, at the lowest point, is $v = \omega L\theta_o$. Applying Newton's second law at the lowest point we have

$$\Sigma F_y = T - mg = mv^2/L$$

Thus $T = m(g + L\omega^2\theta_o^2)$.

CHAPTER 16

Question 1.
(a) With $k = 2\pi/\lambda = 10\pi$ rad/m, and $v = (F/\mu)^{1/2} = 20$ m/s, we have $\omega = vk = 200\pi$ rad/s. The standard expression for a wave traveling along the +x axis is

$$y = A \sin(kx - \omega t + \phi) = (5\text{x}10^{-3}\text{ m}) \sin(10\pi x - 200\pi t + \phi)\text{ m}$$

We are told that $y = 0$ at $x = 0$ and $t = 0$, so $\sin\phi = 0$, which means $\phi = 0, \pi$. To decide which is appropriate we consider the particle velocity,

$$\partial y/\partial t = -\omega A \cos(kx - \omega t + \phi)$$

We are told that this is positive at $x = 0$, $t = 0$, thus $\cos\phi < 0$ and so we pick $\phi = \pi$.
(b) The average power transmission is

$$P_{av} = 1/2\ \mu(\omega A)^2\ v$$

$$= 1/2\ (25\text{x}10^{-3}\text{ kg/m})(\pi)^2\ 20\text{ m/s} = 2.47\text{ W}$$

Question 2.
The frequency of the fundamental mode is ($\mu = M/L$)

$$f = v/2L = 0.5(F/LM)^{1/2}$$

where $v = (F/\mu)$. The ratio of the two tensions is

$$F_D/F_A = (f_D/f_A)^2\ (M_D/M_A)(L_D/L_A)$$

$$= (4/3)^2\ (4/5)(5/6) = 1.19$$

CHAPTER 17

Question 1.

(a) An open pipe has all harmonics, whereas a closed pipe has only the odd harmonics. We must test the ratio of the consecutive frequencies to decide. Note that 1925 Hz/1375 Hz = 1.4. Thus,

$$1.4 = (n + 1)/n$$

leads to n = 2.5. Since n must be an integer, the pipe is not open.

$$1.4 = (n + 2)/n$$

leads to n = 5. The pipe is closed.

(b) Two conseutive frequencies are nv/4L and (n + 2)v/4L, so the difference is v/2L - which inour case is 550 Hz. The fundamental frequency is v/4L = 275 Hz.

Question 2.

The ratio of the transmitted intensity to the incident intensity is $I_2/I_1 = 0.05$. The change in intensity level (in effect, I_1, is the "reference" value)

$$\beta = 10 \log(I_2/I_1)$$

$$= 10 \log(0.05) = -13 \text{ dB}$$

Question 3.

If v is the speed of sound and v_s is the speed of the source,

$$f_A = vf_o/(v - v_s)$$

$$f_R = vf_o/(v + v_s)$$

We need to eliminate v and v_s. From these equations we find

$$f_o/f_A = 1 - v_s/v$$

$$f_o/f_B = 1 + v_s/v$$

Add these to find $f_o(1/f_A + 1/f_B) = 2$, which leads to

$$f_o = 2f_Af_B/(f_A + f_B)$$

Question 4.

(a) The speed of the waves is v = ω/k = 333 m/s. The displacement amplitude is

$$s_o = p_o/\rho\omega v$$

$$= (2\text{x}10^{-2} \text{ Pa})/(1.29 \text{ kg/m}^3)(600 \text{ rad/s})(333 \text{ m/s}) = 7.76\text{x}10^{-8} \text{ m}$$

(b) The intensity is given by

$$I = p_o^2/2\rho v$$

$$= (2\text{x}10^{-2}\ \text{Pa})^2/2(1.29\ \text{kg/m}^3)(333\ \text{m/s}) = 4.66\text{x}10^{-7}\ \text{W/m}^2$$

CHAPTER 18

Question 1.

The change in volume of the gasoline is $\Delta V_g = \beta V \Delta T$. The change in volume of the glass container is $\Delta V_c = 3\alpha V \Delta T$. The amount of liquid that spills out is

$$\delta V = (\beta - 3\alpha)V\Delta T$$

$$= (92.3\text{x}10^{-5}\ ^\circ\text{C}^{-1})(0.6\ \text{L})(30\ ^\circ\text{C}) = 1.66\text{x}10^{-2}\ \text{L}$$

Question 2.

(a) From PV = nRT, we have that

$$P_1V_1/T_1 = P_2V_2/T_2$$

Thus

$$P_2 = (T_2/T_1)(V_1/V_2)P_1$$

$$= (308\ \text{K}/293\ \text{K})(40\ \text{L}/25\ \text{L})(150\ \text{kPa}) = 252\ \text{kPa}.$$

(b) From PV = NkT, we have

$$N/V = P/kT$$

$$= (10^{-8}\ \text{Pa})/(1.38\text{x}10^{-23}\ \text{J/K})(273\ \text{K}) = 2.65\text{x}10^{12}/\text{m}^3.$$

Question 3.

Since $L_2 = L_1(1 + \alpha\ \Delta T)$, the ratio of the new period to the old period is

$$t_2/t_1 = (1 + \alpha\ \Delta T)^{1/2} \approx 1 + \alpha\ \Delta T/2$$

The fractional change in the period is

$$(t_2 - t_1)/t_1 = \alpha\ \Delta T/2 = 10^{-4}$$

CHAPTER 19

Question 1.

Since the net heat transfer is zero

$$m_1c_1(T_f - T_1) + m_2c_2(T_f - T_2) = 0$$

$$(0.5\ \text{kg})(4190\ \text{J/kg.K})(T_f - 20\ °\text{C}) + (0.8\ \text{kg})(450\ \text{J/kg.K})(T_f - 100\ °\text{C}) = 0$$

We find $T_f = 31.7$ °C.

Question 2.

Since PV = nRT, we have

$$T_1 = PV_1/nR = 301\ \text{K}; \qquad T_2 = PV_2/nR = 421\ \text{K}$$

The heat input required for this temperature change is

$$Q = nC_p(T_2 - T_1)$$

$$= (2\ \text{mol})(29\ \text{J/mol.K})(120\ \text{K}) = 6.96\ \text{kJ}$$

The work done by the gas is

$$W = P(V_2 - V_1) = 2\ \text{kJ}$$

From the first law, the change in internal energy is

$$\Delta U = Q - W = 4.86\ \text{kJ}$$

Question 3.

Using the ideal gas law, PV = nRT, the equation for an adiabatic curve, PV^γ = constant, may be written as $TV^{\gamma-1}$ = constant. Thus,

$$T_2 = (V_1/V_2)^{\gamma-1}\ T_1$$

$$= (5/7)^{0.4}\ (301\ \text{K}) = 263\ \text{K}$$

The internal energy of an ideal gas depends only on temperature:

$$\Delta U = nC_v(T_2 - T_1)$$

$$= (2\ \text{mol})(20.7\ \text{J/mol.K})(263\ \text{K} - 301\ \text{K}) = -1.57\ \text{kJ}$$

Compare this result with the calculation in the previous test exercise.

CHAPTER 20

Question 1.

(a) With T = 273 K, the average kinetic energy per molecule is

$$K_{av} = 3kT/2$$

$$= 5.65\times10^{-21}\ \text{J}$$

(b) The rms speed of the molecules is

$$v_{rms} = (3RT/M)^{1/2}$$

where $M = 28\times10^{-3}$ kg/mol. At 273 K, we find v_{rms} = 493 m/s.

(c) The mass density is given by

$$\rho = 3P/v_{rms}^2 = 1.23\ \text{kg/m}^3$$

(d) Since ρ = mass/volume, where mass = nM = 56×10^{-3} kg, the volume is

$$V = \text{mass}/\rho = 4.55\times10^{-2}\ \text{m}^3$$

CHAPTER 21

Question 1.

The heat extracted from the ice is

$$Q_C = mL = 668\ \text{kJ}$$

For a refrigerator the COP = Q_c/W, so the work required is

$$W = Q_C/COP = 668\ \text{kJ}/4 = 167\ \text{kJ}$$

The heat deposited into the room is $Q_H = Q_C + W$ = 835 kJ.

Question 2.

Since there is no net heat transfer,

$$m_w c_w\,(T_f - 90\ °C) + m_j c_j\,(T_f - 20\ °C) = 0$$

$$(0.1\ \text{kg})(4190\ \text{J/kg.K})(T_f - 90) + (0.12\ \text{kg})(385\ \text{J/kg.K})(T_f - 20) = 0$$

Thus T_f = 83 °C = 356 K. The change in entropy is given by

$$\Delta S = \int mc\,dT/T = mc\ \ln(T_f/T_i)$$

For the water ΔS_w = -33.9 J/K, and for the jar ΔS_j = 65.7 J/K. The total change is +31.8 J/K.

Question 3.
The heat absorbed at constant pressure is $Q = nC_p \, \Delta T$, so

$$T_2 - T_1 = Q/nC_p = 57.7 \text{ K}$$

At constant pressure, the chnage in entropy is

$$\Delta S = \int nC_p \, dt/T = nC_p \ln(T_f/T_i)$$

$$= (1 \text{ mol})(20.8 \text{ J/mol.K}) \ln(331 \text{ K}/273 \text{ K}) = 3.37 \text{ J/K}.$$

CHAPTER 22
Question 1.
(a) The net force on q must be zero, that is, $\mathbf{F}_1 + \mathbf{F}_2 = 0$. In order for this condition to be satisifed, q must be placed on the line joining Q_1 and Q_2. It must be also be closer to the smaller charge (Why?). Thus q must be placed to the right of Q_1 at some distance x. The magnitudes of the forces must be equal, that is, $F_1 = F_2$:

$$k|qQ_1|/x^2 = k|qQ_2|/(d - x)^2$$

Cross multiplying and taking the squareroot we find:

$$x - d = \pm (3)^{1/2}x$$

Since x must be positive, we find x = 0.366d. The sign of q does not matter.
(b) In order for the net force on each of the charges to be zero, q must be negative. Considering the magnitudes of the forces on Q_1, we have

$$k|qQ_1|/x^2 = k|Q_1Q_2|/d^2$$

Using x = 0.366d, we find q = -0.804 μC. (Check that the force on Q_2 is also zero.)

Question 2.
The directions of the forces are shown in Fig. 22.4. From the given coordinates we find that Q_1 is $r_1 = \sqrt{5}$cm from the origin and Q_2 is $r_2 = \sqrt{10}$cm from the origin. Thus, the magnitudes of the forces on Q_3 are

$$F_{31} = k|Q_3Q_1|/r_1^2 = 1.08 \text{ N};$$

$$F_{32} = k|Q_3Q_2|/r_2^2 = 1.35 \text{ N}$$

From the coordinates we infer that $\tan\alpha = 1/2$, so $\alpha = 26.6°$ and $\tan\beta = 3$, so $\beta = 71.6°$. The components of $\mathbf{F}_3$ are

$$F_{3x} = -F_{31} \cos\alpha - F_{32} \cos\beta = -1.39 \text{ N}$$

$$F_{3y} = -F_{31} \sin\alpha + F_{32} \sin\beta = 0.$$

Thus $\mathbf{F}_3 = -1.39\mathbf{i} + 0.8\mathbf{j}$ N

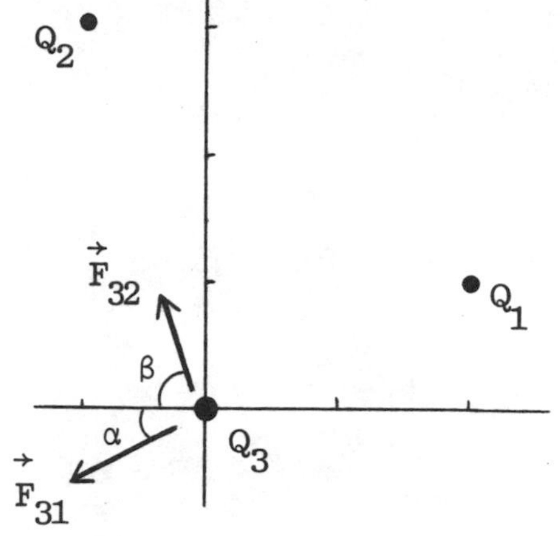

FIGURE 22.4

CHAPTER 23
Question 1.
(a) The directions of the fields due to Q_1 and Q_2 are shown in Fig. 23.14. Their magnitudes are

$$E_1 = kQ_1/r_1^2 = 14.1 \text{ N/C}$$

$$E_2 = kQ_2/r_2^2 = 14.4 \text{ N/C}$$

The components of the resultant field are

$$E_x = -E_1 + E_2 \cos 36.9° = -2.58 \text{ N/C}$$

$$E_y = 0 + E_2 \sin\theta = 8.64 \text{ N/C}$$

Thus $\mathbf{E} = -2.58\mathbf{i} + 8.64\mathbf{j}$ N/C.
(b) Nothing, the field due to Q_1 and Q_2 is not affected by q.

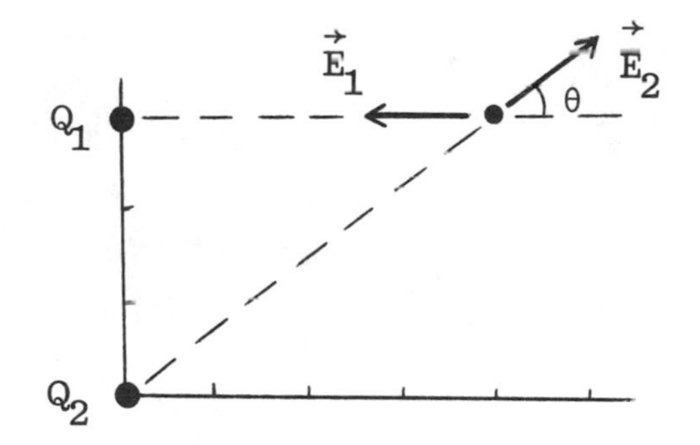

FIGURE 23.14

Question 2.
The forces acting on the particle are shown in Fig. 23.15.

$$\Sigma F_x = qE - T \sin\theta = 0$$

$$\Sigma F_y = T \cos\theta - mg = 0$$

Taking the ratio of these equations, we find $\tan\theta = qE/mg = 0.163$, thus $\theta = 9.3°$.

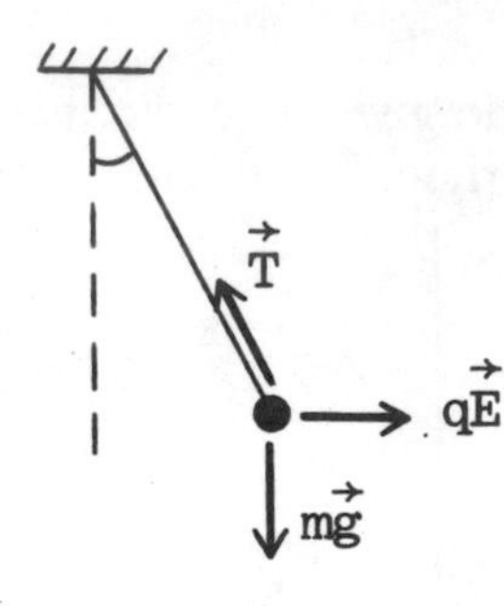

FIGURE 23.15

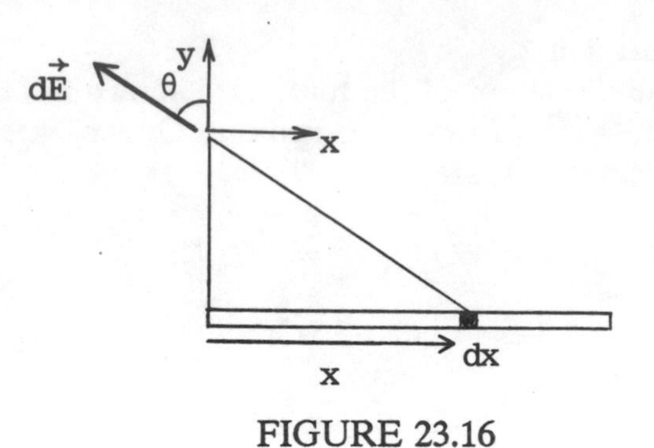

FIGURE 23.16

Question 3.
The field contributed by an arbitrary element with charge $dq = \lambda\ dx$, is

$$dE = k(\lambda\ dx)/r^2$$

where $r^2 = x^2 + b^2$. The x component is

$$dE_x = dE \sin\theta$$

where $\sin\theta = x/b$, see Fig. 23.15. (We ignore the negative sign of the x component.) Thus,

$$E_x = \int dE_x = k\,\lambda \int_0^L x\ dx/(x^2 + b^2)^{3/2}$$

Using the substitution $u = x^2 + b^2$, the integrand is $\int du/2u^{3/2}$.

$$E_x = k\,\lambda\,[-(x^2 + b^2)^{-1/2}]_0^L$$

$$= k\lambda[1/b - 1/(b^2 + L^2)^{1/2}]$$

CHAPTER 24
Question 1.
The charge on the surface of a sphere with a surface charge density is $Q = \sigma A = \sigma(4\pi r^2)$.

Inner: $Q_i = 3\sigma(4\pi)R^2 = 12\sigma\pi R^2$

Outer: $Q_o = -2\sigma(4\pi)(2R)^2 = -32\sigma\pi R^2$

The field between the spheres is determined solely by the charge on the inner sphere, whereas for $r > 2R$ both spheres contribute. The gaussian surface in each case is a sphere.

$(R < r < 2R)$ $$E(4\pi r^2) = Q_i/\epsilon_o$$

$$E = 3\sigma R^2/\epsilon_o r^2$$

$(r > 2R)$ $$E(4\pi r^2) = (Q_i + Q_o)/\epsilon_o$$

$$E = -5\sigma R^2/\epsilon_o r^2$$

Question 2.
(a) The Gaussian surface is a cylinder of radius r and length L.

$(r < R)$ $$E(2\pi rL) = -\lambda L/\epsilon_o$$

$$E = -\lambda/2\pi\epsilon_o r = -2k\lambda/r$$

Outside the cable the charges on both the wire and the sheath contribute; the net linear charge density is 2λ.

$(r > R)$ $$E(2\pi rL) = 2\lambda L$$

$$E = 2\lambda/2\pi\epsilon_o r = 4k\lambda/r$$

(b) Since the lines between the central wire and the sheath must start and end on charges, the net linear charge density on the inside surface of the sheath must be λ.

CHAPTER 25

Question 1.
(a) Note that potential is a scalar, so $V(x) = 2kQ/r = 2kQ/(a^2 + x^2)^{1/2}$

(b) $E_x = -\partial V/\partial x = 2kQx/(a^2 + x^2)^{3/2}$

Question 2.
From the conservation of energy, $\Delta K = -q\Delta V$:

$$1/2\ mv_f^2 - 1/2\ mv_i^2 = -(-e)(-14\ V)$$

$$v_f^2 = v_i^2 - 4.92x10^{12}$$

Thus $v_f = 3.33x10^6$ m/s.

Question 3.
The potential due to an infinitesimal element of charge is $dV = kdq/r$. In this case all the elements are at the same distance $r = (a^2 + y^2)^{1/2}$ from the point in question. Thus the integral $\int dV$ reduces to $(k/r) \int dq = kQ/r$. So,

$$V = kQ/(a^2 + y^2)^{1/2}$$

CHAPTER 26

Question 1.

For the capacitors in parallel, the equivalent capacitance is $C_{23} = C_2 + C_3 = 7.4\ \mu F$. This is in series with C_1, so

$$1/C_{eq} = 1/C_1 + 1/C_{23} = 1/4\ \mu F + 1/7.4\ \mu F$$

Thus $C_{eq} = 2.6\ \mu F$. The magnitude of the charge on C_1 and on the pair $C_2 + C_3$ is

$$Q_1 = Q_2 + Q_3 = (2.6\ \mu F)(15\ V) = 39\ \mu C$$

The potential difference across C_1 is $V_1 = Q_1/C_1 = 9.75$ V, thus the potential difference across the other two is $V_2 = V_3 = 5.25$ V. The charge on C_3 is $Q_3 = C_3V_3 = 26.3\ \mu C$. Finally $Q_2 = 12.7\ \mu C$.

Question 2.

The initial potential difference and charge on C_1 are $V_1 = 24$ V and $Q_1 = C_1V_1 = 120\ \mu C$. When S_2 is closed, charge flows from C_1 to C_2 till their potential difference are equal, that is, $V_1' = V_2'$:

$$Q_1'/C_1 = Q_2'/C_2 \qquad \text{(i)}$$

Thus $Q_1' = (5/3)Q_2'$. The original charge on C_1 is shared by the capacitors, so

$$Q_1' + Q_2' = 120\ \mu C \qquad \text{(ii)}$$

Using (i) in (ii) we find $Q_1' = 75\ \mu C$ and $Q_2' = 45\ \mu C$. Using $U = Q^2/2C$, the final energies are $U_1' = 0.56$ mJ; $U_2' = 0.34$ mJ.

CHAPTER 27

Question 1.

(a) The resistivity is given by $\rho = RA/L = (0.3\ \Omega)(\pi r^2)/5\ m = 6.79 \times 10^{-8}\ \Omega.m$

(b) The current density is $J = I/A = I/\pi r^2 = 2.21 \times 10^6\ A/m^2$

(c) The electric field is $E = \rho J = 0.15$ V/m.

(d) The power dissipation is $P = I^2R = 1.88$ W

Question 2.

We know that $R = R_o(1 + \alpha\ \Delta T)$, thus

$$\Delta T = (R - R_o)/\alpha R_o = 182\ C^o$$

so, $T_2 = 202$ °C.

CHAPTER 28

Question 1.

Since $P = I^2R$, we have $I = (P/R)^{1/2} = 2.24$ A. The current is given by $I = \mathscr{E}/(r + R)$, where r is the internal resistance. Thus, $r = \mathscr{E}/I - R = 0.36\ \Omega$.

Question 2.

(a) Let us trace each loop in the clockwise sense.

Left loop: $\mathscr{E}_1 - I_1R_1 + \mathscr{E}_2 - I_2R_2 = 0$

Right loop: $+I_2R_2 - \mathscr{E}_2 - I_3R_3 + \mathscr{E}_3 = 0$

(b) Substituting the given values into the loop equations we find

$$7 - 2I_1 - 3I_2 = 0 \qquad \text{(i)}$$

$$3I_2 - I_3 + 9 = 0 \qquad \text{(ii)}$$

Next we use the junction rule to substitute $I_1 = I_2 + I_3$ into (i):

$$7 - 5I_2 - 2I_3 = 0 \qquad \text{(iii)}$$

2(ii) - (iii) yields $I_2 = -1$ A. From (ii) we find $I_3 = 6$ A. Finally, $I_1 = 5$ A.

Question 3.

(a) $Q = Q_o(1 - e^{-t/RC})$; $I = I_o\, e^{-t/RC}$ where $Q_o = C\mathscr{E}$ and $I_o = \mathscr{E}/R$.

(b) With the given values RC = 10 s. The power loss at t = 5 s is

$$P = I^2R = I_o^2R \exp(-2t/RC) = I_o^2R\, e^{-1} = 0.037 \text{ W}$$

(c) At t = RC, $Q = Q_o(1 - e^{-1}) = 0.632Q_o = 6.32\text{x}10^{-3}$ C. The stored energy is

$$U = Q^2/2C = 0.2 \text{ J}$$

CHAPTER 29

Question 1.

(a) From the top view in Fig. 29.8 we see that $\ell_1 = a \sin37°\mathbf{i} + a \cos37°\mathbf{j}$. Thus,

$$\mathbf{F}_1 = I\boldsymbol{\ell}_1 \times \mathbf{B}$$

$$= (3 \text{ A})(0.12\mathbf{i} + 0.16\mathbf{j} \text{ m})\times(0.5\mathbf{j} \text{ T}) = 0.18\mathbf{k} \text{ N}$$

We can say that $\mathbf{F}_3 = -\mathbf{F}_1 = -0.18\mathbf{k}$ N.

Since ℓ_2 = ck = 0.5**k** m, the force on this side is

$$\mathbf{F}_2 = I\boldsymbol{\ell}_2 \times$$

$$= (3\text{ A})(0.5\mathbf{k}\text{ m})\times(0.5\mathbf{j}\text{ T}) = -0.75\mathbf{i}\text{ N}$$

Then $\mathbf{F}_4 = -\mathbf{F}_2 = +0.75\mathbf{i}$ N.
(b) We could use the forces on the vertical sides to calculate the torque. Instead, let us use the magnetic moment.

$$\mu = NIA = (1)(3\text{ A})(0.1\text{ m}^2) = 0.3\text{ A.m}^2$$

The magnitude of the torque is

$$\tau = \mu B \sin\theta = \mu B \sin 53° = 0.12\text{ N.m}$$

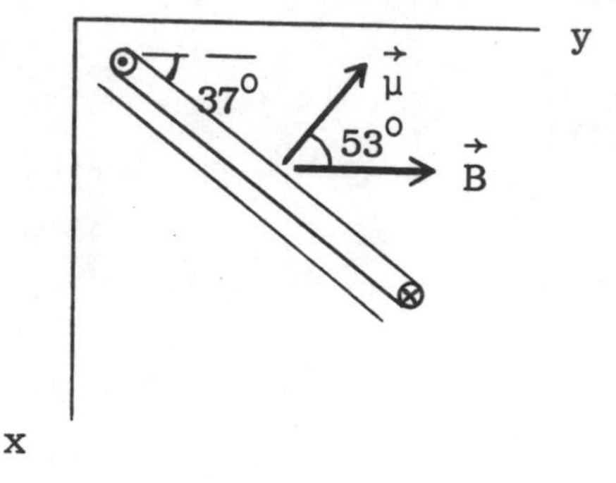

FIGURE 29.8

The sense of the torque is to align $\boldsymbol{\mu}$ along **B**. The vector torque is $\boldsymbol{\tau} = -0.12\mathbf{k}$ N.m. One could also have found the components of $\boldsymbol{\mu}$:

$$\boldsymbol{\mu} = -\mu\sin 53°\mathbf{i} + \mu\cos 53°\mathbf{j} = -0.24\mathbf{i} + 0.18\mathbf{j}\text{ A.m}^2$$

Then,

$$\boldsymbol{\tau} = \boldsymbol{\mu} \times \mathbf{B} = (-0.24\mathbf{i} + 0.18\mathbf{j})\times(0.5\mathbf{j}) = -0.12\mathbf{k}\text{ N.m}$$

Question 2.
We need **F** = q(**E** + **vxB**) = 0, that is, **E** = **-vxB**. Considering only the unit vectors, we have

$$-\mathbf{j} = \mathbf{i} \times ?$$

The only possibility is ? = **k**. The magnitude of the magnetic field is B = E/v = 0.1 T. Thus **B** = 0.1**k** T. The sign of the charge does not matter. However, it is a good idea just to check the directions of the electric and magnetic forces with the right hand rule (keeping in mind the sign of the charge).

Question 3.
(a) From F = ma, we have qvB = mv^2/r, thus the cyclotron angular frequency is $\Omega = v/r = qB/m$. The period is

$$T = 2\pi/\Omega = 2\pi m/qB$$

$$= 1.19\times10^{-8}\text{ s.}$$

(b) The linear momentum is p = mv = qrB = 3.84×10^{-23} kg.m/s

(c) The kinetic energy is K = $p^2/2m$ = $(qrB)^2/2m$ = 8.09×10^{-16} J = 0.06 keV.

CHAPTER 30

Question 1.

(a) The directions of the fields are shown in Fig. 30.13. Note that in each case B is perpendicular to the line from the wire to the point. The magnitudes are

$$B_1 = \mu_o I_1/2\pi r_1$$

$$= (2\text{x}10^{-7}\ \text{T.A/m})(8\ \text{A})/(0.08\ \text{m}) = 2\text{x}10^{-5}\ \text{T}$$

Similarly $B_2 = 4\text{x}10^{-5}$ T. The components of the resultant field

$$B_x = B_1 \cos37° + B_2 \cos53° = 4.8\text{x}10^{-5}\ \text{T}$$

$$B_y = B_1 \sin37° - B_2 \sin53° = -2.0\text{x}10^{-5}\ \text{T}$$

$$\mathbf{B} = (4.8\mathbf{i} - 2.0\mathbf{j})\text{x}10^{-5}\ \text{T}$$

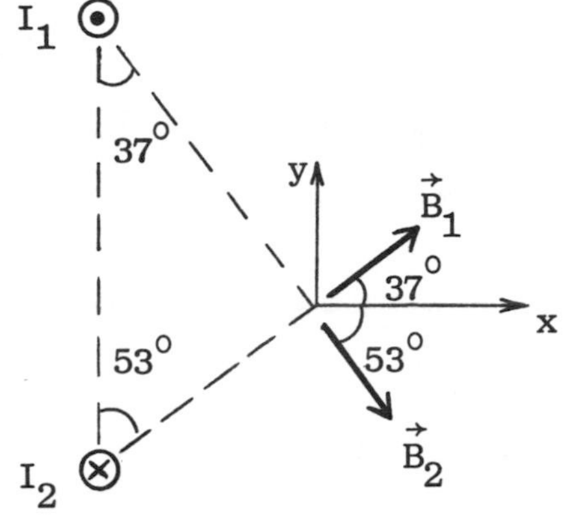

FIGURE 30.13

(b) The force is $F_{12} = I_1 \ell_1 B_1$ so the force per unit length is

$$F/\ell = \mu_o I_1 I_2/2\pi d$$

where d = 0.1 m. Using the given values we find $F/\ell = 1.92\text{x}10^{-4}$ N/m. The force is upward.

Question 2.

(a) The B of the wire is directed along -y for points on the +x axis. If **v** is parallel to **B**, the force is zero.

(b) $\mathbf{F} = e(-v\mathbf{i})\text{x}(-B\mathbf{j}) = evB\mathbf{k}$.

Question 3.

The sections of wire that "point" directly to P do not contribute anything to the field at P. In the text it is shown that for a finite wire

$$B = (\mu_o I/4\pi R)(\sin\alpha_1 + \sin\alpha_2)$$

where α_1 and α_2 are the angles shown in Fig. 30.14 (You should be able to derive this from the expression for d**B**.) In the present case, $\alpha_1 = 0$ and $\tan\alpha_2 = \ell/R = 2$, so $\alpha_2 = 63.4°$. The field is

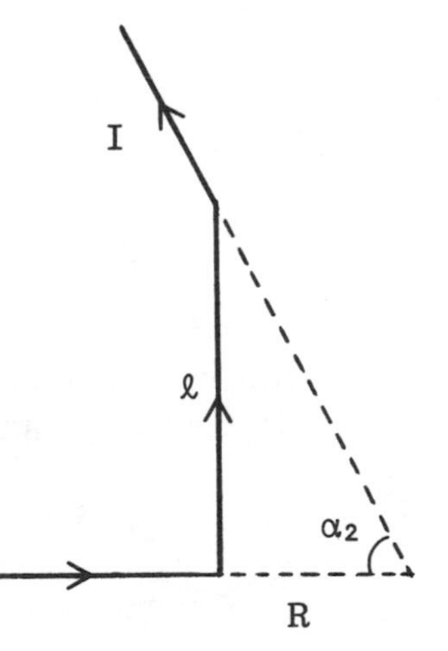

$$B = (10^{-7}\ \text{T.A/m})(10\ \text{A}) \sin63.4°/(1.5\ \text{m}) = 5.96\text{x}10^{-7}\ \text{T}$$

FIGURE 30.14

Question 4.
The fraction of the current flowing through a circle of radius r is determined by the ratio of the areas:

$$I' = I(r^2 - a^2)/(R^2 - a^2)$$

Applying Ampere's law to a circular loop of radius r,

$$B(2\pi r) = \mu_o I(r^2 - a^2)/(R^2 - a^2)$$

CHAPTER 31

Question 1.
The circumference of the circle is $2\pi r$ = 94.24 cm. Thus one side of the square is 23.6 cm. The change in flux arises from the change in area, so the average induced emf is

$$\mathscr{E} = -\Delta\phi/\Delta t = -B\Delta A/\Delta t = -B(L^2 - \pi r^2)/\Delta t$$

$$= -(0.3\ \text{T})(5.57\text{x}10^{-2}\ \text{m}^2 - 7.07\text{x}10^{-2}\ \text{m}^2)/(0.05\ \text{s}) = 0.09\ \text{V}$$

The positive sign means that the induced magnetic field tries to reinforce the original field (the flux decreases since the area decreases).

Question 2.
(a) The flux through the loop ϕ = BLx is decreasing, and the magntiude of the induced emf is $\mathscr{E}$ = BLv = 7.5 V. Since ϕ is decreasing the induced magnetic field tries to reinforce the external field. Thus the induced current flows clockwise around the loop, specifically, it is downward through the rod. The magnetic force on this current in the rod is

$$F = ILB = (\mathscr{E}/R)LB = 1.8\ \text{N}$$

and is directed to the right (opposite to v).
(b) The mechanical power needed to keep the rod moving is P = Fv = 45 W

(c) The electrical power dissipated in the rod is $P = I^2R$ = 45 W.

Question 3.
(a) Since the field produced by the wire is not uniform, we must first find an expression for the flux through an arbitrary element, an infinitesimal strip of width dx:

$$d\phi = B\ dA = (\mu_o I/2\pi x)(c\ dx)$$

The flux through the whole loop is

$$\phi = (\mu_o Ic/2\pi) \int dx/x = (\mu_o Ic/2\pi)\ \ln[(a + b)/a]$$

(b) $\mathscr{E} = -d\phi/dt = -(\mu_o c/2\pi)(6t + 7)\ln[(a + b)/a]$

CHAPTER 32

Question 1.

(a) Since $\mathcal{E} = -L\, dI/dt$, we find

$$L = \mathcal{E}/(dI/dt) = 1.5\times10^{-2}\ \text{V}/(21\ \text{A/s}) = 0.714\ \text{mH}$$

(b) The energy stored is

$$U = 1/2\ LI^2 = 0.5L(15.5\ \text{A})^2 = 85.8\ \text{mJ}$$

Question 2.

The natural frequency is $f_o = 1/2\pi(LC)^{1/2}$, therefore

$$C = 1/4\pi^2 L f_o^{\,2} = 6.6\ \mu\text{F}$$

Question 3.

(a) The current increases according to $I = I_o\ [1 - \exp(-t/\tau)]$, where $\tau = L/R$. Since $I = 0.3I_o$, we find

$$\exp(-t/\tau) = 0.7$$

$$-Rt/L = \ln(0.7)$$

thus,

$$R = -\ L \ln(0.7)/t = 1.5\ \Omega$$

(b) The self-induced emf is $\mathcal{E} = -L\, dI/dt$, so the power supplied to the inductor is

$$P_L = L\ I\ dI/dt = L\ I_o(1 - e^{-t/\tau}).\ I_o e^{-t/\tau}$$

$$= LI_o^{\,2}(0.3)(0.7) = 0.21LI_o^{\,2}$$

Since $I_o = \mathcal{E}/R = 20$ A, we find $P_L = 4$ W.

CHAPTER 33

Question 1.

(a) The angular frequency is $\omega = 2\pi f = 754$ rad/s, thus $X_L = \omega L = 150.8\ \Omega$, and $X_C = 1/\omega C = 26.5\ \Omega$. The impedance is

$$Z = [R^2 + (\omega L - 1/\omega C)^2]^{1/2} = 148\ \Omega$$

(b) The phase angle is given by

$$\tan\phi = (\omega L - 1/\omega C)/R = 1.55$$

thus $\phi = 57.2°$.

(c) The rms current through the circuit is

$$I = V/Z = 6.76 \text{ A}$$

(d) The rms potential difference across C is

$$V_C = IX_C = 179 \text{ V}$$

The peak potential difference across C is $v_{oC} = (2)^{1/2}V_C = 253$ V.

(e) The rms power supplied by the source is dissipated in the resistor:

$$P = I^2R = 3.66 \text{ kW}$$

(f) The natural (resonance) frequency is

$$f_o = 1/2\pi(LC)^{1/2} = 50.3 \text{ Hz}$$

(g) At f_o, $Z = R$, thus the rms current is

$$I_{max} = V/R = 12.5 \text{ A}$$

Question 2.
From the phasor diagram for an RLC series circuit, we see that

$$V_R/V = R/Z = \cos\phi$$

The impedance of the circuit is $Z = V/I = 12\ \Omega$. Thus $\cos\phi = 7/12$ and $\phi = \pm 54.3°$. Since $\omega L < 1/\omega C$, we choose $\phi = -53.4°$.

CHAPTER 34

Question 1.
(a) The wave propagates in the -y direction, i.e. **S** = -S**j** and we know that **E** = E**i**. Using **S** = **E x B**/μ_o, we consider just the directions -**j** = **i x** ?. We see that ? = **k**. The peak value of the magnetic field is $B_o = E_o/c = 4\text{x}10^{-10}$ T. Thus,

$$B_z = 4\text{x}10^{-10} \sin(1.05\text{x}10^7 y + 3.15\text{x}10^{15} t) \text{ T}$$

(b) The average intensity is

$$S_{av} = E_oB_o/2\mu_o = 2.71 \text{ W/m}^2$$

Question 2.
When an electromagnetic wave of intensity S is completely absorbed, the radiation pressure is S/c. If it is completely reflected, the radiation pressure is 2S/c. In the present problem 0.4S is absorbed and 0.6S is reflected. Thus, the radiation pressure is

$$P = 0.4S/c + 2(0.6S/c) = 1.6S/c = 1.07\text{x}10^{-7} \text{ N/m}^2$$

CHAPTER 35

Question 1.

The reflected ray will also be at 50° to the normal or 40° to the surface, see Fig. 35.12. Since the refracted ray is perpendicular to the reflected ray, the refracted ray must be at 50° to the surface. This means that the angle of refraction is 40°. Applying Snell's law:

$$1 \sin 50^\circ = n \sin 40^\circ$$

Thus $n = \sin 50^\circ / \sin 40^\circ = 1.19$

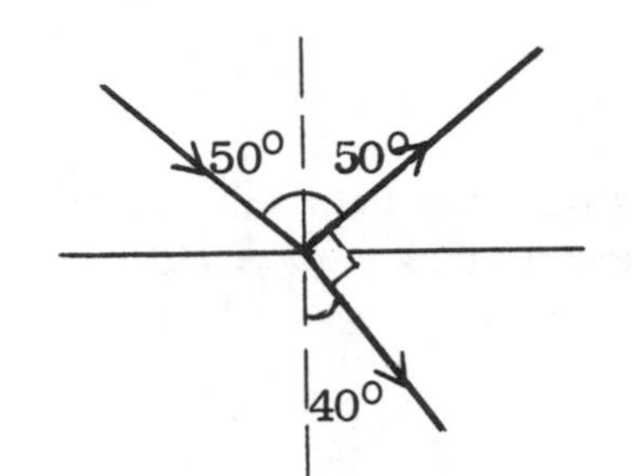

FIGURE 35.12

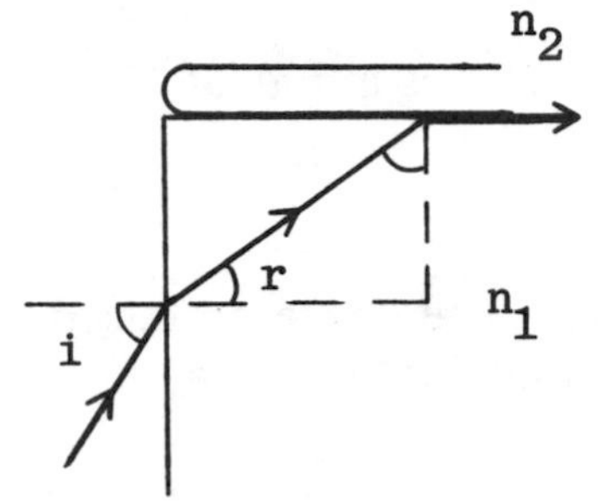

FIGURE 35.13

Question 2.

The critical angle for total internal reflection at the glass-water interface is given by

$$n_1 \sin\theta_c = n_2 \sin 90^\circ \qquad \text{(i)}$$

Since the angle of refaction $r = 90^\circ - \theta_c$, see Fig. 35.13, Snell's law at the glass-air boundary is

$$\sin i = n_1 \sin r = n_1 \cos\theta_c \qquad \text{(ii)}$$

Square (i) and (ii) and add them to find (note $\sin^2\theta_c + \cos^2\theta_c = 1$)

$$n_1^2 = n_2^2 + \sin^2 i$$

Thus, $\sin i = (n_1^2 - n_2^2)^{1/2}$, which leads to $i = 43.4^\circ$.

Question 3.

(a) We are given $m = -q/p = -0.5$. (How do we know that m must be negative?) Since $q = 0.5p$, the lens formula is

$$1/p + 1/0.5p = 1/f$$

With $f = R/2 = 12$ cm, we find $p = 36$ cm.

(b) In this case $-q/p = +0.5$, since the reduced image in a convex mirror is upright. Thus $q = -0.5p$, and

$$1/p - 1/0.5p = 1/f$$

where $f = R/2 = 12$ cm. We find $p = 12$ cm.

CHAPTER 36

Question 1.

(a) For the first lens in Fig. 36.6

$$1/60 \text{ cm} + 1/q_1 = 1/20 \text{ cm}$$

so $q_1 = 30$ cm. The image produced by the first lens acts as the object for the second. Since the lenses are separated by 18 cm, the object distance for the second lens is $p_2 = -12$ cm (a virtual object).

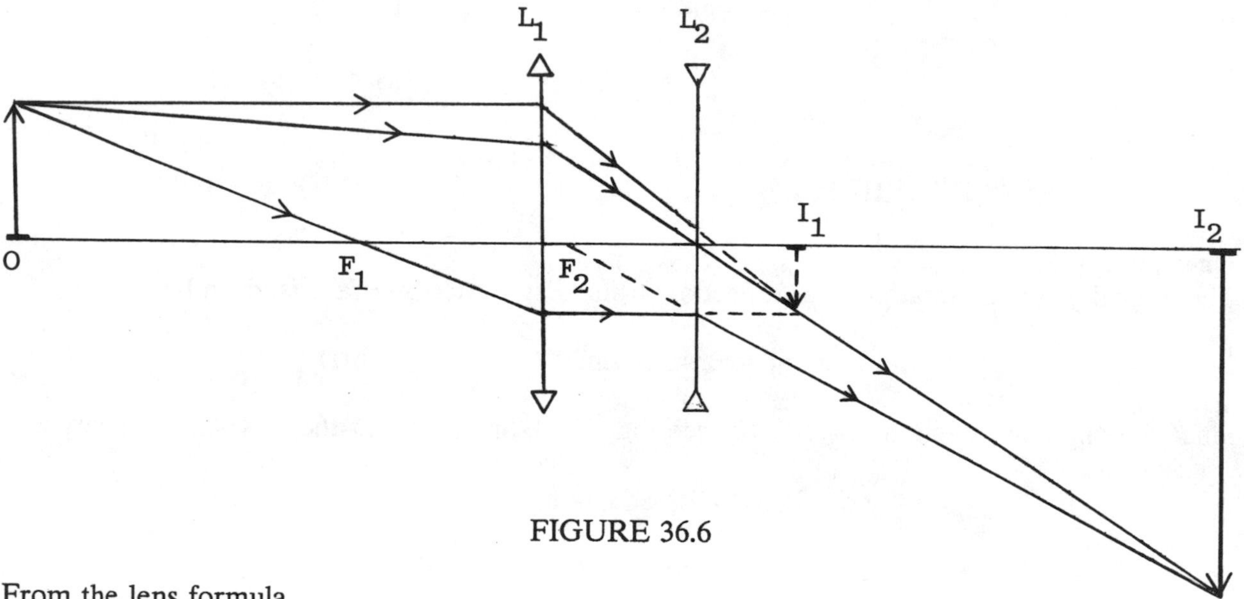

FIGURE 36.6

From the lens formula

$$-1/12 \text{ cm} + 1/q_2 = -1/15 \text{ cm}$$

so $q_2 = 60$ cm.

(b) The final linear magnification is the product of the individual magnifications:

$$m_T = m_1\, m_2 = (-q_1/p_1)(-q_2/p_2) = -2.5$$

Question 2.

(a) We know f_E and $q_E = -50$ cm. From $1/p_E + 1/q_E = 1/f_E$, we find $p_E = 50/11$ cm. The distance between the lenses is $d = \ell + f_O + f_E = 21$ cm, thus $q_O = d - p_E = 181/11$ cm. From $1/p_O + 1/q_O = 1/f_O$, we find $p_O = 181/170 = 1.065$ cm

(b) The angular magnification is $M = -(q_O/p_O)(25/p_E) = -85$

Question 3.
(a) When the glasses are worn, light coming from infinity appears to diverge from the far point of the eye. Thus,

$$1/\infty + 1/FP = -1/2.5 \text{ m}$$

Thus the far point is FP = 0.4 m.
(b) An object at the near point of the unaided eye appears to be at 30 cm when the glasses are worn. Thus,

$$1/0.3 \text{ m} + 1/NP = -1/2.5 \text{ m}$$

Thus the near point (without glasses) is NP = 0.268 m. Diverging lenses cause the near point to move farther from the eye.

CHAPTER 37

Question 1.
The condition for bright fringes is $d \sin\theta = m\lambda$. For small angles, $\sin\theta \approx \tan\theta = y/L$, thus

$$d = m\lambda L/y$$

With the given values we find d = 1.31 mm.

Question 2.
(a) The rays reflected at the top and bottom surface of the oil film undergo a π phase change. Thus, the condition for destructive interference is

$$2t = (m + 1/2)\lambda/n$$

where m = 0, 1, 2, ...
(b) The values of m for two given wavelengths differ by one, so

$$(m + 1/2)588 = (m + 3/2)420$$

This is easily solved and yields m = 2. Thus the minimum thickness of the film is t = (2.5)(588 nm)/2(1.4) = 525 nm.

CHAPTER 38

Question 1.
(a) The 6 th bright fringe of the interference pattern ($d \sin\theta = 6\lambda$) is at the same location as the first diffraction minimum ($a \sin\theta = \lambda$). Thus, a = d/6 = 0.07 mm.
(b) The principal peaks of a diffraction grating are located where $d \sin\theta = m\lambda$. With the information for blue, $d \sin 11° = 5 \times 10^{-7}$ m, thus $d = 2.62 \times 10^{-6}$ m. The location of the red 1 st order red line is given by $d \sin\theta = \lambda$, thus $\sin\theta = \lambda/d = 7 \times 10^{-7} \text{ m}/2.62 \times 10^{-6} \text{ m} = 0.267$, which yields $\theta = 15.5°$.

Question 2.
According to Rayleigh's criterion the critical angle is given by

$$\theta_c = 1.22\lambda/a$$

The angle $\theta_c = d/L$, where d is the separation between sources and L is the distance from the eye to the sources. Thus,

$$L = d/\theta_c = ad/1.22\lambda = 12.6 \text{ km}$$

Question 3.
The intensity of a single-slit diffraction pattern is given by

$$I = I_o \sin^2(\alpha/2)/(\alpha/2)^2$$

where $\alpha = 2\pi a \sin\theta/\lambda$. We are given $\sin\theta \approx = 3\text{x}10^{-3} \text{ m}/2.8 \text{ m} = 1.07\text{x}10^{-3}$, thus

$$\alpha = 2\pi(2\text{x}10^{-4} \text{ m})(1.07\text{x}10^{-3})/(6\text{x}10^{-7} \text{ m}) = 2.24 \text{ rad.}$$

The intensity is

$$I = I_o \sin^2(1.12 \text{ rad})/(1.12 \text{ rad})^2 = 0.646I_o.$$

CHAPTER 39

Question 1.
(a) The person on the platform makes both measurements at the same position and so he measures the proper time interval $T_o = 5\ \mu s$. The observers in the train frame measure

$$T = \gamma T_o = (5/3)(5\ \mu s) = 8.33\ \mu s.$$

(b) The length of the train in the platform frame is

$$L = vT_o = (0.8c)(5\ \mu s) = 1.2 \text{ km.}$$

(c) The observers in the train measure the proper length,

$$L_o = \gamma L = 2 \text{ km.}$$

Question 2.
The velocity of A relative to B is

$$v_{AB} = (v_{AE} + v_{EB})/(1 + v_{AE}v_{EB}/c^2)$$

$$= (0.6c - 0.8c)/(1 - 0.48) = -0.38c$$

Question 3.
(a) Since $E = K + m_oc^2$, we have $K = 3m_oc^2$
(b) $E = \gamma m_oc^2$, so $\gamma = 4$, and so

$$\gamma^2 = 16 = 1/(1 - \beta^2)$$

Find $\beta = 0.97$, or $v = 2.91x10^8$ m/s.

Question 4.
Since $\lambda = c/f$, the Doppler effect may be written for a source that is receding as

$$\lambda = [(c + v)/(c - v)]^{1/2} \lambda_o$$

$$= [(1 + \beta)/(1 - \beta)]^{1/2} \lambda_o$$

where $\beta = v/c$. The fractional shift in wavelength is

$$(\lambda - \lambda_o)/\lambda_o = [(1 + \beta)/(1 - \beta)]^{1/2} - 1 = 0.1$$

Thus,

$$(1 + \beta)/(1 - \beta) = 1.1^2$$

which leads to $\beta = 0.095$, or $v = 2.85x10^7$ m/s.

CHAPTER 40
Question 1.
(a) From the photoelectric equation we have $eV_o = hc/\lambda - \phi$, thus

$$\phi = hc/\lambda - eV_o$$

$$= 9.16x10^{-19} \text{ J} - 3.2x10^{-19} \text{ J} = 5.96x10^{-19} \text{ J} = 3.73 \text{ eV}$$

(b) The maximum speed of the photoelectrons is given by

$$1/2 \text{ m } v_{max}^2 = eV_o$$

thus $v_{max} = (2eV_o/m)^{1/2} = 8.38x10^5$ m/s.

Question 2.
From the conservation of energy

$$K = hc(1/\lambda - 1/\lambda')$$

which may be written as $1/\lambda' = 1/\lambda - K/hc$, where $K = 1.92x10^{-18}$ J. thus $1/\lambda' = 5x10^9 \text{ m}^{-1} - 1.85x10^7 \text{ m}^{-1}$. We find $\lambda' = 2.0074x10^{-10}$ m. The equation for the change in wavelength in the Compton effect is

$$\Delta\lambda = \lambda_c(1 - \cos\theta)$$

where $\Delta\lambda = 0.74$ pm and $\lambda_c = 2.43$ pm, thus $\cos\theta = 0.695$, and $\theta = 45.9°$.

Question 3.
The energy levels are $E_1 = -54.4$ eV, $E_2 = -13.6$ eV, $E_3 = -6.04$ eV, and $E_4 = -3.4$ eV. Since $E_4 - E_1 > 50$ eV we need consider no higher levels. There are three possible lines.

$$f_1 = (E_3 - E_1)/h = 1.17\text{x}10^{16}\text{ Hz}$$

$$f_2 = (E_2 - E_1)/h = 9.86\text{x}10^{15}\text{ Hz}$$

$$f_3 = (E_3 - E_2)/h = 1.83\text{x}10^{15}\text{ Hz.}$$

Question 4.
(a) The intensity of the beam is

$$I = P/A = E/A\Delta t = nhf/A\Delta t$$

so, with $\Delta t = 1$ s,

$$n/\Delta t = P/hf = (10^{-3}\text{ W})(6.328\text{x}10^{-7}\text{ m})/hc(\Delta t) = 3.18\text{x}10^{15}\text{ m}^{-2}\text{s}^{-1}$$

(b) The linear momentum is given by

$$p = E/c = hf/c = h/\lambda$$

$$= 1.05\text{x}10^{-27}\text{ kg.m/s}$$

Since $n/\Delta t = 3.18\text{x}10^{15}$ per m^2 per second, the linear momentum transported per m^2 per second is $3.33\text{x}10^{-12}$ N/m^2.

CHAPTER 41
Question 1.
The linear momentum is $p = (2mK)^{1/2} = (2meV)^{1/2}$, so

$$\lambda = h/p = h/(2meV)^{1/2} = 3.87\text{x}10^{-11}\text{ m}$$

Question 2.
(a) From $\Delta x\ \Delta p \approx h$, we find $\Delta x = h/m\Delta v = 1.32\text{x}10^{-13}$ m

(b) According to the HUP we have $\Delta E\ \Delta t \approx h$, but $\Delta E = h\Delta f$, thus

$$\Delta f \approx 1/\Delta t = 2\text{x}10^{7}\text{ Hz}$$

Question 3.
The energy levels for a particle trapped in an infinite well are

$$E = n^2h^2/8mL^2 \qquad n = 1, 2, 3, \ldots$$

For an electron we find

$$E_1 = 2.68\text{x}10^{-18}\text{ J}; \quad E_2 = 1.07\text{x}10^{-17}\text{ J}; \quad E_3 = 2.41\text{x}10^{-17}\text{ J}$$

CHAPTER 42

Question 1.
(a) The magnitude of the angular momentum is

$$L = [\ell(\ell + 1)]^{1/2}\hbar = (12)^{1/2}\hbar$$

(b) The component of L along an external field is

$$L_z = m_\ell\hbar = 2\hbar$$

(c) Since $L_z = L\cos\theta$, the angle between L_z and L is given by

$$\cos\theta = L_z/L = 2/3.46$$

Thus $\theta = 54.7°$.

(d) The component of the magnetic moment along the field is

$$\mu_z = -eL_z/2m = -2eh2m = -eh/m$$

We find $\mu_z = -1.85\text{x}10^{-23}$ J/T.

Question 2.
If we let $x = r/r_o$, we have

$$P_{2s}(x) = (x^2/8r_o)(2 - x)^2 e^{-x}$$

We find $P(0.764) = 0.0519/r_o$, and $P(5.24) = 0.191/r_o$.

CHAPTER 43

Question 1.
(a) This is positron decay, in which the atomic number decreases by one but the mass number is unchanged. Thus the missing particles are 19F and a neutrino.

(b) For β^- decay, the atomic number increases by one, so the missing particles are ^{64}Zn and an antineutrino.

Question 2.
(a) The initial number of atoms or nuclei is

$$N_o = mN_A/M$$

$$= (0.02\ g)(6.022x10^{23}\ atoms/mol)/(60\ g/mol)$$

$$= 2.01x10^{20}\ atoms$$

The initial activity (decay rate) is

$$R_o = \lambda N_o = 0.693\ N_o/T_{1/2} = 7.67x10^{11}\ Bq.$$

(b) We are given $R/R_o = \exp(-\lambda t) = 0.3$, thus

$$t = -\ln(0.3)T_{1/2}/0.693 = 9.21\ y$$

Question 3.
The energy released is $Q = \Delta m.c^2$ or

$$Q = [1.008665\ u + 6.015123\ u - 3.016050\ u - 4.002603\ u](931.5\ MeV/u)$$

$$= 4.78\ MeV$$

NOTES

NOTES

NOTES

NOTES

NOTES

NOTES

NOTES

NOTES

NOTES